Future of Engineeirng

Legacy by Design

by Felicity Furey

Foreword

Engineering has always been a discipline of profound consequence. From the first bridges that carried people across rivers to the networks and digital systems that connect our modern world, engineers have shaped not only the environments we inhabit, but the way we live, move, interact, and imagine possibility. Yet today, our profession stands at a pivotal moment. The challenges facing humanity—climate change, resource constraints and environmental decay, social fragmentation, mistrust in institutions, rapid technological acceleration, and widening inequity—demand more from engineering than precision and performance alone.

They demand intention, courage, and a renewed sense of responsibility for the world our work creates.

For decades, engineering has been guided by values forged in a different era: efficiency, optimisation, control, and delivery. These principles helped build the world we know, but they are no longer sufficient for the world we need. As you will read in these pages, technical

success can unfortunately co-exist with human or environmental failure. Projects can meet every standard and still miss the mark for the communities they are meant to serve. Systems can function perfectly on paper yet erode trust, wellbeing, or connection in practice. This tension—between what we can build and what we should build—is at the heart of the engineering profession today.

That is why this book matters so deeply.

The Future of Engineering: Legacy by Design is not simply a reflection on the profession; it is an invitation, a challenge, and a roadmap. Felicity Furey brings to this work what few can: the rare combination of technical expertise, leadership across complex systems, deep human insight, and lived experience at every level of engineering—from design and delivery to strategy, policy, and education. Felicity has long been a voice pushing the profession forward, championing inclusive, community-centred engineering and empowering thousands of young people through initiatives such as Power of Engineering and WeAspire. She has always understood that the true power of engineering lies not just in what we create, but in why and for whom we create it.

In these chapters, she shines a light on the "hidden DNA" of engineering—the assumptions, cultures, and inherited habits that shape our decisions more than we realise. She tells stories that feel uncomfortably familiar: technically perfect designs that fail real people; processes that optimise for speed but sacrifice connection; systems that work for the average user but exclude those who fall outside the norm. And she shows, with clarity and compassion, that

these patterns are not failures of skill, but failures of intention.

What makes this book truly hopeful is that it does not leave us in critique. It offers a future- facing framework rooted in human wellbeing, regenerative thinking, and long-term legacy. It gives engineers, leaders, and decision-makers practical tools to widen their lens, challenge the brief, design with care, and reconnect with the deeper purpose of our profession. Through stories from around the world—from Medellín to Melbourne, from Vienna to Brisbane—Felicity demonstrates that engineering can be a wonderful force for healing, connection, equity, and regeneration when we choose to design that way.

This book arrives at a time when our profession needs exactly this kind of leadership. Across industries, engineers are beginning to recognise that the projects we deliver today shape the society our children and grandchildren will inherit. We are no longer simply problem-solvers; we are stewards of the systems that support life. The future will require engineers who can hold complexity, who can work with communities, who can think across generations, and who understand that values are not separate from engineering—they are embedded in every design.

I feel fortunate to have worked as an engineer for most of my career at Arup, a global engineering consultancy that has embedded core values such as social usefulness, total design and humane dealings in everything it does. Furthermore, the firm is trust-owned for the benefit of all members past and present, and society at large. Like Felicity, the founder, Ove Arup, truly understood the importance of 'legacy' and the firm he created whilst not perfect has designed many built environment projects over 80 years of operation that are cherished by society. Engineering can be both beautiful and kind.

Reading this book, you cannot help but feel both challenged and energised. Challenged by the responsibility we carry and energized by the possibility that lies ahead when we broaden our definition of success. Felicity reminds us that engineering is not neutral. It never has been. It is an inherently human act—one that shapes behaviour, opportunity, identity, and connection.

The future of engineering will not be defined only by new technologies or tools, but by the values we choose to build into them. This book helps us make those choices consciously, courageously, and with care.

It is my hope that as you turn these pages, you feel what I felt: a renewed sense of purpose, a deeper understanding of our impact, and a belief in what engineering can become when we design not just for today, but for tomorrow—and for the generations beyond.

This is a book for engineers who want to leave a legacy.

A legacy we can be proud of.

A legacy built by design.

Peter Bailey

Trustee Director, Arup Former Engineer, Principal and Group Board Director, Arup Group

Table of Contents

About the Author

Felicity Furey is an award-winning engineer, author, and international keynote speaker redefining engineering as a human-centred, regenerative force for good. She is one of Australia's leading voices on the future of engineering, leadership, and systems change.

Across her career, Felicity has helped deliver major infrastructure projects, advised organisations navigating complexity, and built movements that shift how people understand what engineering is for. Her work challenges traditional frameworks and expands the role of engineers beyond technical delivery, placing people, environment, leadership, and long-term legacy at the centre of design and decision-making.

Felicity is the co-founder of Power of Engineering, which inspires young people - particularly girls and regional students - to see engineering as a meaningful career. She is also the co-founder of WeAspire, a leadership advisory that has supported thousands of emerging leaders across engineering and infrastructure. She has worked with organisations including Qantas, Engineers Australia, Arup, Toyota, AECOM, BHP, and Boeing.

Through compelling storytelling and practical frameworks, Felicity helps leaders design systems that people trust, cultures where people thrive, and futures built with intention and care so the legacy we build serves people, place and future generations.

To the Engineers of the Future

Dear engineers of the future, one hundred years from now, the ones I will never meet,

When this book was written, the world felt uncertain.

We had built extraordinary systems - fast, efficient, powerful - but many were disconnected from the people and places they were meant to serve. We optimised for speed, cost, and performance, yet struggled with trust, equity, and care. We were very good at solving problems, and still learning how to ask better questions.

This book was written in that moment.

At first, only a small number of engineers took its ideas seriously. Most were doing what they had been trained to do - delivering projects, meeting standards, complying with briefs, and doing their best within the systems they had inherited.

But a few paused. A few questioned assumptions. A few chose to think from a different place.

That was enough to begin.

Thank you for what you did next.

Thank you for speaking up in meetings when it would have been easier to stay quiet. For widening the conversation when it tried to narrow again. For questioning briefs, standards, and definitions of success that no longer served people or the planet.

You changed how engineering was taught.

You embedded different values into standards.

You designed with care for people, Country, and future generations.

You didn't do this through grand gestures or heroic moments, but through small, deliberate decisions - one meeting, one design, one question at a time.

Because of you, the world feels calmer. People are more connected to each other and to nature. Long-term thinking became normal, not exceptional.

If the places you now inhabit are more liveable, more just, and more resilient, know this: it did not happen by accident. Engineering reclaimed its deeper purpose - not just to build faster or bigger, but to steward wisely and design with care.

This book did not change the world on its own.

You did.

With gratitude, for what you carried forward and what you chose to leave behind,

Felicity Furey

Part I:
The Hidden DNA

Engineering rarely announces its values out loud. They sit hidden beneath the surface, embedded in standards, assumptions, and habits that feel so normal they stop being visible. This part of the book is about bringing that hidden DNA into view.

First, we need to understand how it has been shaped. Part I explores the inherited values that underpin modern engineering: efficiency, control, optimisation, and certainty. These values helped build the modern world, but they were forged in a different era, for different problems, and they now shape outcomes in ways we don't always intend or notice.

The chapters that follow invite you to look beneath "best practice", and to question what has silently been normalised. This is not about blame or rejection of the past. It is about awareness. Once we can see the hidden DNA of engineering, we gain the ability to decide what we carry forward, and what we are ready to redesign.

Chapter 1:
The Questions That Started It All

Every engineer has moments that unsettle them. Moments where the project is technically sound, the drawings are complete, and the boxes are ticked, yet something still doesn't feel right. This chapter begins there - not with answers, but with the questions that refuse to go away.

Before we can talk about solutions, frameworks, or the future of engineering, we need to understand how we got here. The assumptions we inherited. The values we absorbed without noticing. The ways "doing things properly" became synonymous with not questioning them at all.

This chapter lays the foundation for the rest of the book. It asks us to pause before we solve, and to look honestly at what engineering has been optimised for, and what has silently been left behind.

When Projects Stop Making Sense

There are moments in every engineer's career that stay with you. For me, they weren't just about what we built. They were about what we didn't stop to question.

I get it. The client gives you a brief and, by nature, engineers are trained to solve problems. So, we hone in on efficiently finding solutions and get stuck into making it work. To our credit, engineers deliver. We have built some of the most extraordinary, life-changing things in our world.

I remember standing on the site of a major tunnel project in Melbourne. The scale was immense. The precision, extraordinary. We were building a piece of infrastructure that would connect the city, reduce congestion, make travel safer, and help people travel around the city smoothly for the next two to three decades. The design incorporated new concepts like climate smart engineering.

It was, by every technical measure, a triumph. Yet as I looked at the stretch of road carving its way through the community, another question surfaced in my mind:

We are connecting the city, but what are we disconnecting?

That question began to echo in everything I saw - in the standards we followed, the assumptions we made, the problems we did not stop to challenge. It became clear to me that engineering wasn't just a discipline of design and delivery; it was a reflection of our values.

In every project I've been involved in - from designing concrete structures on water treatment plants to steel portal-frame buildings, new roads, and upgrades of existing ones; from mine rehabilitation projects and leading a team to create a safe dam in regional Australia to major infrastructure projects and critical infrastructure; even helping create a world first engineering degree - each organisation involved brought their own values. Often, those values competed.

There were the people who would operate and maintain the system or the infrastructure. The government who advocated for what had been promised to the community. The designers upholding standards and speed limits for highways. The contractors looking for efficiencies in construction. I have also seen individuals push for their own personal values in projects, beyond what the organisation wanted.

And I have seen many more values embedded in our projects, many of which were inherited from another time. Ones that are so ingrained in us that we can't even see they are there.

The Inherited DNA of Engineering

Engineering today has inherited a set of values that were primarily forged in the early 20th century, during the Industrial Revolution, when the world was obsessed with productivity, efficiency, and control. These principles helped build the modern world, but they were designed for a different era - one focused on extraction, not regeneration; on uniformity, not diversity.

Back then, engineers like Frederick Winslow Taylor transformed factories with the idea of scientific management.[1] He studied every motion, every second, and every process until he found what he believed to be the "one best way". It revolutionised industry. Productivity soared. Taylor became known as the father of efficiency and paved the way for the rules of management.[2]

But his vision wasn't just a method. It was a mindset. It treated people and systems as components to be optimised rather than parts of a living whole. Those same values found their way into our design standards, our procedures, our education, and the very DNA of engineering.

We've been trained to measure success in precision, output, and control. Slowly, these values have permeated our world and worn away at other values until efficiency is often all that is left.

And yet, when I look at the challenges we face today, from climate change to social disconnection, I can't help but wonder: are these still the right measures of success?

What about relationships, legacy, care, and connection? How do we measure those in a project?

Designing for the Past, Not the Future

When we look around at how we design today, we can still see the fingerprints of that efficient industrial mindset everywhere.

Take car safety. For decades, seatbelts and crash tests were based on the "average male" body. When women were finally considered, the female crash-test dummy was just a scaled-down version of a man which, as it turns out, is not a true representation of female anatomy.[3] Studies have shown women are significantly more likely than men to be seriously injured in comparable frontal crashes.[4] Only recently has crash testing begun to better account for female anatomy and the position that many women drive in. Women often sit more forward to see over the bonnet, which is considered "out of position" in some testing assumptions.[5] The engineering works perfectly - for the person it was designed for.

Or consider air conditioning systems. Many HVAC comfort standards are based on average male metabolic rates.[6] Research published in Nature Climate Change found this benchmark often makes offices several degrees too cold for women, who can have different metabolic needs. Again, technically successful, but functionally inequitable. The "standard" user is not neutral; it's often a reflection of who was designing at the time.

What these examples reveal is something I did not see as an engineer until I began speaking about biases in design on International Women's Day and researching further. It is not a conversation I had often until writing this book with other engineers.

Design is not always neutral. Most standards, test protocols, and "average user" assumptions carry the values, priorities, and blind spots of the people who created them.

This happens in other scientific pursuits too, like medicine.

For decades, medical research, including the development and testing of vaccines, was conducted almost exclusively on male bodies.[7] Women were routinely excluded from clinical trials, largely because hormonal cycles were considered "too complex" to account for. That decision simplified testing, reduced variables, and made analysis easier. From a narrow scientific perspective, it made sense. From a systems and outcomes perspective for 51% of the population, it created a profound bias.

Flu vaccines are a clear example. Research shows that women often mount a stronger immune response to the influenza vaccine than men, experiencing both greater protection and higher rates of side effects. Yet dosage, formulation, and testing protocols were historically based on male physiology.[8] While clinical trials today include women, many dosing and formulation assumptions still reflect earlier male-centred research. The result? A "successful" vaccine that is safe and effective, but not optimised for over half the population. The science works. Just not equally.

This wasn't a failure of science. It was a failure of who was considered worth designing for.

The same pattern also shows up in places we don't usually think of as political or cultural, like military equipment. For years, standard-issue army backpacks were designed around male body proportions: shoulder width, torso length, hip structure, and centre of gravity. When women entered combat roles in greater numbers, injury rates climbed.[9] Packs sat incorrectly on the body. Load distribution increased strain on the lower back and hips. Fatigue set in faster. The equipment met its specifications. It carried the load. It passed the test. But it silently transferred risk onto the bodies that didn't match the assumed norm.

Only when injury data became impossible to ignore did redesign begin - not because the original design was "wrong", but because it had been right for a very specific body.

Bias in engineering rarely looks like malice. It often is masked as efficiency. It looks like simplifying assumptions, reducing variables, and designing for the "average user."

But the "average user" is almost never average. It is a proxy - usually male, usually able-bodied, usually reflective of who was in the room when the standard was written.

Seatbelts, air conditioning, medical treatments, military equipment There are many more examples. The book Invisible Women catalogues how widespread this is. These systems all technically succeed. They meet requirements and pass tests that were designed for them. But who designed them? They do what they were designed to do - for the person they were designed for.

Bias doesn't sit at the end of the design process. It sits in the brief, the assumptions, and the standards we inherit and rarely question.

Once those assumptions are encoded into design standards, they become invisible. They feel objective. They feel scientific. They feel settled, until someone's body, experience, or safety tells a different story.

This is how values become infrastructure. Engineers, whether we realise it or not, are designing far more than systems. We are designing who those systems work for.

Even our roads tell this story of inherited priorities and values. In practice, speed and throughput often dominate how these standards are applied. To keep roads "safe", we clear trees and remove roadside furniture, creating sterile corridors that move cars quickly but strip away shade, biodiversity, and human connection.[10] Roads that are efficient, yes, but soulless - built for movement, not meaning.

And the suburbs? Across Australia, we continue to design vast housing estates that look efficient on paper but fail in purpose. They're built fast, at scale, and for maximum profit. Asphalt streets without footpaths. No trees. Black roofs that trap heat. Limited access to public transport. Congested streets just to exit the estate. Communities are left hotter, more isolated, and dependent on cars for even the shortest trips.[11] Research shows these new suburbs can be up to six degrees hotter than established, leafy neighbourhoods.[12] We're designing heat islands and disconnection, then calling it progress.

There are countless more examples, from designs that are dangerous to everyday items that only consider a narrow slice of the population while ignoring everyone else.

It's the same pattern, over and over: technical success, human failure. The product of values that prize efficiency over empathy, and control over care.

When Access Came Late

We have seen slow and marginal progress to improve our design standards.

One of the most cited examples of inclusive design is the humble kerb cut - the small ramps built into footpaths at street crossings. They were not introduced for convenience or aesthetics. They were fought for by wheelchair users in the disability rights movement, often through direct action. In the 1960s and 70s, activists in the United States literally took to the streets, jackhammers in hand, cutting ramps into kerbs so they could move through their cities independently.[13]

What's remarkable is what happened next. Once kerb cuts became standard, it became obvious they benefited far more people than originally imagined: parents with prams, people using mobility aids, delivery workers with trolleys, cyclists, travellers with suitcases, older adults, and anyone moving through the city with a temporary injury.

The ramp designed for disability turned out to be a better design for everyone.

Kerb cuts are now a foundational element of urban design, but they only exist because a group of people were finally recognised as worthy of consideration in the system. This wasn't a technological breakthrough. It was a values breakthrough - a shift in who the city was for.

Another powerful example comes from communication design. Closed captions were originally developed to support people who are deaf or hard of hearing.[14] For years, they were treated as a niche accessibility feature - optional, secondary, and often poorly implemented.

Today, closed captions are everywhere. Not just on television, but on social media, video conferencing platforms, and digital learning environments. And their use has exploded not because disability suddenly became more common, but because people realised captions support a much broader range of human needs: non-native language speakers, people in noisy environments, neurodivergent users who process information visually, and anyone who wants to engage with content more flexibly.

What began as an accessibility accommodation became a performance enhancement.The design didn't change because the technology improved. The technology was always capable.

It changed because designers finally paid attention to how people actually experience the world, rather than how the system assumed they should.

A few years ago, I heard a story about a new station design in Brisbane. The design team planned that after the station had been constructed, they would get vision impaired people and people with disabilities to experience it and make recommendations for improvements based on their lived experience. Some excellent designers suggested something different: incorporate that feedback during the design phase. Thankfully, they did. This saved a huge amount of money compared with retrofitting a brand-new station that was not fit for purpose.

This wasn't because engineers and designers didn't care. It's because they were following a system that told them what mattered: throughput, efficiency, flow. When disability rights and universal design movements began challenging those assumptions, we saw how powerful it was to redesign with inclusion in mind. It showed us that better design begins when we expand who we design for.

Seeing the Pattern

There are pockets of progress, isolated incidents and small cases where people have stopped to think and question the rules, the standards, and the "we have always done it this way" processes. But it is often the exception rather than the rule.

Every day, dutiful engineers pick up the design standards and get on with it, solving problems and applying engineering judgment case by case.

I've been that engineer. Why would I question it, especially when I genuinely love efficiency? There are a lot of very smart people who brought us this far. Engineering has created phenomenal results when it comes to safety, and we have built some of the most incredible things in our world.

However, today, designing for compliance, efficiency, and safety isn't enough. On their own, these are the wrong values for today's world, and the future we are designing for.

Because when we look around us - at the headlines, the failures, and the heartbreaks - we see what happens when systems are optimised for the wrong outcomes.

Case Study: Dieselgate

Take Dieselgate, for example. In 2015, the world learned that Volkswagen Group had installed "defeat devices" in its diesel vehicles - software that could detect when the car was undergoing an emissions test, switch into a low-emissions mode for the test, then revert to normal operation on the road, producing far higher pollution. The scale was staggering, affecting more than 10 million vehicles worldwide, and hundreds of thousands in the US alone.[15]

From an engineering perspective, things became even more telling. Engineers and software developers at Volkswagen, as well as parts suppliers, were asked to deliver diesel engines that met increasingly stringent emissions standards while maintaining performance, fuel economy, and brand promises of "clean diesel". But meeting those demands proved technically very difficult under the existing constraints. Faced with the pressure, the decision was made: rather than redesign the technology properly, a workaround was created. The defeat device was not an accident. It was an intentional choice.

Then came the cover-up, and a systematic failure of trust. Internal warnings reportedly existed (for example, suppliers had raised concerns), but the corporate response was slow and defensive. Once the scheme became public, Volkswagen's credibility collapsed. Regulators and the public realised that the company's claims of "clean diesel" were hollow, and that engineering had prioritised test compliance and marketing over real-world emissions, health, and integrity.

The human and social consequences were real. Studies estimate that thousands of premature deaths may be attributable to the excess pollution caused by the manipulated engines. Meanwhile, Volkswagen faced massive financial penalties, recalls, and class-action lawsuits across many jurisdictions, and a profound loss of trust in the brand and industry.[16]

For our profession, Dieselgate is a sobering mirror. It reveals what happens when values like speed, cost, performance, and image dominate, and values like legacy, care, and truth are sidelined. It shows engineering brilliance used to optimise a test result rather than to serve society. And it demonstrates how fragile trust is, and how hard it can be to rebuild once broken.

When I consider this case as an engineer, I think about the upstream decisions. Who set the brief? What assumptions were accepted uncritically? And when the moment came to ask, "Is this really what the design is for?" someone didn't raise the flag early enough.

Case Study: Optus 000 Outage

In the early hours of Thursday 18 September 2025, a "routine" network change at Optus triggered a failure that blocked some customers from reaching Australia's emergency number, 000, for around 13 hours across parts of WA, SA, NT, and far-west NSW. Normal calls kept working. The fault sat inside the separate emergency-calling path. Early reporting pointed to a firewall upgrade that incorrectly blocked emergency traffic.

Early counts identified ~600 failed Triple-Zero calls. Authorities later confirmed at least three deaths where callers were unable to reach 000. Some reports place the toll at four while investigations continue.[17]

Optus has said a human error occurred during a scheduled firewall change and that established protocols were not followed. In other words, the change management process (design, pre-checks, deployment, validation, and rollback criteria) broke down at the most critical layer: the emergency route. This is an engineering story as much as a governance one. Separation of critical services. Fail-open versus fail-safe design. Targeted monitoring for emergency call traffic. Automatic escalation when anomalies appear. When those controls are weak, a single misstep in a firewall ruleset can sever the lifeline.

The technical failure was compounded by communication and escalation failures. According to timelines reported by the Australian Broadcasting Corporation and others, customers began warning Optus that 000 calls were failing hours before authorities were fully briefed. Police in WA and SA later said information trickled in piecemeal, delaying coordinated welfare checks. ACMA opened a formal investigation into Optus' compliance with emergency call regulations, underscoring that reliability and rapid notification for 000 are fundamental obligations for any carrier.[18]

Trust, already fragile after Optus' 2023 nationwide outage that also affected 000, eroded further.[19] Reuters reported three alleged deaths linked to the incident and consumer groups and media questioned whether Optus had invested in the right safeguards, whether offshore support and internal processes appreciated the gravity of 000 failures, and why mandatory escalation didn't trigger fast enough. When the public learns that a "routine upgrade" can take out emergency calls, "to standard" and "on budget" start to sound like the wrong metrics entirely.[20]

In the weeks and months after, Optus faced a second smaller 000 disruption in NSW and other locations, keeping the story in the headlines. This also deepened the legitimacy problem: if the lifeline isn't reliable, what else can't we trust?

For engineers, the lesson is stark: emergency pathways require different values. Redundancy proven under load. Guardrails that prevent miscoded rules from ever touching critical routes. Monitoring that alarms not just on traffic volume, but on emergency call success rates specifically. For leaders, it's a reminder that investment choices and incident communication are as much "engineering" as the code on the firewall.

Case Study: Qantas and the Cost of an Aging Fleet

For decades, Qantas was synonymous with safety, "the Spirit of Australia", and the airline that had never lost a jet to a crash. That reputation was built on a culture of engineering excellence and meticulous maintenance.[21] But in recent years, cracks - both literal and cultural - have begun to show.

Qantas became one of the world's most profitable airlines during and after the pandemic, - but it did so by cutting into operations and outsourcing key ground and maintenance services. Engineers warned that the "Qantas standard" they had built their careers on was being diluted. The new CEO inherited not just a balance sheet challenge, but a legitimacy one: how do you rebuild trust in a company whose engineering excellence once was its moral compass?[22]

In early 2024, the Australian Transport Safety Bureau released reports outlining several maintenance lapses across Qantas operations.[23] In one instance, an engineer's inspection missed a tool that had been left inside the engine cowling of an Airbus A380.[24] The aircraft flew for nearly 300 hours before it was discovered. In another, a series of cascading procedural errors exposed gaps in quality assurance. While none of these incidents caused harm, they revealed a shift in what was being valued: cost control over craftsmanship, and turnaround time over thoroughness.

These issues emerged alongside a broader pattern: the airline's fleet renewal had stalled. Many of Qantas's Boeing 737s - the backbone

Post-Grenfell, the UK banned combustible materials on the external walls of new high-rise residential buildings (2018), subsequently extending the ban to more building types in 2022. Yet remediation has been slow and costly, with government and leaseholders bearing much of the burden while many responsible firms have faced limited financial penalties to date.

This has prolonged the crisis of trust: residents in hundreds of buildings are still living with legacy risks, and the public continues to question an ecosystem that certified, procured, and installed combustible systems at scale.[44, 45, 46]

So, what does this reveal about engineering's DNA? Grenfell is a human tragedy first and always.

It is also an opportunity for our profession to reflect. Engineers didn't just draw details; we helped write the rules, choose the products, run the tests, sign off the installations, deliver the project, and have input into the emergency operational procedures.

The catastrophe emerged where standards, procurement, and operations intersected with values: appearance over performance, savings over safety, compliance over care. "Stay put" was a rational doctrine only if the building's fire strategy actually worked; the cladding and installation choices made that assumption false. When lives depend on our designs and manuals, questioning the brief isn't optional. It's the job.

The most devastating lesson of Grenfell is not that systems failed. It's that people trusted those systems and the people that were responsible for them. Our legitimacy rests on never letting that trust be misplaced again.

The Weight of What We Build

When I read these stories - Grenfell, Dieselgate, Optus - I don't see villains. I see systems, processes, and designs that have forgotten their purpose. I see versions of myself earlier in my career, proud of getting the job done, not yet realising the silent cost of unexamined briefs, beliefs, and values. These aren't failures of skill; they're failures of intention. And they remind me that we, as engineers, must design not just systems, but values.

When you line up all these examples - from seatbelts to suburbs, from Dieselgate to Grenfell - a pattern emerges. They're not isolated oversights; they're symptoms of the same inherited DNA. Different industries, designs, and standards, but the same values: we've been designing for the average rather than the actual, for speed over connection, for productivity over people. Our systems are brilliant at delivering technical outcomes but often blind to social and environmental consequences.

As engineers we're trained to be this way for good reason. It works, technically.

We are also trained to be risk-averse, and for good reason. The ability to anticipate, calculate, and mitigate risk is at the heart of what makes engineering safe and why people trust us. If we didn't go to the nth degree on calculations and verification of designs, mitigating the risk of technical failure, our bridges wouldn't stand, our tunnels wouldn't hold, our phones wouldn't ring, our lights wouldn't turn on, our water wouldn't be safe to drink, and our systems wouldn't run. But over time, our understanding of risk has become tightly defined around technical, financial, and schedule parameters - time, cost, quality, safety - and rarely extends beyond them.

We design to eliminate technical failure, but not always moral failure. We assess the probability of collapse, but not the possibility of disconnection. What's missing is a broader view of risk - one that includes the human and social risks embedded in every design decision: the risk of eroding trust, of diminishing community, of leaving future generations with systems that no longer serve them. True engineering risk isn't just about what might break; it's about who might be broken by what we build.

Every profession has its culture. In engineering, that culture rewards certainty, precision, and delivery, but it rarely rewards reflection, questioning, or care. We celebrate the problem-solver, not the problem-framer.

That culture shapes everything: what gets measured, what gets funded, what gets praised. Until we evolve the culture of engineering itself, the systems we build will continue to mirror the same narrow set of values.

And so I began to ask myself and my teams these questions:

- Whose body was the design made for?
- Whose place and whose connection was sacrificed in the name of "efficiency"?

- Do we care more about the technical solution than the people we are serving?

These questions sit at the heart of what I now call the Future of Engineering. They are an invitation to pause before we problem-solve, to question before we calculate, to reconnect before we construct.

We have seen that engineering isn't always neutral. Every design carries intention and hidden values, spoken or unspoken, and everything that engineers do and design is, at the end of the day, for people. It is the true reason who engineers exist: to make the world better for people. If we don't choose our values and intentions consciously, we end up repeating the old ones ingrained in our DNA by default.

What if the future of engineering isn't in solving faster, but in caring deeper?

What if the true measure of progress isn't what we build, but what we connect and regenerate?

This is why the future of engineering matters so much for every single person. It's not about abandoning the technical excellence that defines our profession; it's about expanding it and incorporating a wider view. It's about asking not just can we build it, but why are we building it, for whom, with whom, and what legacy?

Engineering for the future begins with awareness: awareness of where our values come from, how they shape our designs, and what they leave behind. Only when we understand that inherited DNA can we start to redesign it. We've built the world once before. We can build it again - this time, with intention, compassion, and care.

Engineering did not begin as optimisation or extraction. It began as stewardship: managing water, shaping shelter, connecting communities, and working with natural systems so people could survive and thrive. Long before standards and spreadsheets, engineers designed with an understanding of place, people, and time. We are not inventing a new way forward; we are remembering one.

This is where the shift begins - not in technology or tools, but in how we think, what we value, and who we include.

To reinvent engineering, we must first reinvent the engineer.

The Takeaway

What emerges from these stories is a pattern, not a series of isolated failures. Across industries, scales, and contexts, the same values keep showing up in our designs, our standards, and our definitions of success.

Engineering, it turns out, is never neutral. Every decision carries intention, whether we name it or not. And when we don't consciously choose our values, we inherit them by default.

If the future of engineering is going to look different, it cannot begin with tools or technology. It must begin with awareness. The next chapter takes that awareness further, challenging one of the profession's most powerful myths: that engineering is value-neutral.

Chapter 2:
From Compliance to Conscious Design

Once you start noticing the values embedded in engineering decisions, it becomes impossible to unsee them. What we measure, what we optimise, and what we ignore all reveal what we care about, even when we insist we are simply being objective.

This chapter picks up where the questions of Chapter 1 leave off. It examines the idea that engineering is neutral, rational, and detached from social consequence, and why that belief no longer serves us.

To move forward, we must understand how deeply values are woven into standards, manuals, procurement processes, and the way we define "good engineering".

The Myth of Value-Neutral Engineering

We often say engineers are objective, that we deal in facts, not feelings. We talk about evidence-based practice, about data, logic, and reason. We tell ourselves that our calculations are neutral, that numbers don't take sides.

The reality is every equation begins with an assumption about what matters.

When we decide what to optimise, what to measure, and what to leave out, we're already making value judgements. When we design a road for speed, we're saying speed matters more than shade, walkability, or community connection. When we specify a building material for lowest cost, we're saying affordability today matters more than durability tomorrow. When we design for one gender rather than considering all, we are choosing what - or who - to leave out.

Those aren't mathematical truths. They're moral choices, embedded in our designs.

Engineering has inherited a powerful myth - the myth of value-neutrality. But neutrality isn't possible when our work shapes how people live, move, breathe, and belong. When everything we design is, or should be, in service of people.

Engineering can often be mistaken for a technical profession, which is a huge misconception and branding issue for engineers. As my favourite lecturer from university, Professor Les Dawes, said in his retirement speech, engineering is actually a socio-technical profession, with the society part deliberately coming first.

The idea of neutrality came from the same era that gave us the scientific method and the Industrial Revolution - an age that celebrated control, mastery, and standardisation. Objectivity was the gold standard: observe without bias and calculate. That mindset built incredible things - railways, bridges, clean water, global communications - and gave us the world we have today.

But it also built systems that extract rather than regenerate, and that still prioritise economic growth over social or ecological wellbeing.

The myth of neutrality has been useful. It has protected engineers from criticism and given us legitimacy. But it's also made us blind to the inherited, hidden DNA and values we carry. We treat design standards as if they're handed down from the heavens, when in fact they're reflections of time and place, and the values and priorities of the people who created them.

When a road design manual says that speed and efficiency are the highest priorities, that's not a law of physics - it's a legacy of 20th-century industrial value. When a procurement policy defines "value for money" purely in dollar terms, it's not a universal rule - it's an economic worldview.

Engineering isn't value-free; it's actually value-laden.

The question is not whether our work reflects values - it's which values it reflects, and whether they're still the right ones for the world we live in now.

Because when we believe we're neutral, we stop questioning the brief. We stop asking who benefits and who bears the cost. We stop seeing the social and environmental ripple effects of what we create.

And that's where the myth becomes dangerous. Objectivity without empathy can become indifference. Neutrality without reflection can become negligence.

The Future of Engineering isn't about choosing between rigour and care, between logic and empathy. It's about bringing them together, recognising that the best engineering is both technically sound and ethically grounded in our world, and as a legacy for future generations.

Because our calculations shape the world, and the world deserves designs built not just with precision, but with intention.

The Hidden DNA: Manuals, Codes, and Standards

The hidden DNA of engineering doesn't just live in our standards or procurement models - it lives in us. We learn it early, through the manuals we study, the systems we work within, and the projects we inherit. At first, we don't even notice it. I never did. We follow the rules, deliver the outcomes, and trust the process.

Personally, this was especially true in my early career as an engineer. In the beginning, I was overwhelmed by the volume of things to learn. It was almost like a new language, with many, many new worlds.

It was a completely foreign environment, where there were few women (just 10% in civil engineering), after attending an all-girls school. Then, when I got into the workplace, the learning curve was steep too, so I felt like I had to do a lot just to tread water. My strategy was to cling onto what seemed most certain. I had to trust the process, the standards, the engineers around me, because if I didn't, it would have been too much to take in on top of everything else.

Somewhere along the way, I had a moment, or several, where the defaults stopped making sense. I know other engineers have experienced this too, for a variety of reasons, when we see that doing things the standard way doesn't always lead to the right result.

For me, those moments came early in my career, working in local government. Council work isn't always seen as the most glamorous side of engineering, but for me, it was life-changing. It was here I learned some of the most profound lessons about standards, value, and risk. They were the projects that taught me to look beneath the surface of "best practice", and start questioning what we really mean by that term.

Seeing Beyond the Standard

When I was a project manager at my local council in the city I grew up in, I was responsible for a portfolio of at least a dozen bikeway projects. They weren't the biggest or flashiest projects; one only had a budget of $50,000. They weren't record-breaking tunnels or major freeways, but they mattered to local communities. Each project connected a missing link in the network, helping people move safely and easily through their neighbourhoods, from local shops to playgrounds, and offering safer connections across the city.

One of the projects I inherited, in the northern suburbs of the city, was already in motion. The detailed design was complete, the drawings finalised, the tender priced and awarded. My role was simply to project manage and deliver construction. This was pretty straight forward and was seen as a "go through the motions" project. Still, the first thing I did was a site walk - something I always make time for, because engineers will know: paper plans never tell the full story.

As I walked the alignment, I noticed a small but curious detail. Along one section of the road, the new shared path would run beside a large empty parcel of land. On the site walk, I noticed the vacant land was house lots,marked out with survey pegs and with services installed, ready for houses to be built. On the drawings and aerial photographs, it simply looked like vacant land that didn't have much use. Between the shared bike path we were about to build and these housing lots, the design showed a low retaining wall - just a step to half a metre high, enough to make the shared bike path have the correct level and slopes. It was true to standard design, keeping the project within the council's land footprint, avoiding encroaching on private land, and following all the standard design rules.

At council, that was the usual approach. We always aimed to keep construction entirely on council land. Entering private property meant extra negotiations, approvals, and paperwork, which often took a long time to resolve and risked delays. The assumption was simple: stay within the boundary, and get the project completed as quickly and simply as possible. It was efficient, predictable, and safe.

But something about this section of the project didn't sit right. When I looked more closely into who owned the lots, I discovered they were earmarked for housing designed for people with disabilities. I imagined what that retaining wall might mean once those homes were built: residents arriving home to a small but unnecessary barrier at the edge of their property, a break between their front doors and the public space they were meant to belong to.

If my memory recalls correctly the retaining wall also went across where the future driveways and footpaths to the front door were going to be located, which meant a step where residents were supposed to drive their cars or access their home. To build the driveway and the footpath to the front door after our project was completed, the land owner would need to knock down the wall.

The design was technically correct for the bikeway - compliant, efficient, and standard. But it wasn't right for the people who would live there or anyone who would live there, disabilities aside.

I started asking questions: could we change the retaining wall to a sloped batter that gently transitioned between the footpath and the properties? It wasn't the usual approach, and it meant redesigning the drawings, reissuing documents, and asking the contractor for a revised price at a very late stage of the project. The timing was inconvenient, as we were only a week or two away from starting construction, and I was 23 - young, new, and aware that this was going to be a pain to deliver on.

But I couldn't let it go. Delivering this outcome that wasn't the best outcome, even if the other option wasn't standard.

I spoke with my boss and explained what I'd found and the approach that I thought we should take - talk to the land owner for their perspective then, if they land agreed, go down the redesign path and do a variation of the contract.

At first, there was hesitation. The contract had already been signed, and changing it meant more work for everyone. There was also pushback that this wasn't the standard approach, and a reminder that we keep the project within council land.

There was a view that altering the design to enter private land would complicate things and slow the project down. In this case, I could see that it would actually be the opposite: a simple change that would improve the outcome and possibly even reduce cost. The batter would be cheaper and easier to build than the retaining wall, and the landowner - in this case, the organisation building the disability housing - would benefit directly.

Eventually, I convinced my boss. I approached the landowner, explained the intent, and they agreed immediately. In fact, the land owner said they hadn't even picked this up when the previous project manager had let them know about the project. They were happy it had now been flagged and quickly agreed this change would be of benefit to them, so we could remove the wall and batter into their land.

Then came the difficult conversations: rework with the design team, contract rework with the procurement team, and discussing the change with the contractor building the project.

What could have been seen as a risk and a major headache for the project - stepping over the property boundary - turned into collaboration and a great outcome for the project, with cheaper construction and a great outcome for the land owner and future residents. It didn't derail the project; it made it work better. It was also a formative time. A new team had been put together to deliver the project, and it cemented a working relationship that showed we could resolve issues collaboratively.

It was a small shift - a few hundred metres of an earth batter replacing a concrete wall - but it changed how those residents could move and live in their space. That project taught me something that's stayed with me ever since: the standards tell us what's acceptable and compliant, but they don't always tell us what the right outcome is.

I know this is a tiny example in the grand scheme of engineering, and there are no doubt much more significant examples of this on mega infrastructure projects. However, I think the small local stories are just as critical, as they too affect and impact people.

Standing up for that decision wasn't easy, especially as a young woman, the only female engineer in the bikeway project management team, and the only female on the construction site. But it showed me that engineering isn't neutral. Every line on a drawing carries a value. And sometimes, all it takes to see the hidden ones is to walk the site, notice the people who aren't yet there, and ask: who is this design really for?

That experience, and many more like it throughout my career, changed how I saw design. I realised that "the standard way" was not the only way, and that sometimes real innovation happens when you question the default, even if it's something seemingly small.

The hidden DNA of engineering doesn't stop at design standards. It also lives in our systems, especially in how we procure and deliver projects. Procurement is where values become contracts and agreements. It's where we decide not just what to build, but why and how.

Redefining Value for Money

When I joined the Major Infrastructure Projects Office at my local council, I thought I knew what value for money meant. Like most young engineers, I believed it was all about the magic numbers: cost benefit ratios, budgets, and timelines, getting things done efficiently and proving you could deliver.

The reality surprised me at council, in a good way.

At the time, the council was undertaking one of the most ambitious programs ever seen in Australian local government - the Road Action! Program. It was a billion-dollar initiative to fast-track fifteen years' worth of road projects into just four.

I was 22 when I started there, less than two years out of university, and I was handed more than $20 million of road projects on the day I arrived. Within a few months, one of my projects wrapped up and I had a birthday, so at 23 years old, I was the Project Manager of $45 million worth of projects. This was my first real step into project management, and it felt like being thrown into the deep end.

It was a very exciting time to be at council. The Major Infrastructure Projects Office was a combination of top performers from across council, fantastic consultants, and new hires. Because of the scale and pace, we had to think creatively about how to deliver a huge amount of work. We were given this license to do things differently, with a focus on delivering great outcomes for the community, aligned with the council's strategy.

It is one thing to say this and another thing to actually do it. To council's credit they did align projects to the strategy and vision, and their procurement wasn't "lowest price wins". What impressed me most was that the council didn't see procurement as a race to the bottom on cost. They saw it as an opportunity to deliver better outcomes for people and the environment.

For every tender I reviewed, we used a multi-criteria analysis. Price was one factor, but so was the capability of the team, their track record with community engagement, their approach to sustainability, and the social or environmental legacy they could deliver. We assigned weightings to each category, and the results often surprised me. Time and time again, the contractor with the lowest price didn't win. The best overall outcome did.

It challenged a belief I hadn't even realised I was carrying: that "value for money" was purely financial. In practice, I saw that value was broader. It meant creating something the community could trust, that wouldn't harm the environment, and that would serve people long after the ribbon-cutting ceremony.

One project that stands out was Gap Creek Road - one of the last unsealed roads in Brisbane, which I was fortunate to be the Project Manager on. The original budget was close to $14 million, and I was handed the project on my very first day at council. The road connected the suburbs of Kenmore Hills and The Gap. Because it was a connector road, the default design assumptions could have pointed toward a higher-speed outcome.

This road went through a state forest, and the area was home to rich native wildlife. The community who lived there was deeply concerned that sealing the road would increase traffic speeds and harm the local fauna.

The easy path would have been to follow the standard: seal the road, create a clear path beside the road, set the speed, and move on. Instead, the team and the community started to question the brief, collaboratively.

The final outcome reduced the operating speed to around 60 km/h and introduced traffic-calming devices to further slow vehicles throughout the alignment through the state forest - an uncommon approach for this kind of local connector road in Australia. We also installed a series of fauna underpasses at key points to allow animals to move safely beneath the road. To preserve as many trees as possible, the project team walked the 5 km stretch of road with the community and an arborist to identify each tree marked for removal.

The design team went above and beyond to modify the alignment to reduce the impact of the road footprint on the forest. Some of these decisions added costs and complexity, but also resulted in savings on the project. Ultimately, this co-created design approach reflected what the community valued: safety, stewardship, co-creation, and co-existence.

There was a point of contention with the community on the speed calming devices, as they needed to be lit for safety. We explored many options and workarounds, but ultimately lighting was installed.

What made this project even more powerful was that, alongside these values-driven decisions, we also found innovative ways to save money and time without compromising on quality. Normally, when constructing a road, you build up the pavement in layers of varying depths to create a strong foundation for the dynamic loading of vehicles travelling along it. Each layer requires separate placement, compaction, and testing to a strict standard. For Gap Creek Road, that meant several layers, including two 300 mm layers for the base and sub-base - a standard design to save money on materials, with the lower layer being a cheaper material.

One of the team proposed an unconventional idea: instead of two separate layers, what if we used a single 600 mm layer of the higher-quality material all the way through? It meant paying more for the base material, but it would create a huge efficiency as the material would only need to be leveled and graded once, not twice. This would halve the required testing, save significant time, and minimise how long we had to close the road.

It was a classic example of innovation and creativity meeting practical wisdom. The design wasn't cheaper because we cut corners; it was cheaper because we thought differently.

For this project, we made the difficult decision to close the road to traffic for several months. While it was going to have a big impact for local road users, this would also mean we wouldn't need to fell more trees to create service roads during construction to let vehicles through. There were a lot of other positive impacts as a result of the closure.

Overall, the project finished earlier, cost less, and the community got a safer, more sustainable road that respected their values. From memory, the final project cost came in at under $10 million - a huge saving against the original budget of $14 million.

Working on that program changed how I think about projects: genuinely involving stakeholders to co-create and create better outcomes, and understanding where the real value lies. I'd assumed councils were slow, bureaucratic, and obsessed with red tape. But what I found was something entirely different: a team of people who cared deeply about getting it right - not just for the budget, but for the people who would live with what we built.

When done well, it's not about extracting the best deal commercially. It's about investing in the best outcome for everyone. It's about seeing the tender not as a transaction, and community consultation not as a check box, but all of this as a design tool - one that encodes our values before a single shovel hits the ground.

That experience reframed how I thought about value too. I realised that "value for money" is only as meaningful as the values you measure it against.

Learning What Risk Really Means

When I started working at council as a Project Manager, I had to unlearn and think about risk in a whole new way. It was the first time I'd stepped beyond pure engineering thinking into project management thinking - and that shift changed how I saw everything.

My training to this point had been a civil/structural engineer degree, work experience on a high-rise construction project while at university, and almost two years in a very technical structural engineering role designing parts for water treatment plants and the resources sector.

As an engineer, risk had always meant technical risk: whether a design would meet code, a structure would hold, or a pavement would perform. But as a Project Manager, I started to see a whole different layer: people, politics, timing, and trust. I learned that some of the biggest risks on projects weren't about design failure at all - they were about stakeholder failure.

I remember one of the early meetings that I had with a local resident adjacent to one of my projects when I was working for council. I felt excited to meet her and share the project that I had the opportunity to lead. In my mind, I was prepared. I had memorised all the facts and figures - how long the project was, how many trees we were planting, and the technical pieces like the amount of concrete being poured.

As I drove to the site with a communication team member with me, I excitedly said, "What do you think the residents are going to ask about?" In my mind, I was thinking: will it be the amount of concrete, the trees to plant, some other design feature? I was convinced the comms person would say a technical project question like that, something that checked off one of the facts I knew.

The very kind communications person held back a laugh and said, "Oh no, the residents will just care about when their driveway will be open and closed." I felt like a deflated helium balloon. I was almost shocked and realised I hadn't even thought about how to answer that question. Luckily, we had time before the meeting for me to prepare and work out when their driveway would be closed.

That was a really eye-opening moment for me. I realised, as someone thinking from an engineering perspective all time, just how far off of a mark I was when it came to engaging with stakeholders and the community. It was a huge wake-up call to realise people didn't care what I cared about; they cared what they cared about. And that moment has never left me.

This brought a whole new lens for me on project risks.

Every project involved engagement with people who could make or break the outcome: residents, local businesses, and large organisations like utilities and infrastructure providers like Energex or Queensland Rail. On paper, we had standard procedures for engagement - set lead times, fixed communication steps, boxes to tick. For example, if we needed to relocate services like power poles or water mains, the rule of thumb was that those works took about twelve months, so we'd engage stakeholders about that far in advance.

But I started to wonder: what if we went earlier? What if, instead of waiting for those formal lead times, we reached out sooner and started the conversations before the timelines told us to?

My colleagues thought I was overthinking it. "Too early, a waste of time," they said. "It'll change anyway." But I decided to give it a go.

And it worked. By engaging earlier, we had smoother conversations, better coordination, far fewer surprises down the track, and I could invest time in building genuine relationships with stakeholders. I learned that communication wasn't an admin task or a box to tick - it was a form of risk management.

One project really brought that lesson home. We were widening a local intersection to improve safety, and to do it we needed to resume land: fourteen houses in total. Two had already been acquired, and I was responsible for negotiating the remaining twelve. I was 23, and the thought of walking into someone's home to tell them the government was resuming their property was terrifying.

My manager at the time, Mary, was incredible. "I'll show you the first one," she said. "Then you can do the others."

We sent letters to the residents, called to arrange meetings, and went out together with the land valuer and a communications officer to explain the process. Those conversations were some of the hardest I've ever had.

The valuer had a smart, empathetic approach. He said, "What if we offer people a price they're happy with - one that's generous and fair, slightly above market value rather than forcing them to fight it out in court?"

He did his job well and researched, looked at comparable house prices in the area, and recommended offering around $10,000 to $20,000 above fair market value.

At first, that seemed counterintuitive to me. We were spending more money than we technically had to. But when we looked at the bigger picture, it made perfect sense. If even one of those residents took the case to court, the legal costs alone could reach $400,000 to $750,000 per house and 6 to 12 months of time to complete the project. This would mean funding the project team for an additional 6 to 12 months, which would cost even more. The project would stall, tensions would rise, and the community's trust would fracture.

We decided to take the risk - or rather, to redefine it.

We offered fair prices. We communicated openly and respectfully. And something remarkable happened: not a single resident took us to court. All twelve properties were resumed peacefully, and as far as I am aware, this was the first time in the entire Road Action! Program that had ever happened.

It was a huge milestone - not just for the project, but for me personally. I realised that the biggest risks are often the ones we can't quantify on a spreadsheet. The risk of losing community trust. The risk of conflict. The risk of treating people like variables and numbers instead of humans.

That project taught me that stakeholder engagement isn't a "soft" part of engineering or a checkbox in a project plan - it's hard risk management. The kind that can save hundreds of thousands of dollars, months of time, and immeasurable goodwill. One that can leave a community better off after a project.

Since then, I've looked at risk differently. It's not just what might break, but who might be broken by what we build - or by how we build it. True engineering risk isn't about avoiding failure; it's also about protecting trust, relationships, connection, care, and legacy.

Looking back, these projects shaped how I think about engineering more than any technical course or design manual ever did.

They taught me that standards can be questioned, that value is multidimensional, and that risk includes the human experience - not just the technical one. They showed me that when we expand what we measure, we expand what we make possible.

That's what I mean when I talk about the Future of Engineering. It's not just about innovation or technology. It's about consciousness: knowing that every decision, every design, every policy, carries values that ripple far beyond the project itself.

Invisible Curriculum: How We're Trained to Think

Before an engineer ever steps onto a project site or signs off on a design, something deeper has already been written into us. It's not in a textbook or a policy manual. It's not even in our design codes. It's invisible: a hidden curriculum of habits, values, and assumptions we absorb long before we realise we're learning them.

Like DNA, this invisible curriculum carries the cultural instructions of our profession. It tells us what to value, what to fear, what to ignore, and what to aspire to. And, like DNA, it gets replicated and passed quietly from one generation of engineers to the next.

What We Learn Before We Know We're Learning It

When I first thought about engineering, I thought it was all about sitting down and doing calculations. In my first year at university, it felt like that was all that mattered: the maths, the physics, the formulas. The people who could look at a page of equations and see the answer instantly seemed like the "real" engineers.

I worked hard, but I struggled. It took so much energy for me to get through those core maths and physics subjects. I began to believe that maybe I wasn't cut out for engineering, that perhaps the way my brain worked - creative, curious, people-oriented - didn't fit the mould.

When I started working, I saw those same unspoken rules play out. The people who were rewarded in my early career weren't necessarily the ones who could build strong teams and relationships or engage meaningfully with the community. It was the ones who were technically brilliant - the ones who could solve the technical problem fastest, and deliver calculation after calculation and report after report.

This wasn't just a workplace issue; it was systemic across school education. Our education system trains us to see the world in binaries - right or wrong, pass or fail, success or mistake. At school and university, we learn to work individually and to prize certainty. In engineering, that culture becomes amplified: if the numbers don't add up, the bridge might fall down. It's serious, of course - but the side effect is that we learn to avoid uncertainty altogether in all aspects of our projects.

The culture silently teaches me: don't speak unless you're sure. Don't question unless you already know the answer. And don't ever admit you don't know.

For me, it went deeper. Often, I was the only woman in the room, and that made me even more reluctant to speak up. I didn't want to be "that woman" who asked a silly question or got something wrong. The invisible curriculum whispered that safety lay in silence.

In some organisations, this conditioning went beyond silence into shame. In one place I worked, there was a whiteboard on which people's mistakes were written for everyone to see. Once, someone had misspelled "consultants" in a report - and the error was posted on the board as "cons-slut-ants." That word stayed there, a public joke - a scar on the wall that said: don't ever get it wrong or you will be humiliated. Unfortunately, I can share many examples of this from my career, where people, including myself, were shamed for making mistakes.

It's hard to feel safe, creative, or connected in a culture like that.

The Cultures We Inherit

Every profession has its unspoken codes - the values you absorb not through lectures or policies, but through what gets rewarded, noticed, and repeated. These unspoken rules are what form engineering's invisible curriculum, the mechanism by which our professional DNA is passed on.

Over time, they've evolved into distinct cultures - shared ways of seeing and behaving that shape how we design, lead, and learn. Each one started for a good reason, but taken too far, they silently limit what engineering can be.

I still remember my first job after university as a structural engineer.

One of my first tasks was to design a simple pad footing for a substation - something I'd never done before. I asked around to see who might know how, and one of the draughtees kindly shared a past drawing with me. It was the perfect starting point: I could see what it looked like, understand the standards, and begin to figure it out.

I was proud of myself. I'd sought help, found a solution in a creative way that was efficient, and learned something new without needing someone to do it for me. But when I showed the design to my manager, he was furious. "You shouldn't copy other people's designs," he said. "You need your own design style as an engineer."

I remember feeling stunned. What I'd thought was initiative was seen as weakness. What I'd thought was collaboration was treated as failure.

A few weeks later, he asked me to design a retaining wall - something I didn't remember doing at university. So, I went to the geotechnical team and asked for help. They sat me down, patiently explained the basics, and I got to work. I completed the design, checked my calculations, and felt confident in the result.

I handed the design to my boss and went on holiday. When I returned from leave, a fellow graduate pulled me aside in secret and said, "Our boss asked me to redesign your wall to check your work and I got the exact same answer. I'm not sure why he wanted me to redo your design, but I thought I should just let you know."

It was confirmation that my design was right, but it didn't feel like a win. Why did my boss go behind my back like that? Didn't he believe me?

That experience taught me more about the culture in some workplaces than any engineering principle ever could. It showed me how easily something can be mistaken for incompetence, how asking for help can be misread as weakness, and how fear can silence learning.

That manager also once told me I'd never be a good engineer because my parents weren't engineers. When I heard this, I knew he wasn't someone to role model my career on. I cried in more meetings in front of him than I care to remember. Eventually, I spoke with HR and they moved me to another office, and I never saw him again.

That experience wasn't just about one manager or one workplace. It was a glimpse into something more. It revealed the culture we've inherited as engineers, one that can show up in some workplaces. It's a culture that prizes precision, knowing, and control. It's the invisible curriculum of our profession, shaping how we think, speak, and even feel long before we realise it's happening.

And while that culture may have helped us build systems of extraordinary reliability, it also built systems of silence - places where people fear mistakes.

This culture wasn't designed by any one person. It evolved out of the responsibility engineers carry to protect lives. But when taken too far, it constrains the very creativity and collaboration that the future now demands.

That's why we need to name it. To see the cultures we've inherited - not to blame them, but to understand them, and to reimagine them.

When we can see the invisible, we can begin to change it.

Before we can reinvent the culture of engineering, we first need to understand the one we've inherited. I know that it isn't the same everywhere. There are a few extraordinary workplaces where collaboration, curiosity, and care are already part of the culture - places where engineers feel safe to question, supported to grow, and trusted to lead. I've worked with some of them, and they give me enormous hope.

But for many of us, the invisible patterns of the profession - the ones passed quietly from generation to generation - still shape how we think, speak, and show up. They're not written in policy manuals or design codes. They live in habits, assumptions, and unspoken rules we absorb without realising it.

These patterns aren't anyone's fault. In fact, most of them were born from good intentions: to keep people safe, to ensure rigour, to deliver with precision. But when they become over-expressed genes in our DNA, and when they start to define what's "right" in every situation, they can hold us back from creativity, inclusion, and innovation.

So, rather than judging these cultural codes, let's bring them into the light. Let's see them clearly enough to choose what we want to carry forward, and what we're ready to evolve.

Because the culture of engineering is not fixed. It's a living system, and each one of us is its authors.

1. The Culture of Certainty

Engineering teaches us to find the right answer. Our entire education system reinforces this.[47] There's a correct solution, and if you calculate it correctly, you're safe. You pass.

That culture of certainty is not just vital; it's essential when you're designing a bridge or a power network that people need to rely on. Lives depend on precision.[48] But it can also make us uncomfortable with ambiguity, conversation, or dissent. It can make us believe that uncertainty equals weakness.

That was certainly my experience - especially early in my career.

The Moment I Stopped Asking

I'll never forget sitting in one of my first engineering design classes. We were designing a timber building. I'd done physics at school but never any engineering subjects, and I didn't know a single engineer personally.

It felt like stepping into a completely new language. Every lecture was full of strange words that seemed to multiply daily - joists, studs, trusses.

One day, after hearing these new terms, I turned to the student next to me and asked, "Hey, what does a joist and a stud actually mean?"

He looked at me, wide-eyed, as if I'd just asked something outrageous. Then he rolled his eyes and said, "Don't you know?"

Well, of course I didn't. That's why I was asking. But he didn't answer. He just turned back to his work and ignored me.

I remember sitting there, stunned and embarrassed. My face felt hot. Was I really an idiot for asking that? At that moment, I decided it was safer not to ask next time.

That single reaction shaped something in me - a lesson that asking wasn't safe, that competence meant always knowing, that questions were a sign of weakness.

And once that belief takes root, it can take years to unlearn.

Years later, I joined a new team as a water engineer and was asked to manage the design of an earth bund dam for water storage. My manager walked me through how it worked, pointing at sketches and talking through layers of clay, compaction ratios, and spillways.

Inside, my mind was racing. I didn't understand half the terms he was using. But instead of asking, I nodded along.

All I could think was, "I should know this by now."

When the meeting ended, he said, "Does that make sense? Great. Off you go."

And I walked out with a notebook full of half-understood scribbles, too afraid to ask anyone else for help - because asking might expose that I didn't belong in this new team.

That's what certainty culture does. It trains us to fear being seen as beginners. It makes learning feel dangerous. It teaches us that "I don't know" is a phrase to be avoided, rather than a doorway to understanding.

Unlearning the Lesson

It took me years - and some great mentors - to realise that uncertainty isn't the opposite of expertise. It's part of it.

The most innovative engineers I know are the ones who ask the best questions. They're curious - not because they don't know, but because they understand how much there still is to discover.

While our work depends on precision, innovation depends on curiosity. When we confuse certainty with safety, we close the very door that leads to innovation.

This is why so many engineers hesitate to speak up. The invisible curriculum whispers: if you're not certain, stay quiet.

But the most transformative ideas - the ones that make the world safer, stronger, and more beautiful - often begin with uncertainty, and with the courage to ask: "What if we're wrong?"

The next generation of engineers deserves a new definition of certainty - one that isn't about knowing everything, but about knowing how to learn.

The best engineers aren't the ones with all the answers. They're the ones brave enough to keep asking questions and leading the best people to discover the answer.

2. The Culture of Risk Aversion

Closely tied to certainty is risk aversion.

We're trained to eliminate risk wherever we can, and for good reason. People's safety and their lives depend on it.[49]

But risk in engineering has come to mean only technical risk: safety, cost, schedule, compliance. We've learned to quantify and control it, to assign numbers and probabilities.[50]

What we haven't learned as deeply is to see the other kinds of risk: the social, environmental, and human ones. The risk of disconnection, apathy, and not listening.

True risk management should ask not only, "Will it stand up?" but "Will it hold up - for people, for place, for generations?"

Yet our inherited DNA tells us to measure what's easiest to count, not what's hardest to care for.

One day I took a risk that had nothing to do with engineering drawings or safety factors.

At the time, I was volunteering with the Women in Engineering Queensland Committee through Engineers Australia. I was asked to attend a meeting with the Queensland Government. Around the table sat women from every corner of STEM (science, technology, engineering, and maths) - women in construction, IT, mining, and science.

We were there to talk about how to get more young people, especially girls, into our professions.

I remember walking into that meeting feeling completely out of my depth in that government boardroom. I'd never done anything like that before, and the committee was trusting me to represent them. I didn't know anyone there, and the conversation wasn't about structures or systems - it was about people, policy, and making change.

This was well outside my lane.

At that stage of my career, I was still the kind of person who found it hard to put my hand up or share an idea, or even announce my name and role in a meeting without going bright red. I was afraid of saying the wrong thing - of what others might think.

But as I listened to the discussion, I felt something shift. I'd read so many reports about the lack of diversity in engineering, so many papers about the "problem", but rarely saw anyone taking action.

Before I could overthink it, I put my hand up and said, "I'll lead this project."

The words came out before I could stop them. And to my surprise, the government official smiled and said, "That sounds great - and we've got some funding to support it."

I walked out of that meeting exhilarated and terrified. I had no idea where to start.

Stepping Into the Unknown

The team of women in STEM and the Women in Engineering Committee for Queensland decided to create a one-day event for high school students to experience engineering - to show them what it really was. But I knew nothing about education, and at that time, STEM wasn't even a common acronym.

The first thing I did was call my old university lecturer, Professor Les Dawes, who I remembered had worked with schools before. I didn't think he'd even remember me, but he did. He was warm and generous, and invited me to meet at QUT to brainstorm ideas.

What I discovered that day was one of the greatest lessons of my career: There's a huge advantage in not knowing how something is "supposed" to be done.

Because I wasn't an expert in education, I was open. Open to ideas, to collaboration, to possibility.

Together, the women in STEM and I decided to work with Year 9 and 10 students, before they chose their senior subjects, and to make the event specifically for girls.

Within three days of promoting the event, even though it was the start of the school year and the event was only a few weeks away, it sold out. By the end of the week, we had double the number of registrations. There was so much interest, we had to put on a second event.

I was shocked - not just by the response, but by how natural it felt to lean into the unknown.

In engineering, I'd been trained to think about what could go wrong and how to mitigate it. But here, because I didn't know what could go wrong, my mind was open to what could go right.

That openness invited others in. Dozens of ideas, workshops, site tours, speakers, and volunteers stepped forward to make it happen

- teachers, engineers, students, sponsors. Together, we created something none of us could have done alone.

That first event became the seed of Power of Engineering - a movement that would go on to inspire tens of thousands of students across Australia.

All because one day, I decided to take a different kind of risk - the risk of saying yes to something uncertain.

What surprised me most wasn't the outcome - it was how much I changed. For the first time, I had a safe space to play, to experiment, and to bring all of myself - my creativity, my people skills, and my leadership - into an engineering context. There was no risk of failure, no fear of getting it "wrong." Just a sense of purpose.

As Power of Engineering grew, so did my confidence to put people, purpose, and care at the centre of problem-solving. I realised I didn't need to know everything. I just needed to bring people together and trust that we could figure it out.

Redefining Risk

That experience changed how I saw risk forever.

Yes, engineering will always be about managing physical safety - that's non-negotiable. But there's another kind of safety that matters too: psychological safety. The safety to try something new, to raise your hand, to be wrong, to learn.

The irony is that our obsession with eliminating risk can sometimes make us riskier in other ways. We miss opportunities, silence innovation, and trade curiosity for compliance.

When we broaden our definition of risk to include the risk of not acting, not asking, not caring, we start to see differently.

The safest systems are not the ones that never fail, but the ones that can adapt. And the safest engineers are not the ones who never make mistakes, but the ones who are open and keep learning from them.

Real resilience isn't built on fear; it's built on courage.

3. The Culture of Neutrality

Engineering carries a proud legacy of objectivity - the belief that we design without bias. We often describe ourselves as "value-neutral", guided by evidence, logic, and data.[51]

But neutrality is not the same as fairness or equity. When we design for the "average user", default to "standard practice", or speak of "optimising" without asking for whom, we reveal the values hidden inside our neutrality.[3]

Some of that evidence, logic, and data was built on biases, because it was designed by a certain group of people, for a certain group of people, at a certain time.[52] Slowly, we're seeing this change, but much of what we build still rests on standards written decades ago.

I first realised how much this culture ran on one of my earliest site visits, not long after graduating from university.

That morning, I'd worn a simple knee-length dress to work, not realising I'd be heading out to an oil refinery later that day. The site required full flame-retardant overalls, so I was handed a pair from the equipment cupboard. The only size available was a men's medium.

Now, I'm not a very tall person - about 160 cm - so when I pulled on those overalls, I quickly realised I didn't have to worry about my dress showing. The crotch hung somewhere just above my knees.

It was comical, but it was also telling.

I wasn't the size of the "standard man" those overalls had been designed for - and I wasn't alone.

At the time, I remember feeling a mix of frustration and disbelief. Safety gear was supposed to protect us, yet it felt unsafe - loose, bulky, easy to trip on. My high-vis vest was so big it turned into a kind of glowing tent that caught on handrails and corners.

Meanwhile, I was secretly jealous of my friend Peta at university. Her mum could sew, so she would take her high-vis shirts home and tailor them - adding darts, shortening the sleeves, chopping off the excess length. Her shirts actually fit. Mine flapped in the wind like a flag.

We'd laugh about it, but underneath the humour was something more serious. The gear we wore every day had been designed for bodies that looked nothing like ours. It wasn't just inconvenient. It was a reminder of who the system had been built for, and who it had forgotten.

Safety wear has improved a lot since then, but the lesson has stayed with me: "standard" is never neutral.

Engineering loves the idea of the "average". We design for the typical user, the median load, the standard condition. It gives us stability and predictability.

But "average" is a myth - and designing for it often means designing for nobody in particular.

When we assume neutrality, we erase the very people who will live, move, and breathe inside what we create.

We measure what's easy to count - strength, cost, efficiency - but forget to measure what truly matters: dignity, access, belonging.

And that's the irony of neutrality. The moment we declare ourselves neutral, we stop seeing the people who fall outside the average.

Designing With, Not For

Since then, I've learned that every design is an ethical act. Every calculation, every assumption, every standard carries a worldview inside it - the invisible fingerprints of the people who wrote it.

The question isn't whether we're neutral. It's whether we're conscious.

The most meaningful projects I've worked on since are the ones that began not with data, but with dialogue - sitting down with the people who would use, live in, or be affected by what we were building.

When we design *with* people instead of *for* them, something shifts. Neutrality gives way to nuance. Assumption gives way to empathy. And "standard practice" becomes living practice grounded in care, not just compliance.

Every choice in engineering - from what we design to what we measure - carries values. Pretending we're neutral doesn't make those values disappear; it just makes them invisible.

4. The Culture of Hierarchy

Then there's hierarchy - a culture that shapes not only how we design, but how we think, act, and behave.[53]

From our first day in the profession, we're taught to respect authority. The client sets the brief, the manager reviews the work, the senior signs it off, and usually there's an extensive verification process. It makes sense - it's how we keep things safe and accountable.

But it can also discourage questioning.

Many engineers I've mentored have told me they've wanted to challenge a design decision or ask "why", but didn't feel it was their place. I've been there too - sitting in meetings where I had a thought that might have made a difference, but the unspoken hierarchy kept me quiet. I'd think: they're more experienced than me; they probably know better.

This culture tells us that our job is to deliver, not to disrupt - especially when there's pressure to deliver for the client, and every minute is a billable hour. I'd often think: don't be difficult. Don't make extra work for everyone.

But progress depends on those who dare to question the brief, who are willing to ask, "Is this still the right way?"

I have worked in several organisations where it was not acceptable to speak with peers in another team. I had to talk to my boss, who would talk to their boss, and then, once I had permission, then I could talk to my peer. This always seemed ridiculous, inefficient, and the opposite of collaboration, yet it is how many organisations operated, with hierarchy and silos embedded in their culture.

When Hierarchy Becomes Habit

Hierarchy was installed early in my career as something sacred. As a young structural engineer, the only person I was allowed to speak to was my boss. I wasn't permitted to call or talk directly to the client. That was considered "out of line".

Our job was to sit quietly, do the calculations, and pass the work up the chain. The message was clear: stay in your lane.

I remember how that culture began to shape me. I'm a naturally outgoing person, but over time, the repeated message made me hesitant. Clients - the very people with the answers to my design questions - started to feel off-limits.

I saw the same dynamic play out across teams. People were scared to talk to clients, scared to ask questions, scared to break the chain of command.

And yet, it made no sense. The clients often had the information we needed to make better decisions, to clarify design assumptions, to solve problems faster.

Still, the rule was the rule: go through your boss.

I can't count how many times I went directly to the person who had the answer in a range of different workplaces, only to be pulled aside later and told off.

"No," my boss would say. "You should have come to me first. I'll talk to their manager, and once we've had that conversation, then you can speak to them."

It felt ridiculous - a process designed to protect authority more than progress.

On reflection, I can understand why some people worked this way. They wanted to control the situation, and make sure the relationship with the client wasn't "ruined".

That lack of direct communication and collaboration slowed projects down, blocked learning, and created silos.

The irony is that engineering is built on interconnection - systems within systems. Yet too often, the way we organise ourselves makes those systems brittle.

Redesigning Authority

Hierarchy in engineering was built with good intentions: to keep people safe, to ensure accountability, to maintain quality. And those intentions still matter.

But there comes a point when hierarchy stops protecting us and starts constraining us. It discourages the very dialogue and diversity of thinking that innovation requires.

When everyone in the room feels safe to ask a question, share an idea, or challenge an assumption, the quality of our work rises. Hierarchy should serve safety, not silence it.

In reality the best designs, and the best decisions, come from collective intelligence, not command and control.

If we want to build systems that are adaptive and resilient, we must build cultures that are too. That means replacing hierarchy with trust, rank with respect, and authority with shared accountability.

One of the most powerful questions in engineering isn't "Who's in charge?" It's "What can we learn from each other?"

5. The Culture of the What and How, Not the Why

Potentially the most embedded strand in our engineering DNA is the culture of the what and the how.[54]

Ask an engineer about their project and they'll describe, often with pride and precision, and with a huge amount of detail, exactly what it is and how it works: the design, the process, the material, the code. I could happily talk for hours about the cleverness of a structure, the elegance of a bridge, or the impactful magic of a new technology. I am that person on holidays dragging friends and family along to marvel at the local engineering.

But what's often missing from our training - and too often from our daily work - is the why.

Why does this matter?
Why will anyone care?
Why are we building this?
Why now, and for whom?

It's the why that connects our work to people and purpose. It's the why that turns projects into legacies. Yet we rarely talk about it. The efficient engineer in me wonders if this is because the why can seem a pointless pursuit and a waste of time. Yet, I think it is the most important thing we do as engineers. For successful projects, we are always solving a why.

However, all too often we forget the why, and we risk building a world that works perfectly on paper and fails for people and planet.

Finding My Own Why

The lesson of why came alive for me early in my career, long before I could name it.

In my first year at university, I almost quit engineering. I couldn't see the point of it all - the endless maths, the physics, the theory. It was very dry and I remember thinking, why am I learning this? What does this actually mean? What is it all for?

At the end of first year, a friend heading into his final year, Cameron Hately, encouraged me to hang in there. "It'll make sense when it becomes real, when you apply it in second year."

And in second year, it did.

We had a subject where we had to design a housing estate. I know, not the most glamorous project, but it was our first real taste of engineering applied to life. We went out to survey land in Pullenvale, Brisbane, to plan roads and decide where the houses would go.

This turned into being a life changing, "ah-ha" moment. I learnt that if you make roads wide and straight, people drive faster. If you design in curves or even just change the line marking to make roads feel narrower, you can actually slow people down.

It was a revelation. I realised that engineers don't just build things. We shape, and can literally change, people's behaviour. Engineers create the world people live in, often without them even realising it. Engineering wasn't just about structures; it was about behaviour and shaping society. We weren't just designing for safety or efficiency; we were creating how people moved, lived, interacted with, and experienced the world.

At this point in my studies, I had found the engineering course extremely challenging. I'd failed first year maths and materials engineering and dropped a subject. Now I had something powerful: an insight into the impact I could have on the world that was worth working for.

For the next three semesters, I overloaded on subjects, taking five each semester to catch up, and I graduated with my peers. I had to get special permission from the Head of School to do so. I have no idea how I convinced him in our meeting, but I am very glad he believed in me.

That moment became the why that carried me through my degree. I realised that engineering could change how people experienced the world - that it was a social force as much as a technical one. I realised that by becoming an engineer, I could change the world.

When attending the retirement party of Professor Les Dawes, I realised that not only did he help start Power of Engineering, but he also taught that second year subject where I had my "ah-ha" moment. As I prepared to speak at his farewell, I reflected on how much his teaching had shaped me. He had not only taught that subject, but more importantly made me feel like I belonged. He often reminded us that engineering was a socio-technical discipline - equally as social, ethical, and human as it is mathematical and physical.

It's amazing how one person can have such an impact on you and rewrite part of your DNA.

The Power of Engineering

When I co-founded the national not-for-profit charity Power of Engineering, I wanted to help others find that same spark - an "ah-ha" moment or their why - especially young people who didn't yet see themselves in engineering. I wanted to change those numbers in the reports I kept reading about the lack of women and diversity, including diversity of thought, in our profession.

I chose the name deliberately: The Power of Engineering. I wanted every student to understand that they have the power to change the world with engineering.

We ran hands-on workshops for students for girls and regional students in Years 9 and 10, and I watched their faces light up when they discovered what engineering really is.

They learned things like how the electricity network powers their homes and how coded text messages work using binary zeros and ones. They built water filters from everyday materials and watched dirty water turn clear. They designed cities for the future. They got a tour of the A380, from first class to changing the tires and inspecting the engines. They visited tunnel management centres and construction sites, seeing the heartbeat of a city from behind the screens - the people and systems keeping others safe.

Attending many of these events over the years, my favourite part was being with the students and seeing the shift happen right there before my eyes. They started to see a why: why doing maths and physics at school mattered, why engineering can be an incredible career, how it can change the world and be a pathway to solving any problem you care about.

It wasn't just about formulas or structures anymore - though those are the building blocks and the tools. It was about people, possibility, and purpose.

At the start of our one-day events, most students told us they'd never considered engineering. By the end, over 70% said they'd changed their minds - that they would now consider a career in engineering.

That transformation wasn't about selling a career. It was about giving them a connection to why engineering matters and changing the diversity of the people in our profession.

Every one of those students was offered the same lightbulb moment I'd had years earlier: the moment when engineering stops being abstract and starts being alive and relevant to the real world.

Why the Why Matters

That's the power of the why. It can change everything.

When we rediscover the purpose behind our work, we don't just perform better. We become more creative, collaborative, motivated, resilient, and courageous. We think more systemically, design more humanely, and remember that the work we do is in service of something - or someone - bigger than ourselves.

Our invisible curriculum has taught us that the what and the how are enough - that if we can calculate and deliver, we've done our job. In actual fact, the why is what keeps us connected - to the people who use what we design, to the communities we serve, and to the legacy we leave behind.

When we lead with why, we build more than projects. We build possibilities for others and ourselves. It is that connection to why and my purpose that has kept me motivated in the most challenging times of my life and my career.

Rewriting the Code

Each of these cultures - certainty, risk aversion, neutrality, hierarchy, what/how thinking, and silence - are not flaws. They were born from good intentions: safety, accountability, rigour, control. But they've become overexpressed genes in our professional DNA.

To evolve, we don't need to erase them. They can still be important and helpful in our work. However, we do need to rebalance them. To complement certainty with curiosity. Risk aversion with courage. Neutrality with empathy and equity. Hierarchy with collaboration. And the what and how with a deeper, enduring why.

That's what it means to start rewriting the DNA of engineering - to build not just structures, but a new way of seeing, thinking, and caring in the work that we do.

What We Need to Unlearn

Every profession has its culture. Engineering naturally trains us to be risk-averse, methodical, and exacting - qualities that are vital when public safety depends on our work. But when that risk aversion extends beyond safety into imagination, compassion, or dialogue, it limits us.

I still see this today with many engineers starting out. They sit quietly in meetings, waiting for permission to speak. They're afraid to say something unless it's perfect, because that's what the culture has taught them. They've inherited the same DNA I did - one that prizes technical certainty over human insight.

But this is changing. More and more engineers are beginning to question the brief, to bring curiosity and empathy into their practice. They're realising that risk isn't just about failure; it's also about missing what matters.

Every step of my own journey - from being a quiet young graduate to leading teams on major infrastructure projects and co-founding multiple businesses - has been a journey of unlearning that inherited DNA. Unlearning that safety comes only from certainty. Unlearning that questions are a sign of weakness. Unlearning that value is measured only in cost or time.

When we unlearn those invisible lessons, we begin to write new ones - lessons that see engineering as a living, evolving practice that includes care, connection, and creativity.

This is how we start to rewrite the DNA of our profession.

Leveraging Language to Change our Hidden DNA

While it might seem like a subtle change, shifting the words we use and thinking about rethinking what we really mean by risk can have a profound impact.

Psychology tells us that the language we speak shapes how we perceive reality.[55]

Economist Keith Chen discovered that Chinese speakers, whose language does not separate past, present, and future tenses, are better at saving money and maintaining their health than English speakers. In Chinese, you say "it's raining" for yesterday, today, or tomorrow, so health and wealth are understood as part of the now. In

English, we say "it rained", "it's raining", and "it will rain" - subtly placing the future at a distance.[56]

Another striking example comes from research into colour perception and language. Cognitive scientists have shown that the words a language gives us don't just describe the world: they actively shape how we perceive it.

In a well-known study led by psychologist Jonathan Winawer, working with researchers including Paul Kay and Terry Regier, Russian and English speakers were asked to distinguish between different shades of blue. In English, we use a single word - blue - to describe a wide range of shades. Russian, however, has two basic colour terms: goluboy for light blue and siniy for dark blue.

What the researchers found was remarkable. Russian speakers were significantly faster and more accurate at distinguishing between shades that crossed the goluboy-siniy boundary than English speakers were. This wasn't about eyesight or intelligence; it was about language. Because Russian speakers had distinct words for these colours, their brains treated them as more distinct categories.

Even more compelling, the effect appeared early. Children learning Russian showed the same perceptual advantage. And when researchers disrupted participants' language processing with a verbal task, the advantage disappeared, suggesting that language was directly involved in how the colours were being perceived, not just labelled afterward.[57]

In other words, Russian speakers don't just talk about blue differently. They see blue differently.

This research helps us understand something profound: perception is not purely objective. It is trained. It is shaped by the categories, distinctions, and assumptions we inherit through language and culture. The words we use, and the ones we don't, train our attention. They determine what we notice, what we separate, and what we collapse into the background.

When we change language, we don't just change conversations. We change what becomes visible. These studies show us that language doesn't just describe the world; it creates it.

And that matters for engineers too because the terms we use in briefs, standards, and design criteria don't just describe reality. They shape what we are able to see, measure, and ultimately design.

When we learn to talk about "efficiency", "throughput", "optimisation", "risk mitigation", or "value for money", those words shape what we see and what we measure. They train us to see what fits the model and to overlook what doesn't.

When we use words like "social license to operate" rather than "legitimacy", we are doing something to someone rather than proving public trust. The same can be said for consultation and co-design. I've often found that consultation means "we have a design and I will tell you about it", while co-design is means collaboratively creating something.

That's why I believe we need to decode the language of engineering - to look at the inherited DNA that lives inside our words before we even begin the conversation. Because if language shapes perception, then our engineering vocabulary is literally shaping the very world we build.

I think it is exciting to consider this in the context of how we can use the power of language to rewire the Future of Engineering.

Mapping the Current DNA

Every time we open a design standard, an engineering report, a procurement template, or a risk register, we're not starting from scratch - we're unfolding a set of inherited values. These values are the genes of engineering's current DNA. They shape how we think, what we prioritise, and what gets left out. To change the future, we need to map these traits so we can recognise them, challenge them, and rewrite them.

Making the Invisible Visible

Consider a camera lens. Zoom in and you will see the rivets, the calculations, the beams. Zoom out and you'll see the city, the people, the lived experience of the structure, who it serves, and who is connected to it. For too long, engineering has trained us to stay zoomed in - to focus on the what and the how. When we widen the lens, we begin to see patterns. We begin to see the DNA.

For example, a freeway widening project might meet every technical spec, hit the budget, and open on time, yet fragment a community and leave local roads congested. A building may pass every code but fail to include people whose bodies don't match the "average" standard. A major infrastructure asset may be delivered, but if the maintenance budget is tiny, the asset ages fast, and the cost of remediation becomes enormous. In each case, the outcome may seem successful by the narrow metrics - but legacy fails.

For too long, engineering has kept the lens fixed on the technical - on precision, compliance, and performance. But the deeper patterns - the human and ethical codes that shape our work, and the broader fabric of our society - are seldom part of our design and delivery systems. When we adjust the lens, we begin to see that engineering's true blueprint isn't only in what we build - it's in how we build, and why.

Traits of the Current DNA

Here are some of the dominant "genes" I've observed in engineering systems, and the consequences that follow when they're unchallenged.

Speed

We're applauded for "fast delivery". The narrative is simple: get it done, move on. But when fast becomes the metric, what is lost? Side roads, bike paths, community consultation, detailed maintenance planning, and community connection often don't make the cut.[58]

Cost minimisation

Lowest upfront cost wins the tender, but the focus on short-term cost often ignores whole-of-life cost and human cost. For example, in many civil infrastructure projects that I worked on at university, we learnt to design something, hand it over, then walk away. Maintenance was an afterthought, if it was even discussed at all. This is the classic "build and forget" model. Research into deferred maintenance shows that putting off upkeep leads to exponentially higher costs and greater risk.[59]

Efficiency and "highest and best use"

Efficiency is the sacred metric, and for me it's often a strong value, even in my personal life. We're taught to optimise for the "highest and best use" of time, land, and capital. We optimise for throughput, for volume, for capacity. Efficiency is praised and rewarded.

But this way of thinking can make anything that doesn't quite fit the formula - community connection, cultural heritage, or environmental care - seem inefficient.

We celebrate optimisation, but forget that what we optimise for defines what we value. Efficiency measured only in technical output doesn't often capture inclusion, resilience, or the long-term wellbeing of people and place. True efficiency should be measured in regeneration per unit of effort, not just productivity per hour.[60]

Control and predictability

Engineering loves certainty. We design to codes and deliver to schedule. Control is our comfort zone. But it also means we sometimes design for the "expected" rather than the "possible". We hedge against failure, but rarely embrace nuance.[61]

Compliance

Meeting standards becomes the objective rather than the starting point. The standard becomes the ceiling, not the floor. When compliance is framed as sufficient, the value of questioning what the standard leaves out gets lost.[62]

Uniformity

We design for the "average user" rather than the diverse actual population. The comfort models of many HVAC systems, for example, are still based on the metabolic rate, body assumptions, and clothing models that originated from studies of adult men. The "average" is never actually average.[3]

Short-termism and isolation

Projects are often viewed in isolation - deliver this project, hand it over, move on to the next. Legacy, community impact, and future generations are rarely measured. The DNA of cycling through projects means long-term connections suffer.[63]

Rationality over empathy

From the first day of university, we're taught that reason is our greatest tool, that logic and objectivity are what make engineering rigorous.

But when rationality becomes our only language, empathy gets edited out of the equation.

We pride ourselves on being the calm, analytical ones - the people who keep emotion out of decision-making. Yet emotion is what connects us to meaning, to community, to the why behind our work.

When we ignore empathy, we design for performance, not for people.[64]

Scalability and replication

We love systems that scale - modular, repeatable, standardised. This thinking has allowed us to industrialise brilliance. But it can also limit our ability to localise, to listen, to design for the specificity of place.

What's efficient at scale is not always effective on Country. A regenerative mindset asks not "Can we replicate this?" but "Should we?"[65]

Perfectionism

We're trained to eliminate errors, not only to get calculations right but to be perfect before release. This keeps people safe, but it can also make us afraid to experiment and therefore innovate.

Perfectionism kills iteration - and without iteration, innovation stalls. We forget that the bridge we're standing on today was once someone else's prototype.

A well-known teaching story, described by David Bayles and Ted Orland in their book *Art & Fear*[66], comes from a photography lecturer who divided his class into two groups. One group would be graded solely on quantity - they were required to produce a large number of photographs over the semester. The other group would be graded on quality - they needed to submit just one carefully crafted image.

At the end of the semester, the highest-quality photographs overwhelmingly came from the quantity group. While the "quality" group spent their time theorising, planning, and trying to get it right, the quantity group learned through doing - experimenting, failing, adjusting, and refining. Their work improved not because they aimed for perfection, but because repetition gave them feedback.

The lesson was simple and confronting: excellence rarely comes from thinking alone. It emerges through iteration. Through practice. Through learning what doesn't work. Progress came not from getting it right the first time, but from being willing to try - again and again.

A Familiar Pattern: The DNA Across Disciplines

Once you start to see this inherited DNA, it becomes almost impossible not to notice it. It shows up everywhere across every field of engineering, like a familiar rhythm running underneath our work. I see it now walking around the streets of any city. Whether it's transport, water, buildings, energy, or digital systems, the same patterns of speed, efficiency, control, and cost repeat themselves, often with remarkably similar consequences.

In transport engineering, the pattern reveals itself in how cities measure success. When a freeway is widened to "increase throughput", it looks like progress. Vehicles move faster; the metrics on efficiency improve. Yet, at the same time, investment might be pulled from local streets, bike paths, and public spaces. The smaller networks - the places where people actually live, walk, and connect - begin to fray.

What appears as technical efficiency on paper often translates to social inefficiency on the ground: communities cut off, congestion simply displaced, and the human experience of the city diminished in service of speed.

Consider this example from local transport policy, described in the book *Invisible Women* by Caroline Criado Perez.[3] In one Scandinavian municipality, snow-ploughing priorities were originally set to clear main roads first - the routes assumed to matter most for getting people to work. Residential streets, footpaths, and cycleways were cleared later.

On paper, the policy made sense: maximise vehicle flow, protect productivity, keep traffic moving. But in practice, it ignored how many people actually travelled on foot, with prams, on bikes, or using public transport. When the council reframed the problem and prioritised clearing local access streets, footpaths, and cycle routes first, something remarkable happened. Slips, trips, and fall injuries dropped by around 50%. The change didn't require new technology or more funding - just a different definition of what success looked like. Once again, the system had been working exactly as designed. It was simply designed around a narrow set of assumptions about whose movement mattered most.

In environmental engineering, the same DNA could take a different form. A water utility, faced with tight budgets and deadlines, might select the cheapest pipe material that meets compliance.

On the surface, it's a rational, defensible decision - cost-effective and on schedule.

But years later, the hidden costs emerge. Maintenance is higher than expected, replacements come sooner, and the system becomes less resilient to shocks. What saved money in the short term costs more in the long run. This is the legacy of a system optimised for delivery, not durability; for completion, not continuity.

In digital and software engineering, the story is similar - only faster. The push to deliver systems to market quickly rewards speed over scrutiny. Accessibility testing, ethical data governance, or cybersecurity reviews are sometimes deferred until after release, if they're done at all. The product ships and the metrics look good, but the invisible risks remain.

What's gained in time is often lost in trust. Here, our inherited DNA of efficiency over reflection plays out in code instead of concrete.

And in energy infrastructure, the same familiar logic reappears again. Renewable projects are often evaluated by their lowest upfront cost per megawatt. It looks like a victory for the bottom line - clean energy delivered at record low cost.

But when workforce development, community benefit, and local procurement are left out of the equation, the social and regenerative potential of the project is lost. We meet the technical goal, but miss the human one. The opportunity to build capability, legacy, and shared ownership vanishes beneath the numbers.

Across all of these disciplines, the pattern is similar. It's the same gene expressing itself in different ways - a core professional instinct to optimise for what is measurable and controllable, often at the expense of what is meaningful and enduring. It is not born of negligence or ill intent, but of habit.

The systems we work within reward delivery, not depth; completion, not connection. Yet, when we learn to look beyond the metrics - to see the DNA beneath the design - we begin to recognise that these choices are not inevitable. They are inherited. And what's inherited can always be reimagined.

Case Study: When Efficiency Wins but Legacy Fails

Sometimes the cracks in our engineering DNA aren't metaphorical - they're literal.

One striking case is Paradise Dam in Queensland. Built in the early to mid 2000s, it was celebrated for its speed and efficiency. Engineers used a relatively new technique called roller-compacted concrete (RCC) - a fast and cost-effective method that allows concrete to be laid in thin layers and compacted with heavy rollers, rather than poured traditionally. It reduced costs and construction time dramatically. At the time, it was hailed as an engineering triumph - innovative, efficient, and affordable.[67]

But less than two decades later, the dam began showing serious structural issues. In 2019, safety concerns about the stability of the dam wall forced the water level to be lowered.[68] Independent investigations later found that the bond between layers of roller-compacted concrete was inadequate. The dam was, in essence, delaminating.[69] The very method that made it efficient to build was now undermining its integrity.

The dam had to be partially decommissioned.[70] Irrigation communities were left uncertain about water security, and the Queensland Government committed billions of dollars to rebuild the dam "the right way". Worse still, no one could be held accountable. The company that designed and built the dam had long since been disbanded. It was a dam built fast, built cheap - and now it must be built twice. A textbook example of short-term efficiency eroding long-term legacy.

The same pattern appears globally. The Longfellow Bridge in Boston, built in 1907, received minimal maintenance for decades. By the time full restoration was needed, the cost had ballooned from an estimated US $81 million to US $306 million. In both cases, the initial projects were celebrated for innovation and cost-saving - until the legacy costs came due.[71]

The lesson is clear. When our metrics are speed, cost, and delivery, we can easily mistake "done" for "complete". Yet legacy, safety, and trust are built over lifetimes, not project schedules.

Where to From Here?

Mapping this current DNA doesn't mean rejecting everything we do well. It means recognising what we still carry forward and asking: are these the traits the world needs now? The build-once-forget-forever model? The fastest job wins? The lowest cost tender wins? The standard user wins?

What if engineering success looked different? What if our measures included questions like "How did people live in this world twenty years later?" or "Who did this design exclude?" or "What connections did it undo or create?"

Engineering for the future begins with awareness.

Spotting the Hidden DNA

Once we learn to see the patterns, we start noticing them everywhere: in meeting rooms, design standards, project briefs, and even in the language we use every day. The old DNA isn't just in the systems; it's in our conversations, our assumptions, our jokes, our definitions of success.

At first, spotting it feels like trying to see the wind. You notice it only by how it moves things. Over time, the signs become unmistakable.

When Something Feels "Off"

I remember working on a water resources project - a dam in regional Queensland. It was a fairly straightforward, standard design of an earth bund dam - one of those projects geotech teams love: hydrology reports, flood modelling, soil analysis, geology profiles from the beginning of time.

When I was handed the project, there wasn't much of a handover. It was just, "Here's the job. The relationship with the client isn't good. Get started." So I did what I'd been trained to do: opened the reports, reviewed the data, studied the drawings. I made a plan on how to get the project finished within the timeframe with the client.

The design was predominantly the dam structure and a large new access road that led into the nearby town. The geotech team had done extensive research, and rightly so. All of it was needed to design and build the dam and the access road. You could lose yourself in the detail of all the background information they had provided in the report and the detailed calculations.

As the project got closer to completion, I started to pull together the technical specifications to go alongside the report. I started to feel this unease. I couldn't quite put my finger on it at first. Eventually, I realised it wasn't about the numbers or the design itself. It was about what wasn't there.

There was no mention of the context of the land we were on. No discussion of who lived downstream, or what might happen to them if the water flow changed.

And not once, in all the reports and meetings, did anyone raise the question of Country.

Whose Country were we on? What did this place mean, beyond its coordinates and the area surrounding it? It was like the dam and road were being designed in space rather than in reality.

It struck me that we were designing something *on* the land, but not *with* it. We were treating the landscape as a technical surface, not as a living system with story, history, and belonging.

That realisation landed heavily. It felt strange to have such a disconnect. It wasn't that anyone was doing the wrong thing technically. All the calculations added up. It was just that the context had gone missing.

We were operating inside the narrow frame of the project's technical DNA: deliver the dam, meet the brief, stay on budget, work out the cost of construction. But the deeper human and environmental DNA - the one that holds connection, responsibility, and care - had been left out.

That's what it feels like when old DNA is at play. Everything seems right on paper, but something in your gut knows something is missing.

At the time, I didn't have the words for it. I just knew that it didn't feel right to keep moving forward as if nothing was missing. The project met every engineering standard, but it didn't meet mine: the standard that asked, who is this really for? What was the point? Were we making the world a better place, or just ticking the boxes for the client? Or worse, having a negative impact? I didn't know the answer.

After completing the project successfully for the client, I found out months later that the dam was too expensive and wasn't going to be built. I felt relieved and couldn't explain why.

This seemingly ordinary and benign project stayed with me. It taught me that the feeling of unease isn't something to ignore. It's information. It's our intuition trying to tell us that the frame is too small, and that something human, cultural, or ethical has been left outside the technical boundaries we're working within.

Now, when I feel that same dissonance, when a project is technically sound but something feels off, I treat it as a signal. A pause point. A clue to go after.

It's the moment to ask questions like: Why are we building this again? Whose voices are missing? Who is connected to this place beyond the client we are working with? What's the story of this place? Are we building for resilience and for the future - or just for delivery?

That feeling - the one engineers are often trained to suppress - is actually one of our most valuable tools. It's how we know when the old DNA is speaking through the system. When we listen to it, we create the space to build something aligned to humanity and the planet - something connected, contextual, and whole.

That inner moment of unease, when something feels technically fine but intuitively wrong, is the first sign that old DNA might be at play. Learning to recognise that feeling and pause before acting is one of the most valuable leadership skills an engineer can develop.

This concept of listening to your intuition could be a whole other book in itself. If you would like to go deeper on this topic, I highly recommend *The Signs* by Dr Tara Swart.

The Language That Reveals the DNA

I've noticed that the most powerful clues to our DNA are often hidden in plain sight, in the everyday phrases we use. Our professional language is full of shorthand for what we truly value. When you start to listen for it, you can hear the old genes speaking through us.

Here are some common examples of phrases that reflect engineering's inherited DNA - each one revealing what's prized and what's missing:

Common Phrase	What It Can Signal	What's Missing or Unspoken
"Fit for purpose"	Compliance with minimum requirements, efficiency	Whose purpose? Who defines "fit"?
"On time, on budget"	Efficiency, control	Value, connection, legacy
"Value for money"	Lowest upfront cost	Long-term benefit, social or environmental value
"Meeting the brief"	Obedience to the client	Courage to challenge assumptions
"Highest and best use"	Economic optimisation	Cultural, ecological, or community value
"Deliverables"	Output over outcome	Impact, lived experience
"Mitigate risk"	Fear of uncertainty, playing it safe	Embracing possibility, adaptive design
"Optimise performance"	Quantifiable efficiency	Human wellbeing, diversity, regeneration
"In scope/ out of scope"	Neat boundaries, silos	Interconnection, complexity
"Stakeholder management"	Control of relationships	Genuine collaboration, shared ownership
"Industry best practice"	Adherence to convention	Curiosity, experimentation, evolution

Each phrase is a small window into our collective mindset, often a linguistic fossil of the Industrial Era.

We don't need to abandon these phrases altogether. But we do need to notice them and listen to what is behind them. Every time we say "on time, on budget", we can also ask, "And for whom did this create value?"

The Framework: Three Perspectives for Seeing the Hidden DNA

Spotting the hidden DNA isn't just about critique. It's about awareness. It's about learning to see differently.

I often describe this as learning to look at your project through three lenses: systems, relationships, and stories.

1. Systems: What's Built In

Every engineering process has embedded values that are often invisible until we question them. We get so used to them that they just become "the way we do things". That is why they can be difficult to change. They're comfortable and seem like the best way of working.

Ask your team:

- What's the intention? Why?
- What and who are we optimising for in this project?
- How are success and failure defined?
- Where are the connections? Who, where, what, and why does this connect to...?
- What assumptions are built into our standards, codes, or risk models?
- Where have we inherited rather than chosen these values?

These questions expose the system-level DNA - the metrics and measures that guide decisions.

2. Relationships: Who Holds Power and Who Has a Voice

Engineering projects are human systems, even when they're made of steel and concrete. Look around your table:

- Whose perspectives are shaping this brief the most and least? You can look at this from a team, organisational, or broader level.
- Who's missing from this conversation?
- Who is feeling safe or unsafe to share their perspectives?
- How are decisions being made - collaboratively or hierarchically?

- When was the last time we paused to listen rather than defend or blame?

Power sits silently in engineering culture. The hidden DNA often expresses itself through hierarchy - where junior engineers, community members, or marginalised voices hesitate to speak. Spotting that imbalance early can change the entire trajectory of a project.

3. Stories: What Narrative Are We Telling?

Every project tells a story about what we value, what we believe progress looks like, and who we think engineering is for.

Ask your team:

- What story are we telling about this project?
- Is it a story of speed and delivery, or of connection and legacy?
- How will this project be remembered in 10, 20, or 50 years?
- What story do we want the community to tell about it?
- How would others tell the story of this project?

When we change the story, we change the system.

How to Use This Framework in Practice

This is a personal reflective exercise for leaders and something that can be woven into the life of a project or even turned into a workshop for teams.

Here's one way to do it:

1. **Gather your team**. Include engineers, designers, project managers, and even clients, direct users, or community representatives.
2. **Map your current DNA**. Ask everyone to identify what they think the "genes" of your project are, such as speed, cost, safety, or legacy. Consider what is really driving the project and how people's behaviours reveal this.
3. **Use the three perspectives.** Spend time exploring systems, relationships, and stories. Each group identifies one hidden assumption in their area.

4. **Surface the conflicts.** Where are there clashes between efficiency and empathy, or between legacy and budget? Where are the competing priorities? Which ones are winning?
5. **Reframe the DNA.** Ask: "If this project were designed for care, regeneration, and legacy, what would shift?"
6. **Document the new measures.** How will you know if you've succeeded - not just technically, but ethically, socially, and environmentally as well?

This can be run as a one-hour reflective session, a design-thinking workshop, or as part of your project kickoff process.

The goal isn't to create a perfect framework. It's to make the invisible visible, and to name the hidden DNA before it decides the outcome for you. Once you bring awareness to it, it is visible and can be highlighted as the project continues.

Seeing the Hidden Values

Spotting the hidden DNA is an act of care. It's how we protect people, planet, and legacy.

Every time we pause to ask why, notice the language we use, listen more deeply to others' stories, or seek meaningful connection and intention, we're rewriting the next strand of engineering's evolution.

The good news is our DNA isn't fixed. It's alive, it's evolving, and we get to decide what traits the next generation of engineers will inherit.

The hardest part about hidden DNA is that it has been in our blind spot for decades, seen only by a small number of people who have seen its hidden reality. It hides inside our standards, our language, even our sense of professionalism. It tells us that "This is just how things are done." It becomes part of who we are as engineers. And for a long time, we believed it - because it works, because it keeps things safe, and because it's what we were taught.

Once you start noticing how often the word "efficiency" replaces "care", or how "value for money" really means "lowest upfront cost", the lens shifts. You start hearing the stories behind the drawings, calculations, and spreadsheets. You begin to see the invisible hands shaping the work - and the gaps left behind.

For me, that shift didn't happen in a single moment. It happened over time as my career went on. In realisations on project sites, in design meetings, and in late-night reflections on conversations from the day. It was the feeling that something important was missing, even when everything looked right.

It's like stepping back from a drawing you've been too close to for too long and finally seeing the whole picture.

When you learn to see the hidden values that drive decisions, shape priorities, and define success, you start to design differently. You start to lead differently.

You begin to ask:

- What if "risk" included community, human, and environmental care, not just safety and cost?
- What if "efficiency" meant sustaining people and places, not just saving time?
- What if "value" was measured in legacy, not just in dollars?

Seeing the hidden DNA then helps you see how this can be expressed as values in ourselves, in teams, in projects, and in organisations. This isn't about blame or guilt. It's about awareness. It's about giving ourselves the language and courage to choose what we pass on.

This is the pivot point for engineering: from following inherited truths to creating intentional ones. From designing for what's efficient to designing what endures.

And that's where we go next: learning to spot value conflicts early, before they calcify into projects, systems, or cultures.

That evolution begins the moment we notice what we've been blind to.

The Takeaway

Recognising that engineering is value-laden is not a criticism of the profession. It is an act of honesty. When we stop hiding behind the myth of neutrality, we regain the ability to choose our values deliberately, rather than inheriting them by default.

But principles alone do not change outcomes. Values only matter when they show up in practice, in the way engineers are trained, rewarded, and allowed to behave. They live not just in codes and contracts, but in meeting rooms, classrooms, and site sheds. In what gets praised. In what gets questioned. In what gets avoided.

To understand how engineering really works, we need to look beyond what we say we value, and examine how those values are embodied day to day. The next chapter turns to practice: the invisible habits, cultural norms, and unspoken rules that shape how engineers think, decide, and act long before a drawing is ever signed.

Chapter 3: From Principles to Practice

It's one thing to recognise that our engineering DNA needs to evolve. It's another to start rewriting it. That's where the real work begins.

Rewriting the DNA of engineering doesn't start with tearing up standards or codes. It begins with seeing where values already live inside them.

Every design decision, procurement choice, and specification clause carries a set of assumptions about what matters most.

Efficiency. Cost. Control. Compliance.

Those assumptions are the language of engineering. They shape everything we do - how we think and how we act. The question is whether they are still the right assumptions for the world we are designing for today.

Values can also live in practice, in the small, deliberate acts of care, curiosity, and creativity that change how projects unfold. When we begin to work from intention rather than inheritance, something shifts. We start building systems that not only perform, but belong, connect, and care.

This chapter shows what becomes possible when values are made visible in real projects. Through examples from cities, inventions, and systems, you'll see how teams have begun to rewrite engineering's DNA. Sometimes this happens through bold moves. Sometimes it begins with a single question that shifts everything.

Each case study reveals how old assumptions are challenged and how new values are put into practice, aligned with the Future Engineer Framework. These are not abstract ideas, but perspectives that can be applied within any project, organisation, or contract.

Together, these examples capture moments where engineers, planners, designers, architects, and communities put people, place, and purpose back at the centre. They are not the only stories of change, but they show what is already possible, and what is already underway.

Writing this chapter overwhelmingly filled me with hope. Hope for our planet, our future, and our legacy.

For a deeper dive into the case studies, and as more are brought to light, they will be featured on the Future of Engineering podcast. These stories are not outliers or fantasies - they are proof points. Evidence that regenerative, community-led, inclusive engineering is not only achievable, but already happening.

They show that when engineers centre care, Country, and connection, outcomes improve for everyone: technically, socially, environmentally, and economically.

These case studies are exemplars from around the world that illuminate what engineering could look like, and what the future of the profession can become.

Case Study: School Corridor Congestion

Redesigning the Problem Itself

Sometimes rewriting the DNA of a design doesn't mean changing the structure. It means changing the question and challenging the brief.

In one project shared by architect Alastair Parvin in his TED talk "Architecture for the People by the People"[72], a school approached his firm with what seemed like a clear, technical problem. Their Victorian-era building had narrow, overcrowded corridors. Between classes, the space became chaotic. Students bottlenecked, teachers struggled to maintain order, and bullying increased.

The school's request was straightforward: "We need you to redesign the building to stop this happening."

The assumption beneath the brief - the hidden DNA - was that architecture equals outcome. Change the building's physical form and change the outcome. The assumption was that to fix a design problem, you must redesign the building.

The school was prepared to invest millions to widen corridors, restructure rooms, and create more circulation space.

For the consultants, proceeding with a major redesign and refurbishment would have meant significant revenue. We are often incentivised to deliver exactly what the client asks for. I've seen this culture play out time and time again, where questioning the brief is silently discouraged.

But Parvin's team paused.

They questioned the brief. They asked. Is this really a spatial problem, or is it a systems problem?

Instead of redrawing walls, they asked a different question. What if the issue wasn't the corridor at all, but the bell?

They replaced the single central bell, which sent the entire school surging into the corridors at once, with a sequence of smaller bells that rang in different places at slightly different times. Students began moving in waves, not floods.

Congestion disappeared. So did much of the bullying.

The total cost was a fraction of a building redesign. It has been suggested the intervention cost about two hundred pounds.

Rewriting the DNA of Traditional Problem Solving

In traditional architectural or engineering DNA, scale is often equated with success. Bigger budgets. Bigger interventions. More visible outputs.

But in this story, success came from doing less. From finding elegance in simplicity and systems thinking.

The project's shift wasn't really about corridors or bells. It was about reprogramming the values behind the design process, moving from outcome-driven thinking to intention-led design.

The architects effectively designed themselves "out of a job", but into something more important - a solution that fulfilled the true intention of the project while saving time, resources, and money.

This example captures what it means to rewrite the DNA of engineering and design: to stop treating the problem statement as fixed, and instead look for the intent beneath. It shows that creative, regenerative design often doesn't require more resources - just more imagination.

If the old engineering DNA prizes efficiency, precision, and predictability, the rewritten DNA prizes curiosity, humility, and human insight. This project didn't need a new standard; it needed a new perspective.

The same shift can happen anywhere: in the layout of a hospital, the scheduling of a bus network, or the operation of a dam. It's not about breaking compliance; it's about questioning the inherited logic that led to the problem in the first place.

Sometimes the smartest engineering solution isn't the one that builds more or showcases technical brilliance. It's the one that builds less, but better.

Reflection Prompt

For the projects, systems and problems you are working on, consider:

- What problem are you solving, and who defined it?
- What assumptions are baked into the brief?
- What's the smallest intervention that could solve this differently?

Case Study: WikiHouse - Access to Design for Anyone, Anywhere

If the story of the school without a bell shows how we can rewrite the DNA within a system, WikiHouse shows what happens when we rewrite the DNA of the system itself.

For centuries, the architecture and construction industries have been built around a simple assumption: big projects require big organisations.

Design, finance, and power have all been concentrated in the hands of the few - those who can afford to commission, design, or construct. It's a pattern mirrored across engineering disciplines: centralised expertise, complex procurement, and high barriers to entry.

When architect Alastair Parvin and his team began to question this logic of the system, they landed on a radical idea. What if the ability to design and build wasn't restricted to a professional elite? What if citizens themselves could be part of shaping their built environment safely, affordably, and collaboratively? Their answer became WikiHouse - an open-source construction system Inspired by the principles of Wikipedia and open-source software.[72]

WikiHouse is an open-source, CNC-cut timber framing system. Components can be fabricated from sheet material and assembled quickly on site, often with a small team, reducing complexity compared to conventional builds. WikiHouse provides free, downloadable blueprints for homes that anyone can adapt, customise, and fabricate using simple materials and accessible digital tools. A user can download the 3D model files, adjust them in free software (like SketchUp), and send them to a CNC cutting machine - a kind of giant, digital jigsaw. The machine cuts the house's structure from sheets of plywood, which can be assembled without bolts, glue, or specialised construction skills.

The designs are shared under open licenses, meaning anyone can improve or modify them and then re-upload their version back to the shared library - just like open-source code.[73] It's architecture not for the 1%, but by and for the 100%.

Rethinking Traditional Construction Methodologies

Traditional construction is built on a DNA of control, scarcity, and scale. Only those with resources can commission. Only those with licenses can design. Only those with machinery can build.

WikiHouse flips this logic.

Its DNA is one of openness, collaboration, and decentralisation. Architecture and engineering are no longer services delivered to people, but languages that anyone can learn, collaborate in, and build with.

By lowering the barriers of time, cost, and technical expertise, WikiHouse reframes what design and engineering mean. It shifts the narrative from "Who can afford this?" to "Who can participate?"

Parvin calls this "the social economy of design". Just as open-source software transformed computing, open-source construction has the potential to transform cities - not as monolithic developments built for profit, but as living, evolving systems built by communities.

Lessons for Engineering

WikiHouse challenges one of the oldest assumptions in the engineering DNA: that complexity must be centralised and is only for technical specialists, not for the people. For generations, engineers have been told that control equals safety. However, this project shows that trust and transparency can be equally powerful.

When design knowledge is shared freely, it doesn't erode expertise. It multiplies it. More perspectives improve design quality, resilience, and relevance. When communities co-design their homes, they don't just create shelter. They create belonging, agency, and long-term resilience.

This shift - from closed to open, from product to process - mirrors the regenerative mindset emerging across disciplines. It's not just about what we build, but how it enables life, learning, and connection.

The Legacy of Open Source Design

WikiHouse has since inspired projects around the world - from post-earthquake housing in Christchurch, New Zealand to favela regeneration initiatives in Rio de Janeiro. Each project adapts the DNA slightly differently, but the underlying code remains: empower people through open design.

The genius of WikiHouse isn't its technology - it's its philosophy. It reminds us that the tools we use are only as ethical as the values they carry.

In the old model, a few built for many. In the new model, the many build for themselves.

Reflection Prompt

For the projects, systems and problems you are working on, consider:

- What would happen if your design knowledge were open-source?
- What would collaboration look like if trust replaced control?

Case Study: James Dyson - Engineering as a Creative, Innovative and Pioneering Act

If the Industrial Revolution hardwired efficiency, output, and scale into the DNA of engineering, James Dyson has spent his life rewriting that code, one invention at a time.

In a world obsessed with margins and market share, Dyson's motivation was never money. It was excellence. His career is a masterclass in what happens when curiosity, creativity, and integrity become the driving values of engineering.

When Dyson invented his first bagless vacuum cleaner, he wasn't setting out to disrupt a global industry. He was simply frustrated. He'd noticed how quickly his household vacuum lost suction, and realised the problem was the bag itself. The bag filled with dust, blocked airflow, and reduced performance.

He believed there had to be a better way. He wanted to create an invention that worked exceptionally well, that could make a positive difference for everyday people.

For four years, Dyson worked in his workshop at home, building and testing a whopping number of prototypes. He meticulously recorded each iteration and what he was testing. Prototype after prototype failed or fell short.

Prototype 5,127 finally worked.

That was the first cyclonic vacuum, using centrifugal force instead of a bag to separate dust from air.

When Dyson tried to sell the design, every major manufacturer turned him down. The existing companies made more profit from selling replacement bags than from selling vacuums. The system worked for them - not for customers. Dyson's idea, however brilliant, threatened their business model.

One major appliance company initially agreed to manufacture the design. They worked closely with Dyson then abandoned the project, only to release a copied version years later. Dyson sued. The case took years, cost millions, and he ultimately won.

To get his product into homes, Dyson first launched in Japan. It worked, but he had a bigger vision.

So he did what all great engineers do: he changed the system.

In 1993, Dyson launched his own company in the UK. His vacuum cleaner became a global success - not because of clever marketing, but because of engineering excellence and brilliant design. He made a product that solved a problem completely.[74]

That's what happens when success is measured by performance and purpose, not profit alone. Dyson applied the same values to his company as to his products. He retained ownership to protect long-term vision. He kept core capabilities in-house. Even marketing remained internal, prioritising creativity and ingenuity from within, aligned to their values. Dyson's book Reinvention is an incredible account of his story, documenting what I think is one of the greatest engineering journeys of our time.

A Return to Engineering's Creative Soul

In traditional manufacturing, the DNA is simple. Maximise efficiency, minimise cost and produce at scale.

Dyson flipped this on its head. His DNA reads differently. Maximise effectiveness, creativity and innovation; invest in craft, design and making beautiful things and produce for longevity.

He refused to compromise engineering quality, design, or ownership for profit. Dyson doesn't outsource core research or rely on planned obsolescence. In Dyson's world, engineers are not a cost centre; they are the beating heart of innovation.

This was seen in another example when the team at Dyson were looking for a small motor to go in their products. When they weren't able to find one good enough for what they needed, they invented their own. They continue to improve the effectiveness of the motor every year, as with all their products, and never settle for "good enough" design.

Dyson's approach rejected the inherited industrial mindset. Where the old DNA rewards standardisation and sameness, Dyson rewards innovation, continuous improvement on already great design, and creativity. He builds products that cost more not because of status, but because they work better, have replaceable parts, and last longer.

Dyson's story reminds us that engineering is, at its heart, a creative discipline - the meeting point of imagination and precision. This spirit once defined the profession before industrialisation narrowed its focus: the engineer as the pioneering inventor who builds not for scale, but for impact. This spirit, I believe, has been lost in the modern world of engineering.

Dyson's engineering DNA is the kind of DNA we need to bring back: invention over optimisation, curiosity over compliance, creativity over convenience.

We can learn a great deal from James Dyson. He teaches us that when we stop optimising purely for speed and profit, we make room for beauty, curiosity, and care and ultimately create better outcomes. His vacuum cleaner isn't just an appliance. It's an icon, a manifesto - proof that function and form, science and art, ethics and performance can co-exist.

He reminds us that great engineering is not about compliance - it's often about courage. It's about the willingness to question why the system works the way it does and the determination to make something better, not just cheaper or more efficient.

Reflection Prompt

For the projects, systems, and problems you are working on, consider: What are you optimising for: efficiency or effectiveness for all?

Case Study: Sydney Harbour Bridge Whole-of-Life Value and Regeneration

The Sydney Harbour Bridge is one of Australia's most iconic engineering achievements, but it is also a powerful example of long-term value, unintended regeneration, and the importance of designing beyond a project's immediate purpose.

When the Sydney Harbour Bridge was constructed in the 1920s and 30s, the now-famous sandstone pylons at each end were never required for the bridge to stand. The steel arch is anchored deep into bedrock by enormous concrete and steel structures hidden from view. The pylons carry no structural load.

And yet, they were deliberately built - monumental, permanent, and unapologetically civic. Their role was not technical necessity, but meaning. They connected the modern steel arch to the sandstone geology and architectural language of the city beneath it.[75, 76]

This was a moment when engineers understood that infrastructure does more than function. It shapes identity. Nearly a century later, those pylons remain - not as structural elements, but as proof that beauty has always had a place in engineering, when we choose to allow it.

And beauty, remarkably, led to purpose. When the Sydney Harbour Tunnel was constructed decades later, engineers needed a large ventilation stack - a structure that typically disrupts a landscape and often becomes an eyesore. Could you imagine what the iconic Sydney Harbour would look like today, with a tunnel ventilation stack designed in the late 80s sticking up between the Opera House and the Sydney Harbour Bridge or in front of Luna Park?

So where did they put it? Instead of having it on public display, they tucked the inside the northern pylon of the Sydney Harbour Bridge. A temporary scaffolding tower, preserved for aesthetic reasons, became one of the city's most practical pieces of infrastructure. Form became function. What was once unnecessary became essential.

This story is a reminder that when engineers design with care - for presence, for beauty, for connection to place - we create possibilities that future generations can build on. We design not only for the immediate task, but for the unknown opportunities that may emerge long after construction is complete.

Consider also the design life of the Sydney Harbour Bridge itself. Modern Australian standards often adopt around 100 years as a typical design life for major bridges.[77] We don't have a definitive published design-life assumption from the 1920s, but the century mark is now close. Historical records don't confirm the exact design assumptions made in the 1920s for the Sydney Harbour Bridge, but it's likely it was planned with a similar timeframe in mind. That milestone is fast approaching.

And yet, the bridge is still vital, still beautiful, and still a defining symbol of a city and a nation. We cannot simply replace it. We must regenerate it, maintain it, and strengthen it beyond what was originally imagined. With sustained stewardship, some commentators and engineers have suggested the bridge could remain serviceable for centuries.

What makes this possible is one of the most rigorous, continuous maintenance programs.

The Sydney Harbour Bridge has been under near-constant maintenance since the day it opened in 1932. Rather than treating maintenance as periodic repair, it is managed as a living system. At any given time, hundreds of thousands of individual steel elements are inspected, monitored, repaired, or protected as part of a rolling program that never truly ends.[78]

One of the most visible examples is the painting regime. The bridge is effectively painted continuously, with teams working through a cycle that takes several years to complete before starting again. This isn't cosmetic. The paint system is a primary corrosion-protection layer, critical to preserving the integrity of the steel structure in a harsh marine environment. Old paint is carefully removed, steel is inspected and repaired where needed, and modern protective coatings are applied - layer by layer, member by member.

Beyond paint, the bridge is subject to detailed structural health monitoring. Engineers routinely inspect rivets, bearings, joints, deck elements, and fatigue-critical components.

Over its life, the bridge has been strengthened and adapted multiple times - not because it was failing, but because the demands placed on it have changed. Traffic loads are far higher than those imagined in the 1930s. New lanes have been added. Rail systems have been upgraded. Services have been threaded through the structure. Each intervention has required careful analysis, reinforcement, and restraint - adding capacity without compromising the whole.

Crucially, this maintenance is not reactive. It is anticipatory. Engineers work decades ahead, planning interventions long before failure is visible. It is this mindset - seeing maintenance as stewardship rather than cost - that underpins estimates suggesting the bridge could last several more centuries under the current regime.

The Sydney Harbour Bridge survives not because it was overbuilt, but because it has been continuously cared for. Its longevity is not an accident of history; it is the outcome of sustained engineering attention, institutional commitment, and an understanding that legacy infrastructure is something you look after, not something you use up.

This is regenerative engineering in practice: strengthening what already exists, extending service beyond original assumptions, and designing maintenance itself as an act of care for future generations.

Think about that. A structure designed for perhaps 100 years may endure for four centuries.

Why? Because it was built with care, with craftsmanship, and a sense of presence and purpose. Because its designers created things that were not strictly necessary, but meaningful. And because its custodians have understood that engineering is never finished - it is an ongoing relationship between people, place, and time.

The Sydney Harbour Bridge is a testament to regenerative thinking before we had language for it. Its value has grown, its functions have expanded, and its lifespan has stretched far beyond its original intent. It reminds us that engineering is not just the act of building - it is the stewardship of legacy.

Legacy and Crafting Stewardship

The Sydney Harbour Bridge teaches us that legacy is not a by-product of overdesign or excess. It is the outcome of intention, care, and stewardship sustained over time.

Its engineers did not know how the city would grow, what technologies would emerge, or what demands would be placed on the structure a century later. But they understood something fundamental: infrastructure would outlive them. That understanding shaped how they designed - not just for load and function, but for dignity, adaptability, and presence.

Equally important is what happened after construction. The bridge's longevity has been earned, not assumed. It has survived because successive generations of engineers, maintainers, and decision-makers chose stewardship over neglect, and long-term value over short-term savings. Maintenance was not treated as a cost to be minimised, but as an essential act of responsibility to the future.

This is what it looks like when the DNA of engineering shifts from build-and-move-on to build-and-care.

The lesson is not that every project should be monumental, or that beauty always requires grandeur. The lesson is that when engineers design with care - for place, for people, for future use - they create assets that can evolve rather than expire. They create systems that grow in value over time.

The Sydney Harbour Bridge reminds us that regenerative engineering is not new. We have done it before. The question is whether we are willing to do it again - deliberately, consistently, and at scale.

Reflection Prompts

As you consider the projects, systems, or assets you are responsible for, reflect on the following:

- Are you optimising for initial delivery, or for long-term stewardship and adaptability?
- How is maintenance treated in your project: as a cost to reduce, or as a strategy for resilience and value creation?
- If someone were to inherit your project 100 years from now, what would they thank you for?

Legacy is not something engineers leave behind accidentally. It is something we design for, - one decision at a time.

Case Study: Designing for the Night - When Safety, Health, and Equity Are Left in the Dark

For most of modern engineering history, the built environment has been designed with an implicit assumption: that it will be used primarily during the day.

Daytime is where we test. Daytime is where we visit sites. Daytime is where standards are written from. And as a result, nighttime often becomes an afterthought - something to be managed rather than designed for.

This has profound consequences.

Across universities, health precincts, transport corridors, private developments, and public spaces, night-time users experience the built environment very differently. Shift workers, hospital staff, cleaners, security guards, students leaving late lectures, delivery workers, and night-time economy participants often move through spaces that were never truly designed with them in mind.

The Problem: Protecting Assets, Not People

Much of our current approach to night-time design is rooted in Crime Prevention Through Environmental Design (CPTED) and security frameworks. While these frameworks are important, they are often applied with a narrow focus: protecting the asset and deterring a potential perpetrator.[79]

In practice, this means design decisions are frequently shaped by the question, "How do we stop something bad from happening here?" rather than "How does it feel to be here at night?"

The result is a focus on the perpetrator - a tiny percentage of users - rather than the lived experience of the overwhelming majority. Spaces may technically meet safety requirements, yet still feel uncomfortable, intimidating, isolating, or unsafe. Enjoyment, wellbeing, and comfort are rarely part of the brief.

Road design in Australia is a clear example of this imbalance. Our standards overwhelmingly prioritise the vehicle: sight distances for drivers, illumination for hazards, and clarity for traffic movement. People are considered secondary. Design focuses on vehicles seeing the road, not people seeing vehicles, and not people seeing one another.[80]

In my own experience working in road design and delivery, these secondary perspectives like pedestrian visibility, social connection, and perception of safety were rarely, if ever, raised in design meetings. Not because anyone didn't care, but because our standards didn't require us to.

What Gets Missed When We Forget the Night

When night-time isn't deliberately designed for, the impacts compound.

We underutilise expensive assets by designing them for daytime use only. We ignore entire sectors of the economy that operate after dark. We overlook night-shift workers, despite relying on them for essential services like healthcare, transport, logistics, and education.

More importantly, it becomes a health and equity issue.

Designs that fail to consider perception of safety, not just actual safety, can increase stress, anxiety, and avoidance behaviours. People change routes, rush through spaces, or stop using them altogether. The environment may be technically safe, but it does not *feel* safe and that distinction matters deeply.

This isn't malicious. It's cultural and entwined in our standards. Night-time design is often forgotten because it sits outside traditional engineering defaults.

Night time design is one that is personal for me. I live a 10 minute walk from the train station, but at night it can feel sketchy. When it's dark, I'll avoid the train if it means a walk home. Ubers don't frequently service my area and taxis can take 30 mins to arrive, which means I could be waiting a long time at the station, which feels unsafe. Sometimes, I stay overnight where I am working, just to avoid this short walk late at night, which means missing out on time with my young family. It's a small thing but it changes my behaviour, and that is the impact of our designs - it shapes behaviour.

A Different Approach: Designing with Lived Experience

One person making a difference in this space is Alice Vincent, Associate Resilience, Security and Risk and part of Arup's Nighttime Design Team NSW/ACT. With a background in security design, Alice has helped shape what she calls night-time design - an approach that reframes darkness not as a risk to be controlled, but as a condition to be designed for.

Alice shared with me how the Nighttime Design team's work highlights several critical gaps in conventional practice:

- We rarely conduct site visits at night.
- We design to standards that are outdated and largely daytime-centric.
- We "value-engineer" night-time considerations out because they are not yet embedded in standards.
- We overlook perception of safety in favour of technical compliance.
- We fail to consider night-time workers as primary users.

Alice's work emphasises that safety is not just about crime statistics; it's about how safe a space feels. In some projects, this has led to warmer lighting tones, improved visibility between people, and design choices that foster comfort rather than surveillance.[81]

Randwick Health Precinct

A powerful example of this thinking in practice is the Randwick Health Precinct in Sydney - a project which Alice was involved in.

As a major 24-hour health campus, the precinct serves clinicians, patients, visitors, and staff at all hours. Rather than treating night-time as an operational inconvenience, the redevelopment deliberately considered the full 24-hour experience of the place.

Through a creative lighting masterplan, the project integrated lighting and art to improve wayfinding, comfort, and perception of safety after dark. Art installations curated by Sophie Forbat were used not just as decoration, but as tools to humanise space, create visual interest, and help people orient themselves emotionally as well as physically.

The result was a precinct that acknowledged night-time users as legitimate, valued occupants, not anomalies to be managed.[82]

What This Reveals About Engineering Values

Night-time design exposes a familiar pattern in engineering: when something is not written into the standards, it is often treated as optional.

Value engineering typically focuses on commercial efficiency. And because night-time considerations are still emerging, they can struggle to survive through the procurement, delivery, and construction phases.

The lesson here is - when we design only for the day, we exclude by default. When we design only for assets, we overlook people. When we design only for actual safety, we ignore perceived safety, and with it wellbeing, equity, and trust.

Designing for the night requires us to expand our definition of value and risk. It asks us to listen to lived experience, to walk sites after dark, and to consider who is moving through our spaces when most of us are asleep.

This is not about adding cost. It is about adding care.

And like so many shifts in engineering, it begins not with new technology but with a new question: Who are we designing for, after dark?

Case Study: Medellín Metrocable - Mobility as Dignity

Once infamous as the world's murder capital, Medellín, Colombia, is now a global symbol of urban transformation. One of the most powerful chapters in that story begins with a cable car.

In the early 2000s, Medellín faced a huge spatial injustice. Informal settlements clung to the steep hillsides surrounding the city, physically and socially disconnected from jobs, education, healthcare, and opportunity. Traditional transport solutions like roads, buses, rail were technically possible, but economically and politically unrealistic. These neighbourhoods were hard to reach, easy to ignore, and had long been excluded from the city's investment priorities.

Rather than forcing a conventional engineering solution onto an unconventional landscape, the city chose to rethink the problem entirely.

In 2004, Medellín launched the Metrocable - an aerial cable-car system designed to connect the city's poorest hillside communities directly into the existing metro network. On the surface, it was a transport project. In practice, it was something much more ambitious: a statement about who the city was for.

The Metrocable dramatically reduced travel times - in some cases from over an hour to minutes - but its true impact lay beyond efficiency. For the first time, residents of informal barrios could reliably access employment, schools, healthcare, and civic life. The system didn't just move people faster; it expanded what was possible in their daily lives.

Crucially, the Metrocable was never delivered as a standalone piece of infrastructure. Each line was paired with library parks, public plazas, schools, playgrounds, and cultural facilities, deliberately embedded within the neighbourhoods it served. Stations were designed as civic anchors, not just transport nodes. High-quality architecture, often by leading Colombian architects, was placed intentionally in areas that had historically received the least investment.[83]

The message was unmistakable: this city values you.

This approach became known as "social urbanism" - the idea that physical infrastructure must be inseparable from social infrastructure, and that engineering decisions carry moral and political weight whether we acknowledge them or not.[84,85]

Engineering is Never Just Technical

The inherited DNA of transport engineering is familiar - move the greatest number of people as efficiently as possible, optimising for speed, capacity, and cost.

The Metrocable reframed mobility not as throughput, but as inclusion. Transport was no longer just about movement - it became a tool for dignity, visibility, and belonging. Engineering success was measured not only in passenger numbers or travel times, but in access, safety, and social connection.

As one planner involved in Medellín's transformation famously noted, "You cannot solve social inequality with transport alone - but you also cannot solve it without access."

This was not engineering charity. It was engineering intelligence - recognising that cities function best when everyone can participate.

The Metrocable did not "fix" Medellín, but it changed the trajectory of the city.

It demonstrated that infrastructure can heal as well as move. That dignity can be designed into systems. That when engineers expand the definition of success, outcomes expand too.

The most important lesson is not that cable cars are the answer for every city. It is that engineering problems are rarely just technical. They are social, political, and ethical - whether we choose to acknowledge that or not.

Medellín shows us what becomes possible when engineers stop asking "How do we move people faster?" and start asking "Whose lives are we connecting?"

This is regenerative engineering at an urban scale: repairing broken connections, restoring trust, and designing systems that strengthen the social fabric over time.

Reflection Prompts

As you consider the systems and projects you are involved in, reflect on the following:

- Who benefits most from your current design decisions, and who remains disconnected?

- What would it mean to treat access, dignity, and inclusion as performance criteria?
- How might redefining "success" change what you choose to build, and where?

Case Study: Vienna's "Female City" - Designing for Everyone by Designing for Women

In Vienna, the question was never how to build more. It was how to build fairer.

By the early 1990s, city officials began to recognise a pattern. Despite decades of technically competent planning, parts of the city simply did not work equally well for everyone. Women, particularly those juggling paid work, care responsibilities, and household labour, experienced the city differently. Their daily journeys were more complex. Their safety concerns were different. Their time was fragmented across multiple destinations rather than a single commute.

Rather than treating these differences as marginal or anecdotal, Vienna chose to take them seriously and treat them as a design input.[86]

In 1991, Vienna became one of the first cities in the world to formally adopt gender mainstreaming in urban planning. This meant systematically analysing how planning decisions affected women and men differently, then redesigning accordingly. One of the most visible outcomes of this approach was the Frauen-Werk-Stadt ("Women-Work-City") housing initiative, launched in the mid-1990s.[87]

The guiding question was simple, but radical: what does a city look like when it is designed from women's lived experience?

The answers reshaped everyday infrastructure. Sidewalks were widened to accommodate prams, wheelchairs, and people walking side by side. Lighting was designed not just for traffic efficiency, but for visibility and personal safety. Public transport stops were located near childcare centres, schools, medical facilities and shops, recognising that many trips involve chaining multiple purposes together.

Importantly, planners mapped and prioritised care work journeys: the unpaid, often invisible trips that sustain daily life like school drop-offs, grocery runs, and visits to elderly relatives. These journeys had rarely featured in traditional transport models, which were built around a single, linear commute to and from work.

The result was not a city "for women only", but a city that worked better for everyone.[88]

Parents reported greater ease of movement. Older residents felt safer. Men benefitted from more usable public space, better lighting, and more intuitive transport connections. What began as gender-sensitive planning revealed itself as human-centred engineering.

Make the World Work Better for Everyone

The inherited DNA of urban design often assumes a neutral "average user" with linear travel patterns and efficiency defined narrowly as speed and throughput.

Vienna challenged these assumptions.

By designing explicitly for women, whose daily experiences exposed gaps in the system, the city revealed the myth of neutrality. There was never a universal user. There were only unacknowledged assumptions about whose needs mattered most.

Vienna rewrote the DNA of urban engineering from neutral design to inclusive design, and from the average user to the actual user.

This was not ideology layered onto engineering. It was better engineering informed by better data.

Vienna's experience demonstrates that equity is not a trade-off against efficiency. It is a pathway to it.

By expanding who was considered in the design brief, the city improved outcomes across the board. The infrastructure became safer, more intuitive, more resilient, and more widely used. Designing for women did not exclude men; it revealed blind spots that had limited everyone.

Vienna reminds us that fairness is not a constraint on good design. It is often the key to unlocking it.

Reflection Prompts

As you consider your own projects and systems, reflect on the following:

- Who is the "default user" in your design, and who does that leave out?
- How might designing for the most complex users improve outcomes for everyone?

Case Study: Vauban, Freiburg - A Child-First Neighbourhood

On the edge of Freiburg, Germany, on the site of a former French military base, the neighbourhood of Vauban rewrote what urban life could mean.

Instead of starting with traffic models, parking ratios, or development yield, planners and residents began with a different question: what would a neighbourhood look like if it were designed first for children?

The answer was radical, and practical.

Developed from the mid-1990s onward, Vauban emerged as a community-led regeneration project shaped by strong citizen participation, environmental ambition, and a deliberate rejection of car-dominated urban design. Streets were designed as social spaces, not traffic corridors. Children could walk, cycle, and play safely close to home. Cars were not banned, but they were deprioritised, kept to the edges of the neighbourhood or stored in shared parking structures.[89]

Daily life moved at human speed.

Housing was designed to stringent energy standards, with many buildings meeting passive house or energy-positive performance. Some generated more energy than they consumed through rooftop solar and district heating systems. Green spaces, shared gardens, and courtyards were embedded throughout the neighbourhood, creating places for informal connection, play, and care.

Crucially, Vauban was not imposed from above. Residents co-designed their homes and streets through building cooperatives and participatory planning processes. This collective ownership fostered strong social ties and a shared responsibility for the neighbourhood's success.

The results were striking. Car ownership rates fell far below city averages.[90] Children gained independence earlier. Traffic injuries dropped. Emissions decreased.[91] Health outcomes improved. Social connection deepened.[92]

Vauban didn't just reduce environmental impact - it changed how people lived together.

Shifting the Design Reference Point

Vauban shifted the centre of gravity from cars to children, from speed to safety, and from individual ownership to shared care. Engineering success was no longer measured by how quickly vehicles moved through space, but by how safely people could live within it.

When children became the design reference point, everything else aligned: slower streets reduced emissions, shared spaces strengthened community, and energy-efficient buildings lowered costs over time. Instead of sacrificing efficiency, it redefined it.

The lesson here is that engineering priorities are always value-laden, whether we name them or not. Vauban made its values of safety, care, connection, and wellbeing explicit and then engineered accordingly.

This shows us that regeneration is not only about fixing environmental damage. It is about repairing relationships between people, place, and systems.

Vauban is proof that engineering can serve life first. When it does, everything else follows naturally.

Reflection Prompts

As you reflect on the projects and places you influence, consider:

- Who is the design reference point in your project - adults, vehicles, profits, or children?
- How might slowing things down create better outcomes overall?

Case Study: Nightingale Housing - Designing for Affordability and Belonging

In Melbourne, Australia, Nightingale Housing is redefining what "value" means in the housing market.

Founded in 2016 by architect Jeremy McLeod, Nightingale emerged from a simple but confronting realisation: the way we design, finance, and deliver housing is not neutral. It encodes values, and too often those values prioritise profit, speculation, and short-term return over affordability, sustainability, and community.

Rather than accepting those constraints as inevitable, Nightingale set out to redesign the system itself.

At its core, Nightingale Housing is a not-for-profit, architect-led development model. It brings together architects, residents, and ethical financiers to deliver multi-residential housing that is fossil-fuel-free, environmentally high-performing, and genuinely affordable - not as a compromise, but as a design outcome.[93]

Each Nightingale building follows a clear set of principles. Developments are typically car-free, deliberately located near public transport and services. Buildings are all-electric, powered by renewable energy. Apartments are designed for long-term living, with generous natural light, cross-ventilation, and durable materials. Instead of private luxury amenities, Nightingale prioritises shared spaces - rooftop gardens, laundries, bike storage, and communal areas that encourage everyday interaction.

What makes Nightingale distinctive, however, is not just what is built, but how.

There are no developer profit margins baked into the project. There is no marketing budget inflating sale prices. Apartments are sold at cost, with transparent pricing. Residents are often selected through an expression of interest process that aligns values with finances. In some cases, resale prices are capped to preserve long-term affordability.

The result is housing that costs less to buy and less to run. It offers more than shelter: it offers belonging.[94]

Nightingale demonstrates that affordability and sustainability are not opposing forces. When the system is redesigned around shared values, they reinforce one another.

Reframing the Intent of Housing

The inherited DNA of the housing market is familiar. Housing as a financial asset, value is measured in price per square metre and profit is maximised through scale, speed, and speculation.

Nightingale reframed housing as infrastructure for life, not a commodity for extraction. Value was no longer measured by margins, but by long-term affordability, environmental performance, and social connection.

In this model, the question shifted from "How much can we sell this for?" to "How well will people live here - now and in the future?"

It was not about doing less; it was about doing differently.

Nightingale Housing shows that the greatest constraints in engineering and design are often systemic, not technical. Nothing about fossil-free buildings, shared amenities, or energy-efficient design is new. What Nightingale changed was the logic of delivery - who gets to decide, who gets to benefit, and what success looks like.

By stripping out speculative profit and realigning incentives, Nightingale made space for better outcomes: lower emissions, stronger communities, and housing that people can actually afford to stay in.

What this tells us is that when engineers and designers engage with the economic and governance systems around their work, rather than treating them as fixed, they can unlock solutions that once seemed impossible.

Nightingale reminds us that regenerative engineering is not only about materials and technology. It is also about restructuring systems so that care, fairness, and long-term value are built in by design.

Reflection Prompts

As you consider the projects and systems you work within, reflect on the following.

- What definition of "value" is driving your project, and who does it serve?
- Which costs are visible in your model, and which are hidden or deferred?

Case Study: Healing by Design - Hospitals That Bring Nature Inside

Around the world, healthcare systems are rediscovering something ancient: that nature heals.

For much of the twentieth century, hospitals were designed as machines for treatment.[95] Efficiency, sterility, and clinical control dominated design decisions. Corridors were long and windowless. Materials were hard and washable. Patients were moved through systems optimised for throughput rather than experience.

The engineering worked. The human experience often did not.

Against this backdrop, Maggie's Centres emerged as a radical alternative.

Founded in the UK in the 1990s by Maggie Keswick Jencks and her husband, architectural theorist Charles Jencks, Maggie's Centres were created in response to Maggie's own experience of cancer care. She believed that while hospitals treated disease, they often failed to care for the whole person - emotionally, psychologically, and socially.[96]

The question behind Maggie's was deceptively simple: what would a cancer care space look like if it were designed around how people feel, not just how they are treated?

The answer reshaped healthcare design.

Each Maggie's Centre is designed by a leading architect, but all follow the same core principles. Buildings are flooded with natural light. Gardens are integral, not decorative. Views of nature are prioritised. Kitchens sit at the heart of each centre - open, domestic, and communal - offering a place for people to gather, talk, or simply sit quietly together.

There are no clinical reception desks. No harsh lighting. No institutional hierarchy. Instead, the architecture communicates safety, dignity, and care.

Importantly, this approach is not based on aesthetics alone. It is grounded in decades of research showing that exposure to nature and well-designed environments measurably improves health outcomes. Studies demonstrate that patients with views of greenery recover faster, require less pain medication, experience lower stress and anxiety, and report higher overall wellbeing than those without access to nature.[97]

Maggie's Centres do not replace hospitals. They sit alongside them, complementing clinical care with human care.

Actively Using the Built Environment to Heal

The inherited DNA of healthcare design prioritises efficiency of treatment, infection control and standardisation and predictability.

This makes sense as these elements are all critically needed for healthcare. These priorities are important, but incomplete. Maggie's Centres converted healthcare environments from places of clinical control to spaces of compassionate support. Success was no longer measured solely by how efficiently patients were processed, but by how supported, calm, and human they felt while navigating illness.

Healing was recognised as more than a biological process. It was acknowledged as one that depends on people's environment from physical to emotional, psychological, and social.

Maggie's Centres demonstrate that environments are never neutral. The spaces we design actively shape how people feel, behave, and recover.

By placing nature, beauty, and humanity at the centre of healthcare environments, especially the physical ones, these projects reveal that efficiency and compassion are not opposites. In fact, environments that support wellbeing often reduce stress, improve recovery, and lower long-term costs.

The lesson extends far beyond hospitals. When engineers and designers treat experience as a legitimate performance metric - alongside safety, cost, and function - systems become more resilient and humane. Care stops being something delivered only by professionals and becomes something embedded in the environment itself.

Maggie's Centres remind us that regenerative engineering is not always about new technology. Sometimes it is about remembering what humans need to heal and designing accordingly.

Reflection Prompts

As you consider the environments and systems you help shape, reflect on the following:

- What experiences does your design create, not just outcomes?
- Where could nature be integrated as a core element rather than an add-on?
- What would change if you designed for people at their most vulnerable?

Case Study: The Snow Ploughing Policy - When "Neutral" Design Isn't Neutral at All

In the winter of 2011, the city of Karlskoga in Sweden made a small administrative change to its snow ploughing policy. The results were anything but small.

Like many cities in cold climates, Karlskoga had long prioritised clearing main roads first after snowfall. The logic seemed sound. Major roads carried commuters to work. Clearing them quickly kept the economy moving. Residential streets, footpaths, and cycling routes came later.

On paper, the policy was efficient, rational, and appeared gender-neutral.

In reality, it wasn't.

City officials began to notice a troubling pattern. During winter months, hospital admissions for slips, trips, and falls spiked - particularly among women, older people, and children. These injuries were not occurring on highways or arterial roads. They were happening on footpaths, local streets, and pedestrian routes - the very places cleared last.

When the city examined travel patterns more closely, they noticed something.

Men were more likely to commute directly by car along main roads. Women were more likely to make trip-chained journeys: walking children to school, pushing prams, accessing public transport, shopping, visiting elderly relatives. These journeys relied heavily on local streets, footpaths, and cycling paths.[98]

The snow-ploughing policy hadn't been neutral at all. It had been designed around a male default travel pattern, without anyone ever naming it as such.

So Karlskoga tried something different: they reversed the priority order. Instead of clearing major roads first, the city prioritised footpaths, cycleways, and local streets - the routes people actually used to move through daily life.[99]

The outcome was immediate and measurable.

Hospital admissions due to winter-related falls dropped by half. The city saved money on healthcare and emergency response. Streets became safer. Mobility improved. And the policy worked better - not just for women, but for everyone.[100]

Nothing new was built. No technology was added. The system was simply rethought.

Karlskoga's snow-ploughing policy exposed how much these assumptions shape outcomes - and how costly they can be when left unexamined.

By treating commuting by car as the default, the system unintentionally transferred risk onto those moving differently: pedestrians, carers, children, older people.

The redesign rewrote that DNA - from neutral-on-paper to equitable-in-practice.

The snow-ploughing case shows how values shape outcomes, even in the most mundane engineering decisions.

No one set out to design a policy that disadvantaged women or increased injuries. But by failing to question whose movement patterns mattered most, the system encoded bias - and paid for it in human and economic terms.

When engineers and policymakers broaden their understanding of users, systems become safer, more efficient, and more just. The redesign didn't slow the city down. It reduced harm, saved money, and improved mobility.

This is Legacy by Design applied to policy to design systems that support life as it is actually lived.

Reflection Prompts

As you reflect on the policies, standards, and systems you work with, consider:

- Who benefits most from the way your system is currently prioritised?
- Whose journeys, risks, or needs are invisible in your data?
- Where could a simple reordering of priorities dramatically improve outcomes?

The Patterns Beneath the Projects, People and Inventions

When we step back from these case studies and look at them together, a pattern begins to emerge.

We see traditional systems designed around convenience and efficiency rather than care. Around speed rather than understanding. Around a narrow definition of success rather than the reality of lived experience. We see old assumptions carried forward and embedded in standards, policies, and tools, long after the world they were built for has changed.

Interestingly, in many of these case studies, change didn't begin with a breakthrough technology. It began with a different way of seeing, thinking, and often including.

Someone noticed who was missing. Someone questioned what had been taken for granted. Someone chose to design for how life is actually lived.

When that shift happened, everything else followed. Systems became safer. Places became more usable. Solutions became more durable. And value increased - not because more was added, but because the right things were prioritised.

This is the hidden DNA of engineering at work.

It shapes what we optimise for. It determines whose needs are considered early and whose are managed later. It decides whether problems are contained - or passed on.

The future of engineering does not depend on abandoning rigour or precision. It depends on expanding them. On recognising that technical excellence can sometimes be incomplete without social awareness, environmental understanding, and moral clarity.

The question facing engineers now is not whether we shape the future. We already do.

The real question is how consciously we choose to do it.

This begins with shifting our DNA and the values they inherently bring, by learning to think differently - not just about what we build, but about why, for whom, and to what end.

That is the work of the Future Engineer.

Thinking like a Future Engineer

Widening the View Without Changing the Standard

When you look across all of these case studies - from Medellín's cable cars to Dyson's vacuum cleaner - a pattern emerges.

Each one rewrote the inherited DNA of engineering not by abandoning standards, but by working creatively within them.

That's the art of the Future of Engineering. It's not about tearing the system down - it's about taking a pause, a step back and widening the lens in which we see and considering a wider perspective courage and care.

Engineers are already trained to see constraints, make unknowns know, create certainty and focus on solving a very specific problem. But what if we flipped this and expanded rather than constraining ourselves.

The "Think Like a Future Engineer" Framework

To rewrite our engineering DNA, we need to expand how we think, from a narrow technical view to a layered, values-rich perspective. This framework gives engineers five frames to apply to any project, decision, system, or design.

1. The Micro Frame - Materials, Moments, and Human Experience

Ask: *Who touches this? Who feels it? Who uses this? Who's included or excluded by how it's made?*

Start with just one person - a use case of one. Imagine their experience.

Then widen the view. List the potential users and their attributes. Think in segments: children, adults, seniors; different languages and cultural backgrounds; varying abilities; neurodiverse ways of thinking; people from rural or regional communities; night shift workers.

It might seem like a lot of work, but it always sparks something hidden. Even including one new perspective in a design is a win.

I learned the value of thinking from a totally different perspective - one that was the opposite of my own - in one of the businesses that I co-founded. When we were hiring, we deliberately looked for people different from us, especially for product development roles.

We'd think about the kind of person we wanted to attract using tools like DISC or CliftonStrengths, then write the job description based on the exact strengths and styles for the type of person we wanted.

On one occasion, we asked my co-founder's husband - who matched the profile we were targeting - to review it. He told us we'd missed the mark and rewrote it dramatically. This revision helped us attract outstanding candidates who were perfect for the role and perfectly unlike us.

That's the micro frame: zooming in to the experience level to see how the design meets real people, real bodies, real senses, and real daily moments.

We see examples of this in our daily lives. The tactile arrow button at a pedestrian crossing, for example, is one small interaction point that opens up a whole street for people with vision impairments or other disabilities.

The micro frame invites us to think beyond technical specifications and ask: how do people feel in this space? Does it stress them or calm them? What sensory experience and emotion are we creating?

2. The Meso Frame - Community and Connection

Ask: *Who's impacted by this project as a group - directly, indirectly, and invisibly?*

This frame zooms out from the individual to the neighbourhood, the workplace, and the community. It asks us to look beyond the immediate asset and consider how it shapes the relationships around it. What is the bigger picture here, beyond the direct solution?

When I was working in a management consulting role, I was seconded to a large project for one of Australia's major banks. The task was enormous: moving 12,000 people across 12 buildings in Sydney into just three. The logistics alone were staggering, from lift movements and loading docks to packing schedules, relocation sequencing, and communications planning.

My role centred on the huge dataset of those 12,000 people. In an organisation that size, people were joining and leaving constantly. The data shifted every week. The original aim was to create a single source of truth, but it quickly became clear that this was impossible. The system was always moving.

So instead of spending endless hours trying to get the data to 100% accuracy, we took a different view.

We set the intention of "mostly correct". We accepted that the data would always be slightly out of date, and instead focused on accuracy within a reasonable range.

Letting go of perfection and focusing on what mattered most to the community of people we were moving made the project flow more easily and more effectively.

We zoomed out and looked at the data set and the workplace more holistically, from a team perspective rather than an individual one.

Some examples of the meso frame in action:

- **Medellín's Metrocable.** Transport infrastructure became social infrastructure, reconnecting communities long isolated by geography and violence.
- **Vienna's "City of Women".** Designing with care work in mind made the entire city safer and more accessible for everyone.
- **Snow ploughing policy in Sweden.** Clearing local streets first instead of highways cut slips, trips, and falls by 50%.

This frame can be built into projects by expanding what we measure. Often we do not know the impact until a new approach is tried, like the Sweden snow ploughing example, simply because we've never done it that way before. We can introduce community-focused metrics and intentions into KPIs: social cohesion, accessibility, belonging, participation, and connection.

It's about designing not just for individuals, but for the invisible networks and relationships that allow communities to thrive.

3. The Macro Frame - Systems and Sustainability

Ask: *What are the upstream and downstream impacts of this project? What cycles, seasons, and spaces am I part of?*

The macro frame widens our view from project to planet. It's where systems thinking and regenerative design live, where water, energy, materials, ecology, culture, and time all intersect. When we work at this scale, we stop seeing our design as a standalone asset and start understanding it as part of a living system.

One of the clearest examples I've seen came from my friends Eileen Deemal-Hall, then CEO of an Indigenous organisation in Far North Queensland, and Timothy Rutherford, who was leading an army division responsible for border patrol in the region.

Tim's team were preparing for the annual arrival of illegal fishing vessels. Historically, the boats came at a predictable time each year, based on weather and tide patterns. But one season, Eileen's rangers noticed something unusual: the sea cucumbers were appearing early.

Her team told Tim, "The boats will come sooner this year."

Tim's team were sceptical, as their data suggested arrivals were still weeks away. But Eileen and her rangers were right. The boats came earlier that year. Why?

The reason was simple and profound. The sea cucumbers were responding to subtle environmental cues long before those changes showed up in any dashboard or forecast. Indigenous knowledge holders could read the system because they were connected to it.

This is the power of the macro frame. It reminds us that engineering doesn't sit outside nature. It operates inside ecological, cultural, and seasonal cycles we cannot see on a drawing or a Gantt chart. When we listen to Country, and to those who hold generational knowledge, we gain access to signals modern engineering often overlooks.

Other macro-scale examples show this same pattern:

- **LMS Energy's biogas work in Australia** turns organic waste into renewable energy, transforming a linear waste stream into a circular, regenerative system.
- **Costa Rica's renewable grid**, running at about 98% on hydro, geothermal, and wind, shows what's possible when a nation invests in long-term systems rather than short-term fixes.

Whether it's a border patrol operation, landfill, or national energy strategy, the lesson is the same: systems reveal themselves when we look beyond the asset and into the cycles around it.

At the feasibility or concept stage, map the flows of energy, materials, water, waste, ecology, people, and time. Then ask: where could circular or regenerative loops exist? Who already understands these seasons and cycles? Who on Country do we need to listen to and learn from?

4. The Personal Frame - Leadership and Intention

Ask: What values am I bringing into this decision? What am I really optimising for? What is the intention?

The personal frame is where engineering shifts from technical judgement to human judgement. It's the space where our training meets our intuition, where our expertise meets our integrity, and where the brief meets the quiet internal question that matters most: is this the right thing to do?

Every engineer faces these moments. They rarely appear in drawings or models, but they shape projects just as profoundly.

One of mine came while I was working for a global consulting firm, leading a feasibility study for a university. Their library was an old, dark building badly in need of renewal, and our job was to explore options for improving it. We pulled together a multidisciplinary team - engineers, cost planners, sustainability specialists, and an architect - and developed five detailed design options.

We ran a multi-criteria analysis, carefully considering daylight, operations, heritage, accessibility, cost, student experience, and sustainability. When we submitted the report, the choice was clear. But the person overseeing the project replied, "I don't want that option. Can you make it this option instead?"

There was no technical justification, and eventually I learned why from my boss: the person asking wasn't a university employee but a contractor. Choosing the option he wanted meant a bigger project for them to manage and a higher chance of another contract.

Their intention wasn't aligned with the university's - or ours. It was a personal one for them.

Of course, we said no and submitted the report based on the assessment and outcomes. But at that moment, I was reminded that engineering is rarely just about the engineering itself. It's about the intentions behind the decisions - ours and everyone else's.

Moments like this aren't always so obvious. Often, the personal frame appears earlier and more subtly. It might be noticing that a detail meets code but will cause maintenance problems later; sensing that a space technically complies but will not feel safe; or pushing back on a timeline that reduces community engagement to a token gesture.

These everyday moments can be viewed through a simple question that one of my mentors asked me when I was faced with a very challenging conflict of interest situation. She said "could you, but should you". This question often stands out to me when I am faced with difficult decisions.

These small moments define us and our projects just as much as the big ones.

These are the decisions that rely on judgement, not just standards.

The personal frame lives exactly here:

- when you feel the pull between what's easy and what's right
- when a stakeholder's intention doesn't align with the community's needs

- when the design is technically sound, but the intention is not
- when your values are being asked to bend

It is the inner work of our profession - the courage to pause and ask, *"What are we really optimising for?" even when the answer is uncomfortable. It's present* in the moment a graduate asks the question they're afraid will expose what they don't know. It's present in the moment a leader decides to slow down to honour community trust. It's present in every decision where care, courage, and clarity outweigh convenience.

Building values check-ins into project gateways helps. Ask:

- Why are we choosing this option?
- Whose interests are driving this decision?
- Does this align with the intention we set for this project?
- What would we choose if we were designing for legacy, not just delivery?

The personal frame is where engineering leadership begins - not in the calculations, but in the courage to stay anchored in intention, even when the path around you wavers.

5. The Legacy Frame - Time, Care, and Consequence

Ask: *What will this look like in 50 years? Who inherits it? What story will they tell about us?*

The legacy frame stretches our thinking beyond the life of a project and into future generations. It's the widest, deepest view we can take as engineers - a view that extends past budgets, KPIs, or election cycles, into the hands of the people who will live with what we build long after we're gone.

Engineering is one of the few professions where our decisions truly outlive us. The bridges, roads, systems, and policies we design all become part of the inherited landscape of a community. With that comes a responsibility that is both humbling and profound.

One of my most powerful lessons on legacy came a few years ago when I was in New Zealand presenting the opening keynote for the IPWEA International Public Works Engineers Conference. My opening keynote from a previous conference had inspired the theme of this conference: inspiring the next generation.

Immediately after my talk, five Gen Z engineers presented lightning talks. I'll never forget one young woman who stood up and said:

"My career will be about maintaining assets that were built 50 or 100 years ago. That's the bulk of my work - maintaining the legacy you've handed us."

That insight landed hard. She was right. Her entire career would be shaped by decisions people made decades before she was born. And in that moment, the real weight of legacy became so clear.

I realised I had never really been taught to think like that. My education, and much of my early career, was based on delivery, not stewardship. Handover marked the end of a project. But in reality, that is where its life begins. At Power of Engineering events, we run an exercise asking students to design a chair that lasts one minute, one hour, one year, or one hundred years. Their designs change completely when time changes the brief.

The same is true in engineering.

When we shift the time horizon, the entire design changes.

We can see legacy in action - both the risks and the opportunities - all around us. One example that illustrates legacy done well is the Melbourne Urban Forest Strategy. In the early 2010s, Melbourne's street trees were dying at alarming rates due to drought and heat. The city didn't just plant replacements; they created a 40-year strategy to completely transform the urban canopy. They planted not just for shade today, but for resilience decades from now - a gift to a future city whose citizens will feel the benefit long after the original planners and engineers have moved on. That is legacy-conscious design: long-term, regenerative, and intentional.

Legacy is not just about grandeur or scale. It lives in the everyday decisions too - the materials we choose, the lifespan we design for, how easy something will be to repair, the adaptability we build in, and whether the asset will nurture people or burden them.

When we look to First Nations knowledge, legacy becomes even clearer. Indigenous cultures design through the frame of time, working with cycles, seasons, kinship, and Country, grounded in an understanding that what is done today shapes life far beyond our own generation.

Many Aboriginal American Nations speak of considering the impact of decisions on the next seven generations.[101] It's a simple, powerful principle: design not just for those who are here now, but for those who will come long after we're gone.

Legacy thinking is difficult. Research shows that our brains are simply not built for long-term thinking. We discount rewards and consequences that lie far in the future because our evolutionary wiring, cognitive limits, and the abstraction of our future selves all conspire against it. This phenomenon, known as temporal discounting, means that many of us prefer a smaller benefit now rather than a larger benefit later, even if the later benefit is objectively better.[102]

In engineering, this presents a clear challenge. Our professional training often emphasises the next milestone, the next specification, the next commissioning date - and less often the 50-year or 100-year horizon. In cognitive terms, the distant future is psychologically "far away" ,which means the value we place on it drops sharply.

That makes designing for legacy, rather than just delivery, an uphill battle. Research confirms that when decision-makers feel less connected to their future selves or perceive the future as uncertain, they discount the future even more steeply.[103]

The good news is all is not lost. By acknowledging this bias, we can deliberately "de-discount" the future and design with a longer horizon. That means reframing what we're optimising for - not just what solves the problem today, but what will serve people decades from now. It means embedding the legacy question at every gateway: what story will future generations tell about us? It means shifting from "deliver this asset" to "inheritance for those who follow".[104, 105]

This shift is especially important in contexts where communities may not have the luxury of long horizons - where socio-economic or environmental instability forces focus on survival now, making legacy harder to prioritise.[106] This is the essence of the legacy frame: to design with the people of tomorrow, not the pressures of today.

A project that lasts 100 years doesn't need to be indestructible; it needs to be cared for, adaptable, and connected to the systems that sustain it. Longevity doesn't mean permanence - it means stewardship. We have seen this with the Sydney Harbour Bridge and its projected life span of 400 years.

For your project, system or challenge, consider asking at key project milestones:

- How will this asset perform in 10, 20, 50, or 100 years?
- Does this project regenerate or deplete?
- Who will maintain it, and have we designed for them?
- What would we want future generations to say about it?

When we design with legacy in mind, we're not just building projects. We are shaping our world for future generations, and we're choosing what kind of world we hand to the young engineer who will one day stand on a stage and say, "This is the legacy they left us."

Using the Framework in Practice

The Think Like a Future Engineer framework isn't another compliance checklist. It's a mindset you can bring to any design meeting, specification, or collaborative stakeholder conversation in a moment. It helps engineers ask better questions, challenge briefs, and pause - the kind of actions that reveal values hiding in plain sight.

I see it as a simple framework that could shift thinking on a project, and therefore the outcomes for decades to come. Here is how you might use it:

Scale	Frame Focus	Design Prompt	Example Lever
Micro	Human experience, sensory and embodied interaction	Who touches this? Who feels it? Who's included or excluded in this moment?	Sensory, accessibility and emotional-experience criteria
Meso	Community, relationships and collective impact	Who is affected - directly, indirectly, invisibly? What's the bigger picture beyond the asset?	Community-focused KPIs (e.g., social cohesion, belonging, equitable access)
Macro	Systems, cycles, Country and sustainability	What are the upstream and downstream impacts? What cycles, seasons or dependencies are we part of? Who holds the knowledge of this system?	Mapping resource and ecological flows; identifying circular or regenerative loops with guidance from Country
Personal	Leadership, intention and values based judgement	What values are driving this decision? Whose interests are being served? What am I really optimising for?	Team "values check-in" at project gateways; intention-setting
Legacy	Time, consequence and intergenerational stewardship	How will this age over 10, 50, or 100 years? Who inherits it? What story will they tell about us seven generations from now?	Legacy Scorecard (longevity, adaptability, maintenance, regeneration, equity)

Views You Can Open Right Now

Even within the most rigid standards, there are creative openings where new frames can live and make a subtle but impactful change on a project. Integrate this framework into existing project structures, or use it to ask the right questions in a meeting or bring these frames to your reviews.

Here are a few starting points:

- **Pre-briefing with the client;** Take the time before the brief is responded to deeply understand the intent
- **Project kick-Off:** Take time to establish values, vision, and intention for projects and have this present in each project meeting like a "safety moment".
- **Specifications:** Add intent-based clauses (e.g., "optimise for inclusion and user diversity").
- **Procurement:** Weight social and environmental outcomes in bid evaluations.
- **KPIs:** Balance performance metrics with community and wellbeing measures.
- **Gateways:** Include an ethical reflection checkpoint at concept and detailed design stages.
- **Reporting:**- Include a "legacy statement" alongside cost and schedule summaries.
- **Dear my future self:** Write a letter to your future self or the future project team at least a year in the future.

Thinking like a future engineer means shifting from builder to steward - someone who not only constructs, but cares for the systems that support life.

You don't have to change the code to change the inherited and hidden DNA. You just have to start writing new values into the spaces the code forgot.

The Takeaway

Once we see how engineering culture is formed, it becomes clear that systems do not just shape outcomes. They shape people. The invisible curriculum does not stop at how engineers design; it influences how they think, relate, lead, and cope under pressure.

When cultures reward certainty over curiosity, speed over reflection, and perfection over learning, they do more than narrow decision-making. They place a silent but constant load on the people working within them. Stress becomes normal. Burnout becomes expected. Disconnection becomes invisible.

If engineering is really about serving people, then we must also consider the impact our systems have on those who design, build, and operate them. The next chapter introduces a different lens, one that brings human wellbeing and nervous system stewardship into engineering practice. It asks what becomes possible when we design systems that support regulation, safety, and care, not just performance, and what responsibility engineers carry for the human experience embedded in every project.

Part II:
The Framework - Three Lenses for Better Outcomes

When I think about how we've been trained to define success in engineering, I often remember the number of times in my career a manager would grab a scrap of paper, draw a triangle, and label the corners "time", "cost", and "quality".

Then they'd smile and say, "Now pick two."

This happened in multiple organisations that I worked for.

It was half-joke, half-truth - a shorthand for the idea that you can never have it all. There are always constraints and an assumption that we have limited choices.

For years, that triangle was in the background, subconsciously shaping the way I thought about projects: what do I have to compromise on, give up, or take out?

The funny thing is, no one ever asked, "How can we have all three - and more?" or "What if connection, wellbeing, and legacy were part of the equation too?" That triangle wasn't wrong; it just wasn't complete.

This way of thinking taught us to think in constraints rather than possibilities - to focus on what we might lose, and what we had to sacrifice and compromise on, rather than what we could create. When we start to ask "how can we...?" questions and others, our brain problem-solves in all new ways.

Engineers are also trained in something very powerful: creating certainty. At one consultancy I worked with, this was the core of the sales pitch. They even rebranded the company to align with this. People work with engineers because we give them certainty. That is what they are buying.

One of the reasons why I have jokingly said I loved being a civil engineer was that you always know one thing in every equation. Generally speaking, you don't want movement in your design, so the equation needs to equal zero - it has to balance. I secretly liked this certainty that I could create in the physical world.

As engineers, we are taught to eliminate uncertainty - but people can feel uncertain and unpredictable. Stakeholders and communities don't behave predictably. They disagree. They challenge the data. They take time. So it's no surprise that in many projects, engagement is treated as a risk to manage, not a resource to design with. It can sometimes be seen in projects as something you do after the design is done - not something that shapes the design itself. Or sometimes a "consultation" happens, which is just really an "informing" session, telling people how it will be done.

But every time we exclude people from the process, we lose something essential. We lose perspective. We lose trust. We lose creativity. The greatest missed efficiencies in engineering don't come from stakeholder engagement - they come from ignoring it, because the people who live with the design often see things the drawings never will. They understand the system in ways data can't.

Rewriting the DNA of engineering means rethinking not just what we build, but who we build with, why we are building it, and how long we expect it to last.

Engineering has always been a discipline of balance - between safety and innovation, cost and performance, ambition and restraint. It makes sense: we can't possibly design for everything, so we need to pick something. But some of the most important dimensions of that balance are often not measured in the way we design and deliver projects: how they affect people's wellbeing, how they interact with the full system natural environment, and how they endure over time.

This is why we need to bring in the Three Lenses Framework to engineer the future.

The Three Lenses for the Future of Engineering

Every discipline has its turning point - a moment when the old logic no longer fits the new world. For engineering, I believe that moment is now.

The systems we inherited were designed for extraction, expansion, and efficiency. They did what they were built to do - they delivered growth. But in doing so, they encoded a narrow idea of success: speed over connection, cost over care, control over curiosity.

We've reached a point where that inherited DNA is no longer enough for our future. The challenges we face - climate change, inequality, disconnection, mistrust - are not purely technical problems. They are human problems that demand human solutions.

Rewriting the DNA of engineering doesn't mean discarding our technical brilliance. It means expanding it. It means bringing new values to the design table - values that measure not just what works, but what matters.

From Efficiency to Empathy

Across the world, it is exciting to see projects are already rewriting the script. Engineers, designers, and communities are showing that we can deliver outcomes that are technically excellent and socially intelligent.

They are doing it not by rewriting the standards, but by reframing the questions - shifting from "Can we build it?" to "Who is this for, what does it give back, and how will it endure?"

To make these shifts visible, we will dive into a simple framework: Three Lenses for Better Decisions. Each lens offers a new way of seeing: one that reconnects engineering with life itself.

The Three Lenses

To navigate this shift to create the Future of Engineering, we need new ways of seeing our work - not new rules, but new perspectives. Rewriting the DNA isn't a single act of reform. It's a daily discipline - of seeing, listening, and choosing differently.

And it starts here: by widening our lens. When we see the system through people, nature, and time - we start to design a future that can sustain all three.

I believe this can be boiled down simply to three lenses:

- **Human Wellbeing** - designing for the nervous system.Every design decision affects how people feel: calm or anxious, safe or excluded, connected or isolated. This lens helps us measure what truly matters - dignity, comfort, accessibility, and inclusion - as carefully as we measure stress loads or energy use.

- **Regenerative Environments** - designing with nature, not against it. This lens moves us beyond sustainability's "do less harm" mindset toward active regeneration. It asks: how can this project restore the systems that support life - water, soil, biodiversity, and energy - instead of depleting them?
- **Generational Legacy** - designing for time. The final lens widens our horizon to 50, 100, or even 500 years. It reminds us that our work will outlive us, and challenges us to build things that can evolve gracefully, adapt intelligently, and remain loved long after they are new.

Together, these three lenses form a new blueprint for the profession to create a world that embodies legacy by design.

The Three Lenses are not rules or checklists. They're not an overhaul of engineering itself. They're ways of seeing our designs and the world - tools for decisions that honour life, place, and legacy. Each lens rebalances to enable future focused engineering in a more connected way than ever before, taking into account the human experience, natural systems, and time.

It's a way of thinking, valuing and acting. It invites us to look at every decision through three connected perspectives - each of which adds a deep layer of connection to people, place, and time. When used together, the Three Lenses help engineers, policymakers, and designers make decisions that are not only technically excellent, but socially, environmentally, and ethically sound and inclusive.

The fundamental principle or north star of this framework is connection: connection to one another, to the ecosystems that sustain us, and to the generations who will follow.

Chapter 4: The Human Wellbeing Lens and Nervous System Stewardship

"Every bridge, every app, every system we design touches someone's body. It can calm them, or it can cost them."[107]

Engineering systems do not just move traffic, distribute energy, or manage risk. They shape how people feel in their bodies as they work, live, and interact with the world around them. Yet the human nervous system is rarely treated as something engineers design for, even though it is affected by every system we create.[108]

This chapter introduces the Human Wellbeing Lens and the idea of nervous system stewardship as an extension of engineering responsibility. When systems are designed for constant urgency, uncertainty, and control, they place chronic stress on the people within them. Over time, this affects judgement, safety, trust, and the ability to make good decisions.

To design better systems, we must first understand how humans experience them. This chapter reframes wellbeing not as an individual resilience problem, but as a systems issue, asking what engineering looks like when we consciously design for regulation, safety, and care alongside performance and efficiency.

The Hidden Engineering of Everyday Life

From the moment you wake up, you are surrounded by engineered systems that shape how your body feels.

You might be woken by a jarring synthetic alarm, bathed in cold blue light from your phone before your body has registered it is morning. The water that comes out of your tap - what was added to it to give it that taste, smell, and colour? Your commute might be a sequence of honking horns, brake lights, and screens. You step into a building with filtered air, fluorescent light, and piped music; it smells faintly of carpet glue and coffee.

None of this is accidental.

Every part of it was engineered.

All of these are the product of design decisions.

And almost none of it was designed with your nervous system in mind.

By the time you arrive at work, you've already experienced the invisible architecture of the modern nervous system: a constant background hum of stimulation, optimisation, and subtle unease.

I remember the very first construction site I worked on - an office high-rise building that had a green star rating. My job was to go through the design features and make sure they were ticking the boxes to get the green star rating.

On one particular day, I was looking into the paint that was going on the walls and confirming it had a low volatile organic compound (VOC). I didn't really know what that meant, so I looked it up. I found out that the reason that paint smells is because of this thing called VOC. The paint we were using didn't smell because it didn't have it in.

It really struck me at 20 years old. I thought, "Hang on a second. Why would we put these things in paint in the first place if they're harmful?"

The First Lens

The first lens of the Three Lenses framework - the Human Wellbeing Lens - is about becoming stewards of the nervous system, because engineering touches it at every scale.

When I walk through a space, I can feel it before I can analyse it. You probably can too. Some streets calm you - tree-lined, shaded, rhythmic. Others set your pulse racing - too loud, too fast, too exposed.

I've realised over time that infrastructure isn't neutral. Every design we create interacts with people's nervous systems. It either calms or stresses, connects or isolates. It's not something we're taught in engineering school - but it's something we all experience every day.

When we design our world, as engineers, there is more we can do to embed human wellbeing into what we build and create.

I've spent most of my career talking about projects in terms of risk, budget, and technical performance. But the biggest risks to success aren't always structural; they're human. The design that confuses people, overwhelms them, or isolates them - even if it's technically perfect - fails its most important test: how it makes people feel.

That's what this first lens is about - Human Wellbeing. It's the part of engineering that doesn't fit neatly in a spreadsheet, yet it shapes whether people trust, enjoy, and thrive in the spaces we build.

Feeling Your Way Through a Building

Another way to consider this experience is when you're walking into a new building for the first time. For me, I always get a kick out of the experience walking into a building for the first time and connecting with that first impression.

If it's a large office building in the city, you might think: how does the foyer make you feel? Are you intimidated or excited? Are there different smells or noises that greet you as you walk in? And as you move through that space, is it easy to find where to go? Do you know how to get to your destination despite being there for the first time?

I'm sure you've had that experience before where you're in a rush for a meeting or a job interview, running slightly late. You scan the room trying to find the lifts or a sign pointing you in the right direction. Sometimes it's effortless; other times it's disorienting.

I often smile when people ask me at events, "Hey, do you know where the bathrooms are?"

If we're in a high-rise building, I'll usually say, "They're probably near the lift core," because that's where they almost always are.

Good design gives you a sense of orientation without needing to think about it. It guides you gently, like walking through a forest and following the worn path ahead. It feels natural, like the space itself is telling you where to go and how to experience it. That's the kind of design that fills your cup.

The same principle applies to many other areas of our world, like road signage and knowing where your exit on the freeway is easily.

Nowadays, we even have the expectation with technology that you can just open an app and know how to use it. When I was at school, we needed training on how to use Microsoft Word and other programs because they weren't as intuitive back then.

And then there's the opposite. The busy arterial road that traps you in stop-start traffic, where noise and fumes leave your shoulders tight before you've even reached your destination. That's engineering too - but it's engineering that disrupts the nervous system rather than supporting it.

The difference between those two experiences is invisible but powerful. It's not about architecture versus infrastructure; it's about how the things we build make people feel.

Walking in the City

I've lived in Brisbane, Sydney, and Melbourne, and I've always found that the "vibe" of a city can be felt in the smallest, simplest elements - even a pedestrian crossing.

When my husband and I lived near North Sydney, we'd spend weekends away in nature. On the drive home, it became a running joke that we'd both feel our bodies tense as we approached the North Sydney exit. We called it the North Sydney Effect. It was as if, before we even got home, the city had already wound us up.

I noticed this same intensity in the pedestrian crossings. In Sydney's CBD, the green person flashes for the absolute minimum amount of time. It goes green... and as soon as the fastest walker is halfway across, boom! Flashing red. The city moves fast, and you have to keep up.

Melbourne feels different. The green light holds a little longer. There's room to breathe. The city seems to move *with* you instead of against you. Walking in the suburbs there, I was often confused by drivers giving way so regularly to pedestrians. It was so unexpected, compared with other cities.

Brisbane takes it even further. Walk down Edward Street at a reasonable pace and you'll catch green after green after green. It's as though the city has synced itself to human rhythm, not just traffic flow. Whenever I walk through Brisbane's CBD, I find myself smiling at the synchronicity, because the environment feels like it's designed for me, not despite me.

These are micro decisions - the timing of a light, a pause in a crossing sequence, a moment of flow - but they shape a macro emotional experience. A city can calm you or agitate you long before you consciously notice why.

It makes me wonder: what else are we designing into people without realising it?

How many tiny decisions in our infrastructure - lighting, crossings, benches, signage, materials - accumulate into a nervous system that feels held... or harried?

Designing for the Nervous System

Every piece of infrastructure and every engineered system changes how people feel. We measure the technical performance of our work - load, flow, cost - but rarely do we stop to look at its physiological performance and human impact. How does it make people feel in their bodies? In their senses? In their daily lives?

We live in a stress epidemic. Urban environments expose residents to a cascade of stressors: relentless noise, overcrowding, insufficient transport or green spaces, and a structural lack of pause.[109, 110]

Long-term physiological consequences follow: elevated cortisol, impaired sleep, higher blood pressure, and increased risk of chronic disease.[111, 112]

Likewise, social disconnection is rising. Research shows that the built environment plays a role in loneliness and isolation. For example, one systematic review found that certain structural elements of the built environment are strongly associated with people's feelings of loneliness or social isolation.[113, 114]

This lens asks: "What does this design do to the people who live, move, and work within it?"

Designing for the nervous system is not about soft values. It's about physiological precision and safe spaces. Noise raises cortisol; daylight balances circadian rhythm. Way-finding clarity lowers cognitive load. Open, calm, coherent spaces that restore people through beauty, comfort, and dignity perform better.

One study found that spatial openness reduces stress through brain and visual stimulation.[115] Hospitals with natural light record faster recovery times.[97] Workplaces with acoustic balance and greenery have reduced sick days and staff turnover.[116, 117] Transport systems that include pause, protection, and sensory balance lower aggression and improve compliance.

And this is not indulgence - it's risk management for people. When we embed the principles of biophilic design - nature-rich environments, daylight, direct or indirect nature interaction - we see measurable outcomes: reduced stress, improved cognitive function, faster recovery, lower heart rate, lower cortisol.[118, 119]

In other words: when we build with the nervous system in mind, we build infrastructure that not only serves function, but which sustains human life.

Over the last century, we have designed the world into a constant state of mild activation - a low-level hum of tension that our nervous systems were never built to hold. We're just... *on*. All the time.

We engineered:

Speed without stillness. We optimised for acceleration - faster commutes, faster transactions, faster logistics, faster everything - but we forgot that the human body has no "fast mode". Our nervous system still runs on the same biological rhythms as our ancestors. We designed environments where everything moves quickly except the one thing that matters most: our ability to regulate.[120, 121]

Light without dark. We flooded our world with LEDs - in our bedrooms, on our streets, in our pockets. Day became 24 hours long. Darkness, which once cued restoration, rest, and repair, has become a rare resource. We engineered illumination, but at the cost of circadian coherence. We don't let the planet sleep, and we don't let ourselves sleep either.[122, 123]

Efficiency without empathy. We streamlined processes, minimised friction, optimised throughput, and called it progress. But in shaving off seconds and squeezing in output, we also stripped out softness - the buffers, pauses, and human moments that make a system feel humane. We built workflows that run flawlessly on paper, but leave the people inside them exhausted.[124, 125]

Control without care. We mastered data, dashboards, sensors, and metrics. We can track flows of cars, water, electricity, capital, and carbon with astounding accuracy. Yet we often fail to track the human load - tension, cognitive overwhelm, sensory fatigue, social disconnection. We engineered systems that know everything about performance, and almost nothing about the people performing within them.[126, 127]

We built for performance, but forgot that people are part of the system. We forgot that human beings aren't just users of infrastructure - we *are* infrastructure.

Our bodies regulate based on the environments we inhabit. Our heart rate mirrors the rhythm of our surroundings. Our senses take in every stimulus - noise, light, speed, complexity - and convert it into biology.

Over time, we have created a built world that demands more from people than people were designed to give. And the cost shows up not in spreadsheets or budgets, but in cortisol levels, sleep patterns, emotional bandwidth, and the rising tide of stress, burnout, and disconnection.

Consider the statistics. A report by Boston Consulting Group found that nearly 48% of workers across eight countries are currently grappling with burnout.[128] Another survey found that, in 2024, 39% of global employees cited burnout and exhaustion as their top workplace challenge.[129] These are not just organisational problems; they are physiological ones.

But the built environment isn't just exacerbating stress. It's contributing to another epidemic: disconnection - from nature, from each other, and from our own bodies. Research shows that human connection to nature has declined by over 60% since 1800, driven by urbanisation, digital overload, and shrinking green-space contact.[130, 131]

Studies also reveal that people with stronger connections to nearby nature report better mental health, lower psychological distress and greater vitality - even when controlling for age, income, and neighbourhood.[132]

In this design era, the result is two-fold: a body under constant activation and a body deprived of restorative rhythm. It's like building a car with the engine revved high but never shifting down a gear - and expecting the driver to be relaxed.

Most of the time, harm isn't intentional. It happens in the margins - in the places we overlook, the decisions we make in haste, the assumptions we inherit. The projects under pressure where the focus is time and budget. A corridor without a window. A bench designed to stop someone resting. A public space that looks "efficient" on paper, but makes people feel exposed or unwelcome. These decisions rarely come from malice. They come from forgetting that bodies feel what drawings can't capture.

But there are also moments in modern times where design has been used deliberately to unsettle the body.

Take the architecture of surveillance - prisons, high-security facilities, even some public squares shaped by the logic of the panopticon. The idea is simple: create the sense of being watched from every angle, even when no one is there.[133]

But the experience is profound. People become hyper-vigilant. Their breath shortens. They learn to scan constantly, to brace. Their nervous systems stay on alert long after the environment changes. That is design shaping physiology - and not in a way that nourishes it.[134]

Or consider hostile urban architecture: spikes on ledges, benches split into awkward angles, railings too steep to lean on, sound frequencies designed to drive teenagers away. These interventions send a message long before they send a person away.[135, 136]

They say: "You are not wanted here."

Research now shows what people have always felt - that exclusion by design creates loneliness, mistrust, micro-stress, and a subtle sense of being "on guard" in your own city.[137, 138]

And then there are the places that were never meant to harm but do so accidentally - hospitals built like mazes, stations without sightlines, office floors that echo and overwhelm, or neighbourhoods that feel unsafe to go for a run in because it's night time. These are not acts of cruelty. They are simply cases where the human body was not at the centre of the brief.

When you start to see these patterns, something shifts. You realise that engineering has always had the power to influence how people feel - to steady the heartbeat or to quicken it; to bring ease or to compound stress; to offer belonging or to push people to the edges of public life. And the most important realisation? We get to choose which way it goes.

This is not about blame - far from it. It's about awareness. It's about remembering that every design, however small, leaves an imprint on someone's nervous system.

A handrail. A tunnel. A crossing light. A bench. A waiting room. A path that rises or falls.

Once you see that design can both harm and heal, you're suddenly holding a kind of silent power - the ability to shape environments that soften people rather than sharpen them. Environments that invite people in. Environments that give people space to breathe. Environments where the body feels safe enough to belong.

This is where the work becomes beautiful again, because the same attention that once created stress can create sanctuary. The same engineering that once overwhelmed can offer calm. The same systems that once demanded vigilance can now support dignity, connection, and ease.

We cannot undo the past. But we can design the future - deliberately, compassionately, and humanely.

That is the choice that sits in our hands. The built environment has become a stressor. But what if it could also become a healer, if we choose to design for the nervous system, not just the network?

This Isn't Just About the Built Environment

This lens reaches far beyond the built environment and infrastructure. Every field of engineering shapes the human nervous system, often invisibly. Here are some examples of what it looks like in everyday life.

Forever Chemicals: Convenience That Stays Forever.

We engineered stain-proof sofas, non-stick pans, and waterproof jackets. We also engineered "forever chemicals" (per- and polyfluoroalkyl substances or PFAS) - compounds that persist in soil, water, and human tissue for generations. Their efficiency at resisting stains came with a hidden cost: chronic inflammation, hormonal disruption and even accelerated biological aging. For example, PFAS exposure has been linked to immune dysfunction, endocrine disruption, and elevated inflammatory markers in exposed populations.[139, 140, 141, 142] We optimised for durability, and ended up designing stress into biology itself.

Ultra-Processed Food: Designing for Convenience, Not Nourishment.

Food technologists perfected the "bliss point" - the precise mix of fat, salt, and sugar that lights up reward centres in the brain. The only place this naturally occurs for humans is in breast milk. In Australia, ultra-processed foods now make up around 42% of total energy intake for adults, and some studies show up to 56% of household energy is derived from ultra-processed food purchases.[143, 144, 145, 146, 147] We built a global diet that keeps our nervous systems in constant pursuit of the next hit. It's engineering brilliance - and physiological chaos.

Notification Architecture: Designed Distraction.

Software engineers designed the red dot, the ping, the streak. Each one delivers a micro-dose of dopamine.[148] Together, they train the body for vigilance. The result? A world of people who never truly switch off.[149] We are addicted to mobile phones and disconnected from the people and the planet. We engineered attention capture - but not attention care. Research on smartphone notification overload links constant partial attention to elevated stress, poorer emotional regulation, and reduced sleep quality.[150]

Open-Plan Offices: Collaboration or Chronic Stimulation?

We designed workspaces for efficiency and density. But constant background noise, visual movement, and interruptions keep our nervous systems in low-level alert. Studies have found that typical open-plan office noise raised participants' negative mood by 25% and caused a 34% increase in sweat response - both clear physiological stress indicators - even after just eight minutes of exposure.[151, 152, 153] We've learned to wear noise-cancelling headphones not for focus, but for survival. When we design for maximum visibility, we often erase the conditions for calm thought. Often it feels strange or scary to sit in silence with our own thoughts.

Mobility Pressure: Time as a Source of Stress.

From airports to freeways, every second is measured. Just-in-time systems save money but shift risk to humans - drivers, pilots, passengers - who absorb the stress of zero slack. In the rush for efficiency, we've removed recovery time. The body doesn't just move; it senses the margin for error, the countdown, the acceleration. When that margin disappears, our nervous system remains in perceptual tension.[120, 154, 155]

Artificial Light: Losing the Night.

We filled the world with artificial light - "safe", bright, constant. But by extending daylight into eternity, we disrupted the cues that regulate sleep, metabolism, and mood.[156, 157] Light once meant safety and rhythm. Now, it often means fatigue and restlessness. Studies of circadian disruption show that exposure to artificial light at night is linked to poorer sleep quality, increased risk of depression, and metabolic dysregulation.[159] For example, external light at night has been associated with higher rates of obesity and diabetes.[158] We engineered light for convenience, but our biology paid the cost.

Alarms and Alerts: The Sound of Stress.

Hospital monitors, reversing trucks, crossing signals, sirens - each necessary, each activating. But together they create a wall of anxiety - a city that never stops shouting. Safety shouldn't have to scream. We could design for calm response instead of chronic startle. I remember being in hospital after the birth of my second baby, and leaving a day earlier than officially recommended because the constant interruptions were frustrating and getting in the way of my rest and recovery.[160]

Wearable Tech and the Anxiety of Awareness. We now outsource self-awareness to sensors: step counters, sleep trackers, heart monitors. They promise insight, but they can create obsession. We become data managers of our own bodies - measuring fatigue instead of feeling it. We designed tools to track health, but not to restore it. Early research notes that continuous self-monitoring can increase anxiety, reduce spontaneous movement of the body, and reduce the body's internal interoceptive awareness.[161]

HVAC Systems: Air That Activates Instead of Calms.

HVAC systems are often designed for cost efficiency and energy targets, not for human physiology. Yet indoor air quality has one of the strongest measurable impacts on cognition, mood, and stress. Poor ventilation and fluctuating temperatures force the nervous system into constant micro-adjustment. One scoping review found strong associations between inadequate IAQ and increased absenteeism, reduced performance, and negative health outcomes.[162] Indoor air can even hold higher concentrations of pollutants than outdoor air, increasing respiratory and cardiovascular stress. We may have engineered climate control, but we also created environments where our bodies have to work harder just to stay regulated.[163]

Green Space Loss: Designing Out Our Natural Regulator.

Our separation from nature is not symbolic - it is biological. Human nervous systems co-regulate with natural environments: trees, water, fractal patterns, birdsong, sunlight, and soil. A large review found that access to green space is strongly associated with improved mental health, especially in lower-income communities. Even brief contact - a view from a window, a pocket park, the presence of trees on a street - reduces loneliness, lowers stress, and enhances wellbeing. When we pave over nature, remove trees, or design cities without quiet green refuges, we remove a core regulator of the human nervous system.[164] We engineered dense cities - and subtract the one element proven to calm them.

Urban Heat Islands: Temperature as a Stress Multiplier.

Cities absorb and retain heat through materials like asphalt and concrete, creating temperatures several degrees higher than surrounding areas. Research links heat exposure to increased cortisol, poorer sleep, impaired cognition, and higher rates of aggression. For people without access to cooling or shade, the nervous system remains in chronic activation.[165]

Algorithmic Overload: When Systems Decide Faster Than We Can Think.

Recommendation engines, AI-driven feeds, and personalised ads operate at a speed the human nervous system was never designed to process. Constant micro-decisions - swipe, scroll, filter, approve, respond - keep the nervous system in a state of subtle vigilance. Algorithms optimise for engagement, not wellbeing, creating cognitive fatigue and emotional volatility.[166]

Has Engineering Forgotten What Makes Us Human?

Neuroscientist Dr Tara Swart calls beauty and creativity "nutrients for the brain".[167] They trigger the chemistry of safety, learning, and belonging - dopamine, oxytocin, serotonin. They're not luxuries, but rather biological necessities.[168]

We need harmony, rhythm, and form to regulate our nervous systems. Our brains are pattern-hungry, constantly scanning for symmetry, balance, proportion, and meaning.[169, 170]

Aesthetic coherence helps us orient. It signals safety. It tells the body: you can rest here.

And this connection to beauty isn't new. It's ancient. For most of human history, cultures have adorned themselves with pattern, colour, rhythm, and ritual.[171] Research shows that artistic expression - dancing, painting, music, carving, weaving, ornament, storytelling - is universal across all human societies.[172] One large anthropological study analysing over 900 cultures found that dance, song, and story are not "extras", but foundational human behaviours tied to bonding, regulation, and wellbeing. Another study from the University of British Columbia found that humans are wired for adornment - jewellery, body paint, ornament - not out of vanity, but out of identity, belonging, and emotional expression.[173]

Beauty has always been a language we speak through our bodies. And for centuries, engineering understood this too, while it might not have been captured in a standard.

For most of human history, we built as if we understood this instinctively. Street lamps once cast a golden glow and intricate patterns of ironwork made even a simple utility feel like a small piece of civic art. Bridges were designed as sculptural landmarks. Water fountains were carved with imagery. Public buildings carried flourishes and features that served no "functional purpose" except the most important one: to uplift the human spirit.

Somewhere along the way, modern engineering silently stripped this away. Function became the only virtue. If it worked and was safe, it was "good enough." If it met the standard, aesthetics were optional. If beauty didn't increase ROI, it was cut.

So the lamppost became a cold metal pole. The park bench became a bolted rectangle. And worse, we took this a step further and began deliberately designing discomfort. What used to be a place to rest, pause, connect, or watch the world go by has, in many cities, become an object of exclusion.

There is a technical term for this: hostile architecture (also called defensive architecture or exclusionary design). This includes benches with central bars so no one can lie down, spikes placed under bridges, metal studs on flat surfaces, sloped seating, and "leaning rails" instead of seats.

These designs don't just remove beauty - they remove dignity. They turn public infrastructure into a form of policing rather than a public good. They send a message: you can stay here, but only on our terms. And in doing so, they reveal how far some parts of modern design have drifted from care.

We optimise for lumen output and maintenance cycles, not delight. We forget that the visual world feeds our mental world. We forget that people regulate to their environments - and that when those environments lack beauty, our systems drift toward dullness, disconnection, and stress.

Research confirms this too. Studies on architectural aesthetics show that environments with organic shapes, warm materials, and coherent patterns activate the brain's networks for reward, empathy, and calm,[174] while sterile or monotonous environments activate regions associated with threat detection and vigilance.[175] A 2024 study even found that simply walking through a visually beautiful space increased emotional stability and reduced cortisol levels.[117]

The cost of losing beauty is measurable:

- People feel less connected to place
- Civic pride decreases[176]
- Vandalism increases in "ugly" or neglected environments
- Stress recovery takes longer
- Mental health outcomes worsen in monotonous streetscapes[177]
- Children's play and exploration decrease in harsh, angular environments.[178]
- Trust in public institutions erodes when the built world feels uncaring

In stripping beauty away, we stripped away coherence, identity, and the subtle message a city can communicate: you belong here.

We have the capacity to build for both.

What if we design water treatment plants that are sculptural, not hidden? Craft bridges that are elegant as well as strong? Give everyday systems - pipes, poles, pylons, stations, substations - a quiet grace? Let infrastructure be more than a utility? What if it could be a companion that supports us not just physically, but safely as well?

Beauty, in engineering terms, is coherence made visible. It's the moment a system becomes more than functional - it becomes humane. When we choose beauty, we choose care. When we choose delight, we choose belonging. When we design with grace, we send a message across steel, timber, and concrete: this place was made for you.

Designing for Cognitive Load, Psychological Safety, and Recovery

Our nervous system doesn't just need safety. It needs restoration. It needs places where it can come down a gear.

Psychologist and performance researcher Dr Adam Fraser talks about "The Third Space" - the transition moments in our day between one role and the next. The train ride between work and home.[179] The walk from the carpark to the front door. Even the pause between meetings. These small gaps are not empty time; they are opportunities to reset, reflect, and choose how we want to show up in what comes next.[180]

Engineering rarely designs for this pause. Yet it's essential.

Most of us are moving through environments that steadily load up our attention, our emotions, and our senses, with very little built-in recovery. When we talk about designing for the nervous system, we can think about three layers: cognitive load, psychological safety, and restorative micro-recovery.

Cognitive Load - How Hard It Is to Work Out What to Do Next

Cognitive load is the mental effort required to understand and act in a situation. Cognitive load theory, developed by John Sweller, shows that our working memory is limited. We can only hold a small amount of information at any one time before we get overloaded.[181, 182]

In practice, this means every confusing sign, every conflicting instruction, every sudden alert, and every poorly sequenced step adds to the load. If you've ever tried to follow complex wayfinding in a hospital while worried about a loved one, or navigated a busy interchange where the signage and announcements don't quite match, you've felt this. Your heart rate goes up. You double-check. You hesitate. You feel silly for "not getting it", even when the design is what's unclear.

Good engineering reduces this invisible tax. It makes navigation feel intuitive.

Think of the difference between an old station where signs have been bolted on over decades, versus a newer system where colours, icons, and sightlines all work together. One requires constant problem-solving; the other lets you move almost without thinking. In nervous system terms, one keeps you in mild alert. The other gives your brain a moment off-duty.

When we design with cognitive load in mind, we're not just making things "user friendly". We're freeing up mental bandwidth for more important things: being present, connecting with others, and noticing what's around us.

Psychological Safety - Can I Be Here Without Feeling Small or At Risk?

Psychological safety is often talked about in the context of teams. Amy Edmondson, who coined the term, describes it as a shared belief that it's safe to speak up, ask questions, and make mistakes without fear of humiliation or punishment.[183, 184]

In physical spaces, we can think about psychological safety in a similar way: can I be here without feeling judged, exposed, overwhelmed, or in danger?

Spaces that invite choice, allow pause, and convey dignity give our bodies permission to relax. Clear sightlines, good lighting, places to sit where you're not in the way, accessible entries that don't single you out, design that doesn't make you feel stupid or slow - all of these send tiny messages of welcome.

When these elements are missing, our nervous system notices. Consider:

- A dark underpass with no escape routes.
- A waiting room where everyone faces a single counter and you feel like you're on display.
- A ticket machine that times out and flashes red when you hesitate.

Nothing overtly terrible has happened, but your body has already decided: this is not a safe place to be myself. You close down, rush, withdraw, or brace.

Engineers might not use the language of "psychological safety" here, but every detail - from kerb height to counter design to the tone of a recorded announcement - either contributes to or erodes it.

Restorative Micro-Recovery - Designing Little Places to Exhale

Even with the best wayfinding and welcoming spaces, life will still be full. That's why our nervous systems need something else as well: recovery.

Environmental psychologists Stephen and Rachel Kaplan call this attention restoration - the idea that spending time in, or even looking at, nature allows our directed attention to rest and recharge.[185] Roger

Ulrich's famous study in the 1980s showed that hospital patients whose rooms looked out onto trees rather than a brick wall recovered faster, needed less pain medication, and had shorter stays.[97]

These findings matter for engineering because they tell us something simple and profound: the environments we create can either drain or restore people, even when they are not the "main event".

Restorative micro-recovery can look like:

- A small patch of greenery at a tram stop
- Daylight and a view from a stairwell
- A quiet corner bench that isn't squeezed next to a bin
- Timber, texture, or warm materials in a sea of hard surfaces
- A short walking path away from the main flow where you can take a breath

Historically, many cultures built this in as a matter of course. Roman baths weren't just for getting clean; they were social and restorative spaces - warm water, conversation, ritual.[186] Cloisters in monasteries and courtyards in traditional homes around the world created enclosed gardens where people could step away from noise and busyness for a few moments. Their design acknowledged that humans need somewhere to "come down" during the day.[187]

Today, we often outsource that role to commercial spaces - the café you escape to between meetings, or the shopping centre atrium that happens to have plants and light. But there is no reason those restorative pockets can't be part of our core infrastructure.

When we think about cognitive load, psychological safety, and micro-recovery together, a new question emerges for engineers: not just "Is this safe?" but "Does this place help people come back to themselves?"

We can design roads that make sense at a glance. Stations that feel welcoming rather than overwhelming. Hospitals where the corridors and waiting areas are part of the healing, not just the rooms. Third spaces - literal and metaphorical - that help people transition from pressure back to presence.

Our nervous systems are always in conversation with the environments we create. When we design for that conversation intentionally - reducing unnecessary load, increasing safety, and building in recovery - we're not just solving technical problems. We're engineering conditions for people to think clearly, connect with each other, and go home at the end of the day with a little more capacity than they started with.

For me personally, I've come to realise that designing for recovery isn't just something I talk about as an engineer. It's something I've had to learn to do for myself the hard way. Often, it happens in the smallest, simplest ways.

One of the practices that helps me most is that whenever I move from one part of my day to another - leaving a meeting, arriving home, walking into a room where my kids are - I pay attention to the moment my hand touches a door handle. That tiny sensation of contact has become a physical reminder that I am crossing a threshold. It's my own "third space" - a micro-pause that helps me slow down, breathe, and choose how I want to step into the next moment.

Doorways work the same way. There's something about physically moving through a frame that invites a reset, if I let it. These small rituals have helped me be calmer and more present - especially with my children, where the stakes for presence always feel higher.

The place where I feel my nervous system settle most reliably is in nature. As I write this, I'm sitting on our back deck, looking out at tall trees as kookaburras call to each other. The cockatoos parade across the sky like they own it. A huge water dragon lizard is sunning itself on our fire pit, completely unbothered by anything. The bush turkeys - dozens of them - wander through as if they're supervising the garden. Nature is my happy place. It always has been, even as a child playing in the creek a short bike ride from our family home.

If I can't be outside, sound helps me get there. I'll listen to thunderstorms or rainforest ambience when I need to drop back into myself. Even watching a video of a slow, crackling fire can reset me. There's something primal and comforting about the flames that speaks directly to the nervous system.

Burnout, stress, and nervous system dysregulation are familiar companions in my life - more familiar than I'd like to admit - so I've had to build my own recovery structures. Not having coffee after a certain time of day. Practising yoga regularly. Journaling when I need it. Remembering to breathe deeply when I need to.

These tiny practices accumulate. They don't eliminate the load, but they stop it from becoming unliveable.

I've also started tuning my environment to support me, just as I advocate for in projects. Moving to the Sunshine Coast has helped more than I could have imagined. Being surrounded by nature every day feels like living inside a restorative design experiment

When I'm travelling or spending time in cities, I notice the difference in my body almost immediately - the pace, the noise, the constant stimulation - so now I plan space around it. I schedule recovery into my calendar the same way I schedule meetings. I allow myself to step away, walk, breathe, or simply sit.

All of this is, in its own small way, engineering. It's designing conditions for myself to thrive while we wait for the designs to catch up. I know not everyone has this available to them, and each person has their own things that regulate them. That is part of the reason that I have written this book - so that as engineers, we might create more of these spaces for people who might not have these connections so easily available to them, or even realise they need them.

The environments we create - out there in the world, and inside our own lives - shape us far more than we realise. When we build in moments of pause, clarity, and restoration, we're not just being nice to ourselves. We're tending to the very systems that allow us to show up as humans - thinking, feeling, connected, and whole.

The Key Dimensions of Engineering for the Nervous System

Once you begin to see through this lens, it becomes impossible to unsee it. Every design decision, no matter how small, affects the human body in some way.

Daylight shifts our hormones and synchronises our circadian rhythm. Noise changes our heart rate and our ability to think clearly. The width of a footpath can determine whether two neighbours pause to chat or pass each other with tension in their shoulders.

None of this is "soft science". It's physiology, and it is as real - and as relevant - as any load calculation or performance curve.

In many ways, every engineer is already designing for the nervous system. We just don't always name it that way and we often aren't intentional about it. If we measured calm, connection, clarity, and confidence with the same rigour that we measure strength, flow, or voltage, we would see a different kind of performance emerge - one that lasts, because it works in partnership with the human body, not against it.

You may ask: aren't these already in the standards? For many engineers, standards are the backbone of our practice. They give us the minimum performance requirements we must meet to keep a building or system safe, functional, and compliant. They protect us from risk, guide our decisions, and help us work efficiently across teams and disciplines.

But standards don't usually capture the whole human experience of a place[188], or the new research in neuroscience - yet.[127, 189] They are, by their nature, reductive. They define what can be measured, enforced, and universally applied. That is their strength - and also their limitation.

Once you start viewing engineering through the lens of the nervous system, you begin to notice the difference between what the standards measure - and what the body feels. We can choose to add to the standard and incorporate the nervous system in our designs.

While these elements may not all fit neatly into formal standards yet, they absolutely belong in our professional judgement as engineers and designers. Not everything that matters can or should be captured in a clause.

Some of the most human parts of engineering live in the decisions we make day-to-day - how we sequence a space, the materials we choose, how we operate and maintain an asset over decades, and the values we bring into a design meeting. These considerations can be woven into design guidelines, operational practices, maintenance regimes, procurement conversations, and the culture of our teams.

Standards give us the floor, but our knowledge, intuition, and responsibility give us the ceiling. The nervous system lens is something we can choose to include, regardless of whether it is yet mandated.

So what are some elements that we could consider in our designs to engineer supportive human wellbeing? Each dimension asks us to tune into how people experience the spaces, systems, objects, or processes we create - not just how they function on paper.

Daylight and Views

Natural light is one of the strongest regulators of the human body. Studies consistently show that exposure to daylight improves mood, sleep, and cognition. Even a simple view - a tree outside a window or a glimpse of the sky from a corridor - can reduce stress and help people recover faster. It is extraordinary how often this is treated as optional, when it shapes so much of how alive we feel inside a place.

What most standards cover:

- Minimum daylight factors or illuminance levels for specific building types
- Glazing ratios and window placement for natural light penetration
- Requirements for artificial lighting to meet minimum lux levels (e.g., Australian Standards 1680 series)
- Energy efficiency considerations (e.g., heat gain through glazing)

What they miss:

Standards can miss quality of light - variation, softness, warmth, or alignment with circadian rhythms. They don't consider:

- Whether people have a view of nature
- How daylight changes over the day
- How light supports mood regulation
- Whether a corridor or workstation receives *enough* natural light to restore the nervous system

Acoustics

Noise is not neutral. It creates strain in the body, even when we're not consciously aware of it. Poor acoustics increase fatigue, raise cortisol, and reduce our ability to focus. Conversely, balanced acoustics - softer materials, good reverberation control, pockets of quiet - invite people to breathe differently. Suddenly, the environment feels generous rather than demanding.

What most standards cover:

- Maximum background noise levels
- Reverberation time requirements
- Wall, floor, and ceiling acoustic ratings (e.g., STC, NRC)
- Noise intrusion limits from external sources

What they miss:

Acoustic standards focus on *audibility*, not emotional ease. Missing elements include:

- The cumulative strain of low-level noise
- How sound interacts with movement (crowds, footsteps, trolleys)
- Tonal qualities (sharp versus soft sounds)
- Whether the space feels peaceful or agitating
- How noise shapes cognitive load and stress

A room can be "within limits" and still be exhausting to be in.

Air Quality and Ventilation

Fresh air affects everything: cognitive performance, alertness, wellbeing, even our immune system. Ventilation systems designed purely for energy efficiency often miss this. When air quality is high, people think better and feel better. When it's not, their bodies compensate - sometimes for years - without anyone realising the nervous system has been carrying the load.

What most standards cover:

- Minimum ventilation rates (litres/second/person)
- Filtration and HVAC requirements
- Thermal comfort ranges
- Pollutant thresholds (CO_2, VOCs, particulate matter)

What they miss:

Standards don't address the lived experience of air:

- Freshness
- Movement and flow
- Stability of temperature across the day
- Sensory comfort (stuffy versus clear)
- How stale air affects cognition, mood, and fatigue

Air can meet the standard and still feel draining.

Wayfinding and Flow

When design is intuitive, people move with confidence. When it isn't, they hesitate, scan, double-check, or rush - all small signs of micro-stress. Flow is not just traffic engineering; it is emotional regulation. A well-designed meeting room, path, station, or building feels like it's guiding you forward rather than testing you.

What most standards cover:

- Signage size and visibility
- Emergency exit requirements
- Minimum corridor widths
- Accessibility slopes and turning radii

What they miss:

Humans navigate with feelings as much as with logic. Missing considerations include:

- Intuitive layout and sequencing
- Predictability of sightlines
- Clutter that increases decision-making
- Stress caused by ambiguity
- Emotional clarity: "I know where I'm going"

Accessibility and Dignity

When spaces work for people of all abilities, they work better for everyone. Accessibility isn't simply about compliance - it's about dignity. It's about the hidden message a space sends: you belong here, and you do not need to apologise for how you move through the world. Design that honours different bodies and different ways of thinking creates spaces that feel honest, humane, and generous.

What most standards cover:

- Ramp gradients and handrail heights
- Lift access and door widths
- Accessible bathroom requirements
- Tactile ground indicators

What they miss:

Compliance doesn't guarantee dignity. Standards rarely consider:

- Sensory accessibility (lighting, noise, transitions)
- Intuitive wayfinding for different thinking styles
- Avoiding spaces that single out or stigmatise people
- Emotional safety: "I belong here without effort"
- Neurodiversity, cognitive load, or invisible disabilities

Perceived Safety

We often assume safety is purely technical - the right lighting level, the right guardrail height, the right surface texture. But perceived safety is equally powerful. The nervous system reacts instantly to openness, sightlines, lighting quality, and crowd flow. A bridge that feels unsafe isn't truly safe. A space where you have to steel yourself to walk through it is already asking too much of the human body.

What most standards cover:

- Lighting lux levels
- Guardrail heights
- Slip resistance

- Physical hazard minimisation
- CCTV placement in some guidelines

What they miss:

Standards don't cover the *felt* experience of safety. Missing elements include:

- Clear sightlines
- Avoidance of dark or enclosed corners
- Predictable transitions
- Warm lighting versus harsh glare
- Whether people feel watched or welcome

Additional Dimensions for Today's Engineering

As engineering disciplines evolve, new dimensions of nervous system design are emerging - ones we don't often name, but which profoundly shape how people feel.

Pace and Pressure

Transport engineering sets the tempo of a city. Faster isn't always better. The timing of signals, the spacing of crosswalks, the rhythm of traffic flow - these set the "heartbeat" of a place.

Cities designed only for vehicle throughput often produce a sense of urgency that the nervous system absorbs long before we notice it. Slower, more human rhythms change everything.

What most standards cover:

- Minimum signal timing
- Platform widths
- Clearance zones
- Ramp lengths and gradients
- Vehicle performance and network efficiency

What they miss:

Most guidance optimises for vehicles, not people. Missing considerations include:

- The stress of tight pedestrian crossing times
- Sensory overload on platforms
- Opportunities to pause or recover
- Emotional clarity in busy transport nodes
- Human tempo versus machine tempo

Cognitive Simplicity

Whether you're designing software, a train timetable, a medical device, or a public space, simplicity reduces stress. A system that is clear, consistent, and predictable allows people to feel competent and calm.

When design is cluttered or complex, people feel small, hesitant, or overwhelmed. This is why "user experience" isn't just a tech term - it's a nervous system term.

Material Warmth

Materials speak. Steel can feel cold or commanding. Timber feels warm and grounding. Stone feels stable. Soft textures invite touch. Hard ones repel it. The palette of a space influences mood, emotional safety, and even how much time people want to spend in it.

What most standards cover:

- Slip and fire resistance
- Durability and hygiene requirements
- Toxicity and material safety
- Maintenance considerations

What they miss:

Standards rarely address:

- Tactile comfort
- Sensory overstimulation (glare, harshness)
- Warmth and emotional resonance
- Whether materials support calmness
- The psychological impact of sterile or monotonous surfaces

Rhythm and Coherence

Spaces with a clear rhythm - repeating forms, proportions, and patterns - allow the brain to make sense of them effortlessly.[190] Coherence helps the body settle.

This is one reason why traditional architecture, with its consistent proportions and ornamentation, is often described as "soothing". It is literally easier for us to process.[174]

Nature and Green Infrastructure

Nature is one of the most powerful regulators of the human nervous system. Even small amounts can have outsized effects - a tree canopy on a hot street, a patch of grass beside a bus stop, or planter boxes on a balcony. Green infrastructure isn't ornamental; it's physiological support.[185]

Micro-Moments for Recovery

Transitions matter. Spaces that allow people to step aside, take a breath, lean on a railing, sit for a moment, or orient themselves physically and emotionally create nervous system resilience.

Recovery doesn't require grand gestures. Often, it only requires a single square metre of considered design.[112]

Key Dimensions in Review

The below table is a summary that you or your team might use to review each dimension.

Dimension	Indicator / Metric	Guiding Question	Example Design Response
Air Quality	$PM_{2.5}$, CO_2, VOC levels; ventilation performance	Can people breathe easily here? Does the air support clarity and comfort?	Natural ventilation, greenery, filtration, improved HVAC sequences
Noise & Vibration	Decibel levels, RT60, vibration thresholds	Can people hear what they need, and avoid what they don't?	Acoustic zoning, soft materials, vibration isolation, quiet HVAC strategies
Light & Circadian Rhythm	Daylight exposure, lux levels, glare indices	Does the lighting support the body's clock and help people feel awake or restfulas needed?	Tunable lighting, daylight access, shading, warm-toned lighting choices
Wayfinding & Cognitive Load	Time to orient, error rates, user stress	Would a first-time user feel confident here?	Intuitive layouts, sightlines, clear sequencing, minimal signage
Chemical Exposure	PFAS-free materials, low-VOC products	What residues are we leaving in people's bodies?	Material transparency, low-toxicity sourcing
Sensory Coherence	Visual and auditory consistency; clutter levels	Does this space feel harmonious or overwhelming?	Unified material palette, decluttering, balanced lighting and acoustics
Psychological Safety	User surveys, behavioural patterns, incident reports	Does this place make people feel safe, respected, and at ease?	Better lighting, clear sightlines, inclusive design, predictable transitions
Restorative Opportunities	% of space for rest, nature, pause	Where can people breathe out?	Green pockets, quiet rooms, shaded seating, micro-rest spaces
Technology Interaction Load	Number/ frequency of alerts; required inputs	Does this technology calm or agitate?	Quiet defaults, batch notifications, simple interfaces
Beauty & Coherence	User delight, material quality, craft	Would someone smile walking through here?	Warm materials, detailed design, integrated art, natural textures
Transition Spaces	Presence and quality of buffers	Does the change of mode feel calm?	Vestibules, gradients in texture and light, gentle sequencing

These Are Not Extras - They Are Engineering

I believe these key dimensions aren't "nice-to-haves". They are foundational to whether a system truly works.

A bridge that meets the code but feels unsafe will be avoided. A hospital that increases stress as people navigate it cannot be called healing. A train station that creates anxiety at peak hour isn't performing, no matter what the throughput numbers say. A digital system that overwhelms people with alerts isn't supporting them, even if it's technically flawless.

Good engineering has always been about designing systems that hold people - physically, emotionally, and socially. When we design for the nervous system, we don't make projects "soft"; we make them robust. We make them places that people can return to again and again without paying a physiological price.

I think that can make our world better for everyone. That, ultimately, is what engineering for the future must become.

The Human by Design Framework

Understanding these key dimensions is one thing, but putting them into practice is another. In a real project, we are often juggling constraints, budgets, approvals, time pressures, and competing priorities. It's easy for wellbeing to slip into the background unless we have a way to hold it in the conversation.

That's why I created the Human by Design Framework - not as another compliance tool, but as a simple, practical guide teams can use to ground discussions, test ideas, and reflect on whether a design supports or strains the people who will use it. Think of it as a project compass: a way to translate the science of the nervous system into everyday engineering decisions, from concept to operations.

As engineers, we naturally gravitate toward what we can quantify: load, flow, energy, cost, compliance. These are essential, and they always will be. But when we begin to consider how a person *feels* inside a space or system, we realise that the human nervous system is also carrying a kind of load - one we rarely measure but profoundly shape.

The Human by Design Framework is an invitation to broaden our technical gaze. It offers a way to notice where a design supports people's wellbeing and where it unintentionally places them under strain. It's a way of making the invisible visible - a framework for conversation, reflection, and recalibration that honours the biology of the people who will live, work, move, heal, learn, and connect within what we create.

Importantly, this is not a call to overhaul the whole system or throw out decades of engineering wisdom. Engineers already do extraordinary work - work that keeps people safe, connected, warm, moving, and alive.

The dimensions we've explored aren't criticisms of what exists; they're simply additions to the toolkit, and ways of broadening our field of view to include what we now understand about people and their nervous systems.

Even small shifts - a clearer sightline, a quieter HVAC system, a softer material, a better crossing sequence - can make a meaningful difference to someone's wellbeing, sense of ease and connection to place.

That's the spirit behind the Human by Design Framework. It's a gentle guide to help teams reflect on how their design might feel for the humans inside it - an invitation to build on what we're already doing well, and bring a little more care, clarity, and calm into the systems we create.

How to Apply the Human by Design Framework

Once you begin to see design through the nervous system lens, the next question becomes: how do we practically apply this on a real project, with real constraints, real deadlines and real complexity?

Rather than replacing standards, this framework aims to complement them. It adds the human context that standards were never designed to hold.

It honours the extraordinary work engineers already do and invites us to expand our field of view - gently, iteratively, and creatively.

Part 1: The Human Journey - Walking through the design, moment by moment, in a real body

This first step is about experiencing the design from the inside. Instead of analysing a plan, we imagine (or physically walk) the journey as a real person:

- A commuter rushing between modes
- A parent with a pram
- A wheelchair user
- An elderly neighbour
- A neurodivergent teenager
- A shift worker arriving at 4am
- Someone living with PTSD or trauma
- Someone who is overwhelmed, grieving, or anxious

The nervous system is always scanning: do I belong here? Am I safe? Am I lost? Am I welcome?

As you walk the journey, ask:

- What does my body notice first?
- Do I find myself bracing, scanning, or hesitating?
- Where do I feel calmer or more supported?
- Would someone carrying trauma feel exposed, triggered, or at ease?
- Does this place soothe or startle?

This step builds embodied empathy - an understanding of engineering from the perspective of the people who will live with it every day.

Part 2: The Human Load Map - Mapping where the system strains or restores the nervous system

After walking the journey, zoom out and create a simple map of where the body works hard and where it rests.

Mark the load points:

- Cognitive load (decision bottlenecks, confusing layouts)
- Sensory load (noise, glare, vibration, crowds)
- Emotional load (feeling exposed, unsafe or watched)
- Physical load (steep ramps, long distances, unstable surfaces)

And then mark the recovery points:

- Greenery
- Daylight
- Quiet pockets
- Intuitive wayfinding
- Ceiling height
- Warm materials
- Predictable transitions
- Shaded seating
- Breathing space

The map makes it visually obvious where people may be overwhelmed, and where a small intervention could make a big difference. This can become a shared tool within your team for designing not just functionally, but physiologically.

Part 3: The Rhythms Board - Designing for the cycles and transitions of real life

Human needs shift throughout the day and across emotional states. The Rhythms Board helps teams consider:

Daily rhythms

- Morning rush versus afternoon fatigue
- Evening safety
- Night shift workers
- Bight-time low visibility

- 24-hour operations
- Energy, alertness, and sensory tolerance cycles

Transition rhythms

- Transitions are often the most stressful part of a journey:
- Thresholds and entryways
- Lifts and stairwells
- Platform edges
- Change-of-mode zones
- The shift from outside to inside (and vice versa)

These are moments where overwhelm spikes - and where good design can lower strain.

Emotional rhythms

People don't enter spaces in a neutral state. They arrive carrying:

- Stress
- Grief
- Panic
- Excitement
- Exhaustion
- Fear
- Distraction
- Joy

This lens asks: does this design support people at their worst moments, not just their best ones, and does it give people dignity, predictability and a sense of being held rather than hurried?

Part 4: The Seasons Lens - Designing for the emotional, biological, and life-stage climates of real people

This final lens recognises that humans move through many "seasons" - not only in emotion, but in age, biology, and circumstance. These seasons profoundly shape nervous system responses and often determine whether a space feels safe, overwhelming, or supportive.

The Seasons Lens includes four categories:

1. Emotional Seasons

People carry inner weather systems of their own: curiosity, fear, calm, overwhelm, joy, anxiety, grief, playfulness, and vigilance. Design asks: does this place welcome people in every emotional season?

2. Seasons of Life

Different ages experience environments in profoundly different ways.

Children (0-12)

Closer to the ground, more sensory-driven, more easily overwhelmed, more reliant on predictability. Design: safe exploration, colour, clarity, rest spots.

Teenagers

Navigating identity, social sensitivity, hormonal fluctuation, and a need for autonomy. Design: dignity, privacy, non-patronising cues, safe gathering spaces.

Adults

Often navigating time pressure, cognitive load, and responsibility. Design: ease, clarity, recovery pockets.

Middle-age

Managing careers, caregiving, and physical changes. Design: good lighting, clear contrast, supportive seating.

Elderly

Sensitive to contrast, glare, noise, and instability; need rest points and warm lighting. Design: gentle pacing, predictable surfaces, railings, clear sightlines.

3. Biological Seasons

These are the internal cycles - hormonal, energetic, and physiological - that shape how people feel in their bodies.

Female Biological Seasons

Menstrual Cycle

Changes in energy, concentration, pain sensitivity, temperature, and sensory tolerance.Design considers: warmth, pace, rest, reduced glare, and noise.

Pregnancy

Needs more seating, stability, toilets, chemical safety, space, and temperature control.

Perimenopause & Menopause

Heat sensitivity, anxiety, sleep disruption, cognitive fog, joint pain. Design asks: does this space overwhelm or support someone experiencing these shifts?

Male Biological Seasons

Men also experience hormonal and physiological transitions:

- Gradual testosterone decline
- Sleep changes
- Stress reactivity shifts
- Hearing sensitivity changes

Design: soft acoustics, warm lighting, clear interfaces, recovery pockets.

Gender Diverse & Nonbinary Bodies

People experiencing hormonal cycles through medication or gender-affirming treatment need:

- Stable temperatures
- Low-glare lighting
- Chemical-safe materials
- Predictable environments

Design asks: does this space support a wide range of human bodies with respect and care?

4. Situational Seasons

These are temporary but powerful states that alter nervous system response.

Examples include:

- Illness or injury
- Trauma, PTSD, or hypervigilance
- Grief
- Burnout
- Sensory overwhelm
- Sleep deprivation
- Carers carrying invisible emotional load
- Time pressure or cultural disorientation

This is where trauma-informed design becomes essential. Design asks: does this place trigger or re-traumatise? Or does it hold people with steadiness and gentleness?

Implementing the Human by Design Framework

In practice, a team might use one, a few, or all parts of the framework to get a new perspective on a project, system, or problem. Each part of the framework could be used to:

1. Feel the design from the body's perspective (Walk the Human Journey)
2. Identify strain and restoration points (Draw the Human Load Map)
3. Consider how needs shift across the day and emotional states (Build the Rhythms Board)
4. Ensure the design works for people at all ages, cycles, vulnerabilities, and transitions (Apply the Seasons Lens)

The Human by Design Framework doesn't replace engineering standards. It enriches them to engineer for everyone. It brings humanity into technical excellence - the signature of the Future of Engineering. It's both rigorous and compassionate, and it creates designs that work not just on paper, but in real human bodies.

Practical Uses on Projects

Here is one way you could use the Human by Design Framework with your team on a project. It could be a new or existing project, or one you are reviewing for lessons learned. You can replace "place" with "system", "problem", "scenario", or whatever is relevant for your context.

Step 1: Select Your User Profiles (5-8 minutes)

Pick a few stakeholders relevant to your project:

- Mobility needs
- Sensory needs
- Emotional needs
- Life-stage needs

Step 2: Walk the Human Journey (10-20 minutes)

For each user, answer the journey questions. Capture insights on sticky notes.

Step 3: Build the Human Load Map (10-15 minutes)

On a floor plan or service map:

- Circle strain points
- Highlight recovery points
- Note where users hesitate or brace

Look for patterns.

Step 4: Complete the Rhythms Board (10-15 minutes)

Ask:

- How does this place behave across the day?
- Where does pressure build?
- Where does pace rise or fall?
- How does emotion shift here?

Step 5: Apply the Seasons Lens (15-20 minutes)
Move through:

- Seasons of life
- Biological seasons
- Situational seasons
- Emotional seasons

Capture ideas, concerns, and gaps.

Step 6: Cluster Themes (10 minutes)
Group your notes into:

- Stress points
- Opportunity points
- Confusion points
- Delight points

This step forms the backbone of actionable design choices.

Step 7: Prioritise Key Design Elements (10-15 minutes)
As a team, select:

- 3-5 high-value improvements
- 1-2 low-cost quick wins
- 1-2 restorative elements
- 1 essential inclusion for vulnerable users

Step 8: Integrate Into Concept or Detailed Design
Add selected elements into project or operational documents:

- Design briefs
- Drawings
- Stakeholder sessions
- Cost planning

- Procurement criteria
- Operations & maintenance plans

Step 9: Revisit the Human by Design Framework

Do this at the concept, detailed design, pre-construction, and post-occupancy stages. Treat it as a living lens, not a one-off workshop.

Where this Framework is Already Working

The Human by Design Framework isn't theoretical - it's already alive in projects in countries all around the world. Here and there, in pockets of brilliance and courage, engineers, designers, and communities are creating spaces that honour how people feel, move, heal, and connect. Seeing these projects gives me enormous hope. They remind me that we're not starting from scratch; we're building on a micro movement that has already begun.

Every one of these examples shows what becomes possible when we widen our lens and design with care, imagination, and humanity. Some of them are small interventions. Others are city-shaping. All of them prove that engineering can be a force for wellbeing - and that it already is.

What inspires me even more is that these shifts aren't contained to one sector. They're happening across industries, disciplines, and countries. The case studies span health, transport, the public realm, mental health and neurodiversity, childhood and aging, trauma recovery, ecology and environmental restoration, projects across both the global north and south, and designs that move fluidly between urban environments, natural landscapes, architecture, systems, and operations.

This diversity tells us something important: human-centred engineering isn't niche. It is emerging everywhere people dare to ask, "How can we design this for humans, and for everyone who will use it?"

As you read these case studies, I hope they spark something in you: inspiration, possibility, and perhaps even a sense of responsibility. Imagine if every project carried this level of thoughtfulness. If every design held the nervous system in mind. If the places we build helped people feel calmer, safer, more connected, and more alive.

These projects are not the end point. They're early signals of what engineering could become, and what it can already be when we lead with humanity.

This is engineering for the future through the lens of humanity: Human by Design.

Case Study: Maggie's Centres (UK) - Human by Design Through Calm, Dignity, and the Architecture of Care

Meet Sophie, a 38-year-old mother of two. Three months ago, she was diagnosed with breast cancer. Between hospital appointments, scans, and treatment decisions, she feels disoriented and fatigued. The hospital's corridors feel stark and clinical; Sophie's nervous system is already on high alert.

One drizzly afternoon, a nurse suggests she drop into the nearby Maggie's Centre. Sophie hesitates - "What do you do there?" - but arrives anyway.

The moment she steps through the door, she notices the warm timber floors, abundant daylight, and a small garden view. A barista offers her tea. No appointment. No reception desk looming. Just someone saying, "Sit down. Tell me how you are."

Sophie relaxes. She stays for an hour.

She visits again the next week with her partner. They talk in a kitchen-style space, not a consultant room.

By the third visit, Sophie finds she's laughing with another visitor. Sophie's shoulders are lower.

She says later, "I didn't know a building could make me breathe again."

Maggie's Centres were founded by Maggie Keswick Jencks and Charles Jencks, after Maggie's own experience of cancer treatment made her aware of the emotional and environmental strain patients face. Her vision: a network of centres adjacent to major NHS cancer hospitals - but intentionally unlike them.

The first centre opened in Edinburgh in 1996. Today, there are over 25 across the UK and internationally.[191] Each centre is designed by a renowned architect, but the brief always begins with: how should a person feel?

Maggie's offers free practical, emotional, and social support to anyone affected by cancer - patients, carers, and families. Hundreds of thousands of sessions are delivered annually. In 2022, Maggie's provided over 300,000 support sessions and welcomed over 30,000 first-time visitors.[192]

According to their own reporting, 97% of visitors said the centre improved their ability to manage stress, and 99% said they felt less alone after visiting.[191] These metrics reveal more than user satisfaction. They show how architecture plus service can shift human physiology and emotional wellbeing.

Maggie's Centres exemplify the Human by Design Framework because they start with the human body and nervous system, not just the building. Rather than asking "What rooms do we need?" the team asks: "What does this body need in this moment of life?"

Their design responds to four essential forces:

- The emotional and sensory overwhelm of diagnosis and treatment
- The cognitive load of navigating medical systems
- The important need for dignity, connection and normalcy
- The variation in human bodies, senses, life stages, and vulnerability

The result is care expressed through architecture - where every detail matters.

How Maggie's Reflects the Human by Design Framework

Part 1: The Human Journey

From the moment someone arrives, the experience shifts. A gentle path, transparent entry, warm materials, and a kitchen-rather-than-reception ease the body's tension. Instead of "check-in", you're invited in.

Key design moves include:

- Elevated timber floors and abundant daylight
- Curved walls and natural textures to reduce threat anticipation
- Garden views and domestic-scale spaces for immediate grounding

Part 2: The Human Load Map

Hospitals often produce micro-stress: beeps, glare, confusing signage. Maggie's reduces load.

- Cognitive load is lowered via intuitive layout and no reception queues
- Sensory load is reduced through soft acoustics, daylight, and natural materials
- Emotional load is addressed by providing privacy, domestic cues, and informal social spaces

The building becomes a buffer between the body and the medical machine.

Part 3: The Rhythms Board

Cancer treatment brings variable energy, emotional states, and physical mobility. Maggie's supports these changes by designing for rhythm:

- Areas for quiet reflection, social connection, and kitchen-style gathering
- Spaces that flex with patient cycles: pre-treatment, post-treatment, family delay
- Outdoor and indoor spaces tuned to morning, afternoon, and evening use

Part 4: The Seasons Lens

The design is sensitive to human seasons of life, biology, emotion, and vulnerability.

- Supports various life stages: partners, children, older patients
- Acknowledges biological seasons: induced menopause, fatigue, sensory sensitivity
- Holds situational seasons: grief, fear, trauma, uncertainty
- Space is not one-size-fits-all - it adapts to human change.

Practical Takeaways for Engineers

- **Design details affect physiology.** A handrail, material, colour or window view can reduce cortisol or enable calm.

- **Domestic cues build safety.** Kitchens, wood, and curves signal value, not processing.
- **Predictability supports cognitive ease.** Simple layouts and unambiguous paths reduce mental strain, especially in high-stress contexts.
- **Nature as medicine.** A tree view or garden light is part of healing.
- **Flexibility over one-mode design.** Design for humans at their best *and* at their most vulnerable.

Maggie's Centres show that legacy isn't just about infrastructure durability; it's about how the spaces we design carry people through their hardest seasons. Engineering that supports bodies, minds and hearts leaves a legacy of dignity and care. When you design for the human nervous system, you are leaving more than a building - you are leaving wellbeing.

Case Study: Tokyo Metro Wayfinding (Japan) - Human by Design Through Cognitive Ease, Predictability, and Sensory Clarity

Daniel is a 27-year-old traveller visiting Tokyo for the first time. He's jet-lagged, barely slept on the flight, and is carrying two bags and a vague sense of panic.

Tokyo Station - one of the busiest, most complex transport interchanges in the world - should overwhelm him. Yet as he steps off the train, he notices something surprising.

The system feels... easy. Clear colour coding. Repeated icons. Directional arrows at eye level and foot level. Station numbers that make navigation possible, even without reading Japanese. Even the lighting feels predictable.

Daniel still moves quickly, but he isn't bracing. He feels guided instead of lost. By the time he reaches his hotel connection, he realises he hasn't panicked once. His body stayed calm. This, he later says, "is the first transport system that actually helped me think."

Tokyo Metro and JR East handle over 40 million passenger journeys a day - one of the highest anywhere in the world. The system is enormous, with layers of interchanges, multi-level stations and dense crowds.[193]

In the early 2000s, user studies revealed a major issue: foreign travellers, older adults, and even local passengers frequently experienced cognitive overload while navigating stations.

The system responded with one of the world's most ambitious wayfinding redesigns.

Rather than adding more signs, Tokyo Metro simplified, clarified, and harmonised - from typography and colours to arrows and lighting. What emerged has become an international benchmark, studied by transport designers globally.

The results have been measurable. Passenger surveys show significant reductions in navigation errors and travel anxiety. Stations with updated wayfinding saw a drop in reported "lost traveller" incidents, and a 30% reduction in platform crowd clustering.

While not measured in cortisol, these behavioural shifts reflect physiological ease. People are moving through complexity with confidence, not confusion.[194]

Tokyo Metro exemplifies the Human by Design Framework because it prioritises cognitive and sensory wellbeing in a high-pressure environment. Instead of assuming people would adapt to complexity, they redesigned complexity around human bodies and brains.

They considered fatigue, language barriers, sensory sensitivity, age, impairment, and emotional states - creating a system where millions can move calmly through intensity.

How Tokyo Metro Reflects the Human by Design Framework

Part 1: The Human Journey

From entry to exit, the system guides the body step by step. Key design moves include:

- Colour-coded lines repeated consistently across floors, walls, and signage
- Station numbering (e.g., G12) to remove language dependence
- Arrows positioned where eyes naturally land during movement
- Predictable lighting levels to reduce visual startle
- Uncluttered decision points to support calm navigation

Part 2: The Human Load Map

Tokyo Metro systematically reduces cognitive and sensory load.

- Cognitive load lowered through consistent icons, colour saturation rules, and station numbering
- Sensory load reduced through controlled acoustics and non-glare surfaces
- Emotional load addressed via intuitive layouts that minimise hesitation

The environment compensates for fatigue, stress, and language barriers.

Part 3: The Rhythms Board

Passenger rhythms in Tokyo are intense - morning crush loads, late-night returns, weekend tourism. Design supports these rhythms by:

- Creating "calm paths" with fewer stimuli
- Spreading passenger flow through clear, redundant routes
- Lighting that adjusts subtly throughout the day
- Zones for standing back to regroup during peak congestion

The system works with human rhythms rather than against them.

Part 4: The Seasons Lens

Wayfinding supports users across diverse life seasons:

- Older adults with mobility or visual decline
- Tourists dealing with jet lag and disorientation
- Neurodivergent passengers who rely on predictability
- Parents with prams needing wide, accessible movement paths
- People who feel overwhelmed by crowds or noise

This is navigation as emotional support.

Practical Takeaways for Engineers

- Reduce cognitive load: fewer decisions per moment
- Use repeatable visual cues - predictability calms

- Design signage from user height, not technical drawings
- Assume fatigue, stress and language barriers
- Lighting, material and acoustics shape emotional safety

Tokyo Metro shows that clarity is a form of care, through engineering that helps the body stay settled inside complexity.

Case Study: Superkilen Park (Denmark) - Human by Design Through Cultural Joy, Belonging, and Community Expression

Fatima is an 11-year-old girl who recently arrived in Denmark with her family. Her new neighbourhood, Nørrebro, is highly multicultural, yet she feels out of place.

One afternoon, she wanders into Superkilen, the public park at the heart of the area. She sees a Moroccan fountain like the one in her grandmother's town. A Somali swing set. A neon sign from Russia. A Japanese octopus playground sculpture.

Suddenly, the place doesn't feel foreign. She feels a flicker of belonging. Her culture is not hidden, but celebrated. This becomes the park where Fatima practises Danish with friends and teaches them games from home. A space that felt overwhelming becomes a space she can call hers.

Superkilen was created through a collaboration between BIG (Bjarke Ingels Group), artist group Superflex, and landscape architects Topotek 1. It was commissioned to address social fragmentation and create shared community identity in one of Copenhagen's most ethnically diverse suburbs.

The team held extensive community workshops, asking local residents what objects from their home cultures they wanted represented. These were then sourced - and in some cases, shipped - from over 60 countries.[195]

Superkilen has delivered measurable outcomes. Studies show increased community use, cross-cultural interaction, and stronger feelings of neighbourhood pride. Local crime and antisocial behaviour decreased after opening. Importantly, the park has become a landmark of multicultural inclusion - a place where identity and architecture meet.[196]

Superkilen is Human by Design because it prioritises emotional belonging, sensory joy, and cultural legitimacy. Instead of smoothing out differences, it elevates them. It treats aesthetics as a tool for social wellbeing and recognises that public space regulates nervous systems - especially for people navigating cultural transition.

How Superkilen Reflects the Human by Design Framework

Part 1: The Human Journey

The park is designed as a gentle unfolding of sensory and cultural experiences. Key design moves include:

- Curved pathways encouraging exploration and discovery
- Vivid colours that uplift rather than overwhelm
- Open sightlines creating perceived safety
- Human-scaled seating and gathering zones

Fatima's body relaxes as the environment signals: "You are welcome."

Part 2: The Human Load Map

Superkilen reduces social and emotional load in a multicultural context.

- Visual cues familiar to diverse cultures
- Play areas that reduce inter-age and inter-cultural tension
- Soft surfaces and shaded pockets to support sensory comfort

The emotional load of displacement is softened by recognition.

Part 3: The Rhythms Board

The park supports community rhythms across the day and year.

- Morning walkers
- After-school play
- Evening social gatherings
- Weekend markets and cultural events

The space adapts to shifting energy, temperature, social patterns, and community needs.

Part 4: The Seasons Lens

Superkilen recognises seasons of identity and belonging.

- For migrants acclimating to a new country
- For teenagers building cultural confidence
- For elders seeking familiarity
- For families navigating transition

It is a space that normalises difference.

Practical Takeaways for Engineers

- Co-design with genuine cultural representation
- Use aesthetics to build emotional safety, not just visual appeal
- Create micro-zones for different sensory and social needs
- Celebrate diversity as a design strength
- Include seating, shade and colour strategically for comfort and confidence

Superkilen shows that public space can be an antidote to alienation. This can be achieved through engineering that builds belonging.

Case Study: Wellington Waterfront (New Zealand) - Human by Design Through Nature, Soft Edges, and Trauma-Informed Urbanism

Moana is a 62-year-old Wellington local who remembers the waterfront of her childhood as industrial, windswept, and unsafe. For decades, she avoided it - especially at night.

Recently, after the waterfront redevelopment, she walks there again. This time, she notices something different: soft timber boardwalks underfoot, sheltered pockets from the famous Wellington wind, gentle lighting that doesn't glare, wider pathways where she never feels rushed.

She stops and looks at the harbour - something she hasn't done in years. She sits on a bench shaped to support aging bodies. She feels safe. She feels welcome. She feels part of her city again.

Wellington's waterfront redevelopment transformed a derelict industrial zone into a sequence of parks, boardwalks, cultural spaces, and public amenities. Led by Wellington Waterfront Ltd, the project aimed to restore connection between the city and harbour. The design integrated Māori cultural narratives, ecological repair, and human-centred placemaking. It is internationally recognised for creating a safe, inclusive, nature-connected public realm.[197]

The outcomes are tangible. Local surveys show significantly increased public use, especially by families, older adults, and young people. Reported perceptions of safety improved dramatically, and the waterfront is now considered one of the most beloved public spaces in the city. These changes are not accidental - they are engineered.[198]

Wellington Waterfront is Human by Design because it acknowledges the nervous system's need for softness, predictability, and recovery. It treats public space not as a corridor but as a place of restoration. It respects cultural memory and creates dignity through design.

How the Wellington Waterfront Reflects the Human by Design Framework

Part 1: The Human Journey

The experience is designed to be calm, intuitive and sensory-rich. Key design moves include:

- Wide, legible pathways with clear views
- Timber materials that feel warm and safe
- Sheltered pockets from Wellington's high winds
- Water views that orient and calm the body

The journey slows the nervous system rather than accelerating it.

Part 2: The Human Load Map

The design reduces sensory and emotional load in subtle but powerful ways.

- Non-glare lighting in the evenings
- Noise-buffering landscape features
- Refuge points for rest and retreat
- Predictable surfacing textures

The waterfront becomes supportive for people who feel vulnerable at night or in crowds.

Part 3: The Rhythms Board

The design supports daily and seasonal rhythms of use.

- Morning joggers
- Lunchtime office workers
- After-school families
- Night-time walkers
- Summer festivals and winter calm

It flexes with the emotional rhythms of the city.

Part 4: The Seasons Lens

The waterfront considers:

- Elders needing stability, seating, and gentle gradients
- Children needing safe exploration
- People healing from trauma or past experiences of unsafe public spaces
- Visitors navigating unfamiliarity
- Cultural seasons expressed through Māori design elements

It acknowledges that public space can either retraumatise or restore.

Practical Takeaways for Engineers

- Use materials that signal warmth and safety
- Prioritise non-glare lighting to reduce nighttime anxiety
- Provide micro-sheltered areas in windy or exposed environments
- Integrate cultural narratives to strengthen belonging
- Design for vulnerable users first

Wellington's waterfront shows that when we design for wellbeing, we rebuild trust between people and place. This is legacy expressed through public space.

Case Study: Neurodiversity-Informed Workplaces (Global) - Human by Design Through Sensory Regulation, Autonomy, and Cognitive Clarity

Priya is a 32-year-old software engineer with ADHD. She loves her work, but most offices exhaust her: open-plan noise, bright overhead lights, constant movement in her peripheral vision, the pressure to multitask.

At her new company in Amsterdam, she walks into the office expecting the same. Instead, she notices something different: quiet zones with soft lighting, small focus pods where she can think without interruption, movement-friendly breakout spaces, and meeting rooms with clear, predictable layouts. The acoustics are gentle. The signage is simple. No one stares when she walks away mid-meeting to regulate - that's what the "reset booths" are for.

By lunchtime, she realises she hasn't masked once. Her nervous system feels steady. She actually looks forward to coming back tomorrow.

Over the last decade, workplaces like the BBC Headquarters in London, Deloitte's "The Edge" in Amsterdam, and Spectrum Workplace in New York have embraced neurodiversity-informed design. Driven by research showing that 15-20% of the population is neurodivergent, and that poor sensory environments significantly reduce productivity, retention, and wellbeing, organisations have begun rethinking their approach to space.

A 2022 study by the British Council for Offices found that sensory-friendly workplaces can increase employee satisfaction by up to 30% and reduce cognitive fatigue, especially among neurodivergent staff. Another global workplace survey found that autistic employees were 200% more likely to stay in jobs when workplace environments supported sensory needs.[199, 200, 201]

Neurodiversity workplaces exemplify the Human by Design Framework because they recognise that humans don't all think, sense, or regulate in the same way. Instead of treating difference as an inconvenience, they design with difference as the starting point - creating spaces where more people can thrive.

How Neurodiversity Workplaces Reflect the Human by Design Framework

Part 1: The Human Journey

These workplaces are designed to guide the body gently into work. Key design moves include:

- Gradual sensory transitions from entry to work zones
- Soft, indirect lighting instead of fluorescent glare
- Predictable spatial layouts to reduce anxiety
- Clear sightlines for a sense of safety and orientation

The journey into the workplace is not overstimulating; it is steadying.

Part 2: The Human Load Map

These workplaces actively lower sensory and cognitive load.

- Noise-dampening materials reduce auditory overwhelm
- Zones for focused work separate from high-energy collaboration areas
- Clear wayfinding reduces decision fatigue
- Materials chosen to reduce visual glare and flicker

The environment supports attention rather than fragmenting it.

Part 3: The Rhythms Board

Neurodivergent energy rhythms are highly variable across the day. Design supports these rhythms with:

- Focus pods for deep-concentration phases
- High-movement collaboration zones for kinetic workers
- Reset rooms for emotional regulation
- Flexible seating and posture variety

The space supports regulation instead of enforcing uniformity.

Part 4: The Seasons Lens

This design considers:

- Emotional seasons such as burnout, overwhelm, or high focus
- Biological seasons like hormonal fluctuations that affect sensory tolerance
- Seasons of life - from early career to pregnancy to midlife changes
- Situational seasons: deadlines, fatigue, personal stress

The workplace supports people as they are, not as they "should be."

Practical Takeaways for Engineers

- Noise is a physiological stressor; design for quiet where it matters
- Provide a diversity of work environments, not one-size-fits-all
- Avoid glare and harsh lighting; use layered, tunable lighting
- Include predictable layouts for cognitive safety
- Create intentional "reset" spaces

Case Study: Montessori Classrooms (Global) - Human by Design Through Autonomy, Sensory Grounding, and Developmental Rhythms

Let's imagine Luka, a curious 4-year-old who finds typical classrooms overwhelming. Loud noises, bright walls, cluttered shelves, and rigid transitions make his small body tense.

When he walks into his Montessori classroom for the first time, he pauses. The light is soft. The room is quiet. Everything is at his height - shelves, tables, handwashing stations. The colours are gentle, mostly natural wood. Nothing feels rushed.

He is invited to pick up an activity and choose where he wants to work. He instinctively gravitates toward a tray with beads and sits at a tiny round table. For the first time in a learning environment, Luka's nervous system feels settled. He stays focused for 20 minutes - a new record for him.

When he has energy he needs to get out, he has the option to play in the playground. When he wants water, he can go to the tap that is his height and fill up his water bottle. When he wants to eat, his lunch is available in the fridge. Rather than being structured by the clock, he can learn to intuitively listen to his body.

Maria Montessori's educational philosophy, developed in the early 1900s, is now practiced in over 140 countries.[202] But what many people don't realise is how much Montessori environments are grounded in design for the nervous system.

Montessori observed that children learn best when environments meet their developmental, sensory, and emotional needs. This has been validated by contemporary neuroscience. Self-directed learning environments increase intrinsic motivation, lower stress hormones, and support cognitive development.[203]

A 2017 study in *Frontiers in Psychology* found that children in Montessori programs had stronger executive function, better social cognition, and more sustained attention.[204] Another study showed improved wellbeing and reduced stress behaviours in Montessori environments, compared to traditional classrooms.

Montessori classrooms exemplify the Human by Design Framework because they honour the sensory, emotional, and biological rhythms of children - designing with, not against, their developmental seasons.[203]

How Montessori Classrooms Reflect the Human by Design Framework

Part 1: The Human Journey

Montessori rooms are designed to feel predictable and safe for little nervous systems. Key design moves include:

- Child-height sightlines and furniture
- Clear layout with minimal clutter
- Gentle transitions between activities
- Nature elements for sensory grounding

Children move at their own pace, and are guided rather than directed.

Part 2: The Human Load Map

Montessori intentionally lowers sensory and cognitive overload.

- No bright, chaotic colours
- Natural materials (wood, glass, pottery) instead of plastic
- Silence or soft soundscapes during work periods
- Clear "home" locations for each material

This reduces stress and supports focus.

Part 3: The Rhythms Board

Children have varied daily rhythms that shape learning. Design supports these rhythms with:

- Areas for movement and rest
- Outdoor-indoor flow
- Work cycles that allow deep focus without interruption
- Spaces for both solitude and collaboration

The environment respects children's emotional and energy rhythms.

Part 4: The Seasons Lens

Montessori recognises:

- Sensitive periods of development
- Emotional seasons (separation anxiety, confidence building)
- Biological needs like sleep, hunger, sensory integration
- Social seasons - parallel play, group play, growing autonomy

Each "season" informs how children experience space.

Practical Takeaways for Engineers

- Lower sensory clutter; children feel it physically
- Use natural materials to ground and calm
- Design for autonomy: reachable, navigable, intuitive layouts
- Provide both active and quiet zones
- Consider developmental seasons, not just age groups

Case Study: Dementia Village - Hogeweyk (Netherlands) - Human by Design Through Familiarity, Autonomy, and the Seasons of Memory

Gerard is an 82-year-old man living with dementia. In traditional aged care, he felt lost and agitated - long corridors, locked doors, institutional lighting, unfamiliar routines. He often tried to "escape" because nothing around him made sense.

When he moves into Hogeweyk - the world's first dementia village - something shifts. He walks along a small street that looks like a neighbourhood from his youth. He passes a hairdresser, a little shop, a café. He feels the sun on his face in the courtyard. A caregiver greets him like a neighbour, not a nurse. Nothing feels like a facility. It feels like life.

Gerard wanders, stops, sits, walks again - freely and safely. His agitation decreases. His confidence returns. He is himself again.

Hogeweyk, created in 2009 in the Netherlands, is internationally recognised for its groundbreaking redesign of dementia care. Instead of institutional aged care, it is structured as a real village: households, streets, a supermarket, a theatre, and outdoor squares.[205, 206]

Studies have shown that residents in Hogeweyk:

- Exhibit significantly lower agitation
- Require fewer medications
- Experience higher quality of life
- Engage in more social interaction compared to traditional dementia facilities.

Hogeweyk has since inspired similar developments in Canada, the US, and Australia.[207]

Hogeweyk exemplifies the Human by Design Framework because it respects the nervous system changes experienced with dementia. Rather than trying to contain or manage symptoms, it designs for autonomy, familiarity, and dignity.

How Hogeweyk Reflects the Human by Design Framework

Part 1: The Human Journey

The journey feels like normal daily life, not care management. Key design moves include:

- Familiar household layouts
- Streets with recognisable cues
- Human-scale architecture
- Gardens and outdoor access without barriers

The journey reduces confusion and restores identity.

Part 2: The Human Load Map

Hogeweyk reduces emotional and cognitive strain.

- No institutional corridors
- Visual cues from everyday life

- Consistent materials and colours that trigger recognition
- Hidden care infrastructure so residents don't feel medicalised

Safety occurs through design, not restriction.

Part 3: The Rhythms Board

Dementia symptoms vary across the day. Design supports these rhythms with:

- Activity spaces for daytime engagement
- Calm zones for afternoon rest (when sundowning can peak)
- Predictable mealtime rhythms
- Outdoor routes that support wandering safely

The environment works with cognitive rhythms, not against them.

Part 4: The Seasons Lens

The Seasons Lens is crucial here because dementia is a changing life season. Hogeweyk considers:

- Memory seasons - what residents recall and what they don't
- Emotional seasons - fear, confusion, joy, nostalgia
- Biological seasons - mobility changes, sleep changes
- Social seasons - family visits, loneliness, need for touch

It honours who people were, who they are, and who they are becoming.

Practical Takeaways for Engineers

- Design for familiarity, not novelty
- Support autonomy through layout, not surveillance
- Hide clinical elements to preserve dignity
- Create safe wandering loops
- Use outdoor spaces as essential infrastructure

Hogeweyk shows that human dignity can be engineered - and that care environments can restore identity, not erase it.

Case Study: The High Line (New York, USA) - Human by Design Through Urban Calm, Nature Immersion, and Community Rhythms

Tomas is a 45-year-old office worker in Manhattan. His days are long, his commute is loud, and his nervous system rarely gets a break.

One afternoon, after a difficult meeting, he decides to walk instead of taking the subway. He climbs the stairs to the High Line - the elevated park built along an old freight rail line. The air shifts. The noise drops. Wild grasses rustle in the breeze. Sunlight filters through nearby buildings, creating patterns on the timber walkway. People stroll at a slower pace.

Tomas feels his breathing deepen. He sits on a bench overlooking the Hudson and realises: "Here, my mind can rest."

This becomes part of his daily ritual - a place where he decompresses before returning home to his family.[208]

The High Line was transformed from a derelict elevated rail track into one of the most celebrated urban parks in the world. Designed by James Corner Field Operations, Diller Scofidio + Renfro and Piet Oudolf, it opened in stages from 2009 onwards. What makes it extraordinary is not only its design but its impact. It changed how cities think about public infrastructure and wellbeing.[209]

Since opening, the High Line attracts around 8 million visitors per year and has spurred over $2 billion in urban regeneration.

But beyond economic outcomes, a wide body of research links access to urban nature with lower stress and improved mood. The High Line is often cited as a powerful example of this effect in practice - a place that invites people to slow down, recover attention, and interact more gently with strangers.[185, 210]

The High Line exemplifies the Human by Design Framework because it acknowledges the nervous system's need for softness, slowness, and sensory restoration in one of the world's most overstimulating environments. It's a living example of how engineering can create moments of human refuge.

How the High Line Reflects the Human by Design Framework

Part 1: The Human Journey

Walking the High Line is a sensory downshift. Key design moves include:

- Slow, meandering paths that encourage wandering
- Abundant planting inspired by wild meadows
- Elevated views that provide psychological distance from street chaos
- Human-scale seating pockets for rest

The journey is a gradual exhale.

Part 2: The Human Load Map

The design reduces urban overload:

- Noise is softened through elevation and vegetation
- Visual clutter disappears, replaced by coherent landscape patterns
- Clear pathways reduce decision fatigue
- Integrated shade and wind breaks support comfort

The High Line offers a rare sensory downshift in a dense city - a place many people experience as restorative, where attention softens and the mind feels clearer.

Part 3: The Rhythms Board

The park aligns with daily and seasonal rhythms:

- Morning runners
- Lunchtime strollers
- Sunset watchers
- Seasonal blooms that change the sensory palette
- Night lighting designed for softness, not spectacle

The High Line moves with the rhythm of the city.

Part 4: The Seasons Lens

The design supports:

- Tourists navigating unfamiliarity
- Elderly walkers needing stability and rest
- People recovering from stress or trauma
- Children exploring safely above the traffic
- Seasonal affective shifts, offering greenery year-round

It is a refuge for all seasons of life.

Practical Takeaways for Engineers

- Elevate or buffer spaces to reduce noise and stress
- Use nature as physiological restoration
- Design for wandering, not just destination
- Create micro-spaces for rest and contemplation
- Use planting to create sensory rhythm

The High Line proves that even in the densest cities, engineering can create breathing space - and breathing space creates wellbeing.

Case Study: South Korea's Healing Forests - Human by Design Through Biophilia, Stress Recovery and Ritualised Nature

Meet Eun-ji, a 29-year-old manager in Seoul experiencing chronic stress. Her doctor suggests "forest therapy". This is a concept she initially finds strange - walking in a forest as medical treatment?

But she joins a guided session at Saneum Healing Forest. As she steps beneath the tall pines, the air feels different. The guide invites her to breathe slowly, touch the bark, listen to the wind, and walk barefoot along a cool moss path.

She feels her heart rate slow. Her chest loosens. She forgets her inbox for the first time in months. By the end of the session, Eun-ji feels lighter.

She returns weekly. She later tells a friend: "It saved me."

South Korea has invested heavily in "healing forests" - purpose-designed nature settings where trained guides support sensory, slow, embodied experiences. The country now has dozens of designated forest-healing and forest-welfare facilities, reflecting a serious commitment to nature-based wellbeing at scale.[211]

Research on forest bathing and forest therapy, including studies conducted in South Korea, have reported short-term improvements in stress-related measures such as cortisol, blood pressure, and anxiety.[212]

Across multiple studies, forest bathing and forest therapy programs have been associated with reductions in stress biomarkers, improvements in blood pressure and heart-rate variability, and reductions in anxiety and rumination.[213] Some studies also report changes in immune-related markers.

The strongest evidence to date supports forest-based interventions as a helpful complement for stress, anxiety, and depressive symptoms, with emerging (but less definitive) research in other conditions.[213]

Healing forests are Human by Design because they embody biophilic engineering at a national scale - intentional, evidence-based, and physiologically grounded.

How Healing Forests Reflect the Human by Design Framework

Part 1: The Human Journey

The journey is sensory, slow, and grounding. Key design moves include:

- Gentle trails with soft ground textures
- Shaded sitting decks for reflection
- Immersive olfactory experiences, from pine and cypress
- Water soundscapes engineered for calm

The journey reopens the body's capacity to rest.

Part 2: The Human Load Map

Healing forests actively lower sensory and cognitive load:

- No artificial noise
- No competing signage
- Slow walking pace
- Predictable wayfinding markers

The forest becomes a physiological reset button.

Part 3: The Rhythms Board

Design supports body rhythms:

- Morning sessions for cortisol alignment
- Slow afternoon walks
- Seasonal trails (spring blossoms, autumn leaves)
- Movement-rest cycles guided by therapists

Nature and body rhythms harmonise.

Part 4: The Seasons Lens

Healing forests support:

- Seasons of burnout
- Seasons of grief
- Seasons of transition
- Biological seasons - hormonal shifts, immune recovery
- Situational seasons - trauma, anxiety, chronic stress

Nature adapts to the emotional weather of life.

Practical Takeaways for Engineers

- Biophilia is not optional - it's medicine
- Slow design is as important as efficient design
- Sensory simplicity reduces cognitive chaos

- Natural materials regulate the nervous system
- Landscapes can serve as infrastructure

South Korea shows that wellbeing can be a national design strategy, not simply an afterthought.

Case Study: Jewel at Changi Airport (Singapore) - Human by Design Through Wonder, Orientation, and Calm in Transit

Rafael is a frequent flyer who secretly dreads airports. He associates them with stress - long queues, harsh lighting, dry air, confusing terminals.

When he transits through Singapore's Changi Airport, he expects the same. But as he walks into Jewel, everything shifts. He hears the soft roar of the world's tallest indoor waterfall. He sees sunlight pouring through the glass canopy. There are trees, walking trails, quiet nooks, and gentle mist. Families are relaxed. Travellers are smiling.

Rafael feels his shoulders drop. For the first time in years, he enjoys being in an airport.

Jewel Changi opened in 2019 as a fusion of indoor forest, retail precinct, and transportation hub. Designed by Safdie Architects, the Rain Vortex is surrounded by a multi-storey forest with over 2,000 trees and 100,000 shrubs.[214]

This isn't just decoration - it's engineered biophilia designed to support the wellbeing of the tens of millions of passengers who move through Changi each year.[215]

Changi Airport is frequently ranked among the world's best airports, and its terminal design deliberately prioritises greenery, natural light, and intuitive wayfinding - features that are widely associated with better passenger experience in high-stress environments.[216]

Jewel Changi is Human by Design because it engineers awe, calm, and clarity in a setting traditionally associated with overwhelm.

How Jewel Changi Reflects the Human by Design Framework

Part 1: The Human Journey

Jewel creates a gentle, welcoming arrival. Key design moves include:

- Gradual transition from terminal to nature
- Massive canopy flooding space with natural light
- Curved paths with intuitive sightlines
- Waterfall as orienting landmark

The journey replaces tension with wonder.

Part 2: The Human Load Map

Travel amplifies cognitive and sensory load. Jewel reduces it with:

- No glare; diffuse natural light
- Cool mist and clean air reduce fatigue
- Clear circulation paths reduce decision stress
- Noise masked by waterfall's calming frequencies

The airport becomes restorative, not draining.

Part 3: The Rhythms Board

Designed for 24-hour passenger rhythms:

- Quiet zones for red-eye travellers
- Family-friendly play zones
- Slow walking paths for decompression
- Daylight alignment for jet lag support

The environment respects biological time zones.

Part 4: The Seasons Lens

Jewel considers:

- Travellers experiencing anxiety or sensory overwhelm
- Families juggling children

- Older travellers with mobility needs
- People moving through seasons of grief, stress, or transition

It is a transit environment designed for humanity.

Practical Takeaways for Engineers

- Use orientation anchors (like waterfalls) for cognitive ease
- Engineer biophilia, not just add plants
- Prioritise natural light and high air quality
- Provide micro-environments for different energy and emotional states
- Design for awe - it resets the nervous system

Jewel Changi proves that even a high-tech, high-pressure transport hub can become a sanctuary - a place that lifts the human spirit instead of exhausting it.

We Have Done This Before

For all our modern sophistication, designing for human wellbeing isn't new. Long before neuroscience, biometrics, or performance indicators, ancient civilisations and First Nations peoples across the world instinctively built environments that calmed, oriented, and connected people. They understood something we risk forgetting: that engineering is not just about structures - it is about shaping the human experience.

Indigenous cultures have always designed with Country, seasons, and community at the centre. They worked with the land's rhythms, not against them, creating systems that supported safety, identity, nourishment, and belonging across generations. Their engineering was inseparable from care.

In many ways, history shows us that we once knew how to design with the body and the earth in mind. We built with rhythm, with the senses, with beauty, with generosity, and with an understanding that what we create becomes our legacy. These ancient examples remind us that supportive design is not a modern invention - it is a human instinct we are finding language for again.

Case Study: Roman Bathhouses - Human by Design Through Ritual, Sensory Regulation, and Shared Restoration

Imagine Livia, a woman living in ancient Rome, stepping into the thermae (Roman Bathouse) after a long day.

The air warms her skin. Voices echo softly under vaulted ceilings. She moves through a sequence of rooms - hot, warm, cool. Each transition slows her breath. She meets neighbours, washes away dust, and rests in the warm light filtering through a clerestory window. By the time she enters the cold plunge, her mind is quieter than when she arrived.

Roman bathhouses were feats of engineering - hypocaust heating systems, acoustically balanced chambers, temperature gradients, and social spaces designed around the body's natural responses.[186] They were not just hygiene facilities; they were public wellbeing infrastructure, widely used as part of daily civic life.[217]

Why This Was Human by Design:

- Temperature changes supported nervous system regulation
- Warm light and steam softened sensory overload
- Architecture encouraged both solitude and social connection
- Ritualised sequences created predictable emotional transitions

Framework Elements Reflected:

- Human Journey: a deliberate sensory sequence calming the body
- Human Load Map: reduced cognitive load through predictable ritual
- Rhythms Board: aligned with daily social and biological rhythms

Practical Takeaway: Design sequences, not just rooms. The transitions between spaces can regulate the nervous system as much as the spaces themselves.

Case Study: The Ancient Greek Agora - Human by Design Through Belonging, Social Rhythm, and Civic Safety

Dorian, a craftsman in ancient Athens, walks into the Agora at sunrise. The light hits the colonnades just as merchants set up their stalls.

He hears voices rising in debate, children running between market stands, neighbours sharing news. The open space feels safe - he can see everything around him.

He takes a seat on a low stone bench, knowing he'll meet someone he knows. This is where his nervous system settles into the rhythm of the city.

The Agora was the heart of Greek public life - a carefully scaled civic space designed for conversation, trade, decision-making, and daily social rituals. Its open sightlines, walkable proportions, and shaded colonnades created psychological safety long before the field had a name.[218]

Why This Was Human by Design:

- Open visibility reduced threat perception
- Walkable scale supported natural social interaction
- Shaded edges created comfort in the Mediterranean heat
- Spatial organisation fostered belonging and civic trust

Framework Elements Reflected:

- Human Journey: gentle arrival with predictable movement patterns
- Human Load Map: low cognitive load through open sightlines and intuitive circulation
- Seasons Lens: supported the social seasons of life - trade, learning, debate, rest

Practical Takeaway: Design public space for visibility and connection. When people feel safe, they participate fully.

Case Study: Machu Picchu (Inca) - Human by Design Through Nature Integration, Orientation, and Emotional Elevation

Imagine Amaru, an Inca farmer, climbing the terraces of Machu Picchu at dawn. Mist lifts from the mountains as the sun aligns perfectly with the Temple of the Sun. The stone pathways curve naturally with the mountain's contours.

Every turn reveals a new view - sky, river, stone, forest. Amaru feels guided not just by markers, but by the land itself.

Machu Picchu was a masterpiece of bioclimatic and landscape-integrated engineering. Every building aligned with solar patterns, water flows, and sacred geography. The terraces stabilised slopes, managed water, and created emotional and agricultural resilience.[219]

Why This Was Human by Design:

- Built forms aligned with natural rhythms that oriented the body
- Views and vistas created emotional grounding
- Integration of land, water and architecture reduced environmental stress
- Geometry supported spiritual and psychological coherence

Framework Elements Reflected:

- Human Journey: movement framed by nature and elevation
- Rhythms Board: alignment with solar, seasonal and agricultural cycles
- Seasons Lens: supported generational knowledge and legacy systems

Practical Takeaway: Design with the landscape, not on top of it. Orientation and nature connection regulate the human body.

Case Study: Angkor Wat (Cambodia)- Human by Design Through Ritual, Orientation, and Sensory Coherence

Saran, a young monk, walks the processional path at dawn. In the dim morning light, he crosses the moat and passes through a series of gateways that gradually narrow his attention.

The temple grounds unfold slowly, revealing symmetry, silence and long shadows.

By the time he reaches the central sanctuary, his mind has shifted into stillness. The architecture has prepared him for contemplation.

Angkor Wat was engineered as both a temple and a cosmological map. Its design encoded celestial alignments, symbolic geometry, and a deliberate spatial sequence that guided emotional and spiritual transformation.[220]

Why This Was Human by Design:

- Processional pathways created structured psychological transitions
- Symmetry and proportion lowered cognitive tension
- Controlled light and shadow regulated emotional state
- Water, stone and vegetation formed a coherent sensory palette

Framework Elements Reflected:

- Human Journey: a ritualised pathway shaping emotional regulation
- Human Load Map: reduced sensory clutter through coherent design
- Rhythms Board: structured around solar and ritual rhythms

Practical Takeaway: Use transition spaces intentionally - they shape how people feel before they enter the heart of a place.

Case Study: Zen Buddhist Temples & Gardens (Japan) - Human by Design Through Sensory Simplicity, Minimalism, and Nervous System Quiet

Haru is a poet visiting a Zen garden for the first time.

He steps through a wooden gate and senses the change instantly: gravel raked into calm patterns, moss catching soft morning light, the faint sound of bamboo tapping in the breeze.

There are no bright colours, no clutter, no demand for his attention.

His breath lengthens. The garden gives him permission to slow down.

Zen temples and gardens were designed as environments for meditation and attentional clarity. Every element - from gravel and timber to the placement of a single stone - was engineered to reduce cognitive noise and support embodied presence.[221]

Why This Was Human by Design:

- Minimal stimuli lower cognitive load
- Organic materials regulate sensory systems
- Asymmetry encourages curiosity without overwhelm
- Soundscapes engineered for calm, not silence

Framework Elements Reflected:

- Human Load Map: extremely low sensory and cognitive demand
- Human Journey: gentle, spacious movement through garden paths
- Rhythms Board: aligns with breath, season and daily ritual

Practical Takeaway: Less stimulus often means more clarity. Use design to create mental spaciousness.

Case Study: Indigenous Australian Knowledge Systems - Human by Design Through Country, Seasonal Cycles, and Ecological Alignment

Imagine a young Yorta Yorta child following their auntie across Country during a seasonal burn.

The air is smoky but gentle. The fire is cool, moving slowly through grasses in a patterned rhythm.[222] Birds circle above. The child watches how plants respond - how the land breathes easier afterwards. They see engineering as part of life, not separate from it.

Indigenous Australian engineering includes sophisticated systems of water management, seasonal fire knowledge, shelter placement, and navigation grounded in an intimate relationship with land, sky and ecology.[223]

It is engineering that listens before it acts.

Why This Was Human by Design:

- Seasonal calendars aligned human activity with ecological rhythms
- Fire management reduced long-term risk and supported biodiversity
- Shelters oriented to wind, sun, and protection for bodies
- Knowledge systems integrated emotional, spiritual, and practical wellbeing

Framework Elements Reflected:

- Seasons Lens: everything structured around seasonal belonging
- Rhythms Board: cycles of land, fire, food, water and culture
- Human Journey: movement guided by story, Country and sky

Practical Takeaway: Design that listens - truly listens - creates resilience and care across generations.

When Architecture Was Designed to Confuse the Human Body

Now that we've explored these ancient examples - places shaped with rhythm, connection, nature, ceremony, and human intention - it's worth looking at a very different kind of design. Most historical engineering aligned with the body and supported people's wellbeing, but not all of it did.

There are moments in history where architecture was engineered deliberately to disorient, confuse, or heighten vigilance. These examples help us see even more clearly how profoundly space influences the nervous system, and how design can either calm or activate the body.

Himeji Castle in Japan is one of the most striking illustrations of this. While breathtakingly beautiful, its internal layout was intentionally crafted to unsettle intruders - a reminder that engineering has always had the power to shape human physiology, whether toward connection or toward confusion.

Here's how one of the most iconic castles in the world used space, form and sensory manipulation to trigger the *opposite* of Human by Design.

Case Study: Himeji Castle (Japan) - When Architecture Was Designed to Confuse the Human Body

Imagine a samurai intruder approaching Himeji Castle in the 1600s. From a distance, the structure looks elegant - almost delicate - rising white against the sky. It seems obvious where the front gate is, and where the main tower should be.

But the moment the intruder steps inside the first courtyard, nothing behaves the way the senses expect. The pathways twist. The angles shift. The ground slopes when it should rise. Suddenly, he's disoriented, his nervous system on high alert. Every step feels uncertain.[224, 225]

Himeji Castle is one of the most remarkable examples of architecture intentionally designed not to soothe the human body, but to unsettle it. While its beauty is undeniable, its engineering was deeply strategic:

the castle was a defensive machine that used sensory misdirection and cognitive overload to protect its occupants.[226]

Here are some of the ways Himeji Castle deliberately disrupted perception and physiology:

1. The "Confusing Paths"

Approaching the main keep, it looks as if you're heading directly toward the central tower - yet every path veers at irregular, unpredictable angles. This induces what environmental psychologists call spatial disorientation, a state where the brain's cognitive map loses coherence. Research shows that when humans cannot form a reliable internal map, the amygdala (the threat-detection centre) becomes more active, increasing vigilance and stress responses.

2. Downward-Sloping Paths That Should Rise

When approaching a fortress, the body expects an uphill climb. Himeji reverses that instinct. Descending when you expect to ascend creates proprioceptive mismatch - a subtle but destabilising physiological confusion. Studies on proprioception and threat detection show that when sensory input contradicts expectations, the body interprets it as potential danger, increasing heart rate and narrowing attention.

3. Hidden Watch Points

The walls conceal arrow slits and drop windows at varying heights. Intruders feel observed, but cannot locate the watchers. Research in environmental threat perception consistently shows that "unseen surveillance" - where individuals feel watched without identifying the source - heightens cortisol and activates the body's defensive postures.

4. Disappearing Sightlines

The castle intentionally blocks the view of the main keep until the very final approach. Humans use long sightlines to orient, stabilise the inner landscape, and down-regulate anxiety. When orientation cues disappear, cognitive load and tension increase.

5. Irregular Spatial Compression

Pathways shrink abruptly into narrow chokepoints, then widen without warning. This rapid alternation mirrors patterns known to increase physiological arousal, similar to how haunted houses or high-stress training environments manipulate space to create unpredictability.

The Engineer's Opportunity

I was walking through Brisbane City on one of those quiet mornings where the light hits the river just right and makes it sparkle. I came across a wayfinding sign - one of those simple pedestrian markers with arrows and distances.

The first two entries made perfect sense: **River Walk - 300 m** and **Goodwill Bridge - 500 m.**

The third sign on the same post stopped me. It said: **QUT - New Farm - 1.2 km.**

I was standing at QUT.

I couldn't help but imagine being someone new to the city - late for an interview, maybe rushing to find their way. They'd glance at that sign and wonder: Am I in the wrong place? Am I supposed to be somewhere else? It seems so small - a line of text, a missing dash, a confusing conjunction - but in that tiny detail lives a micro-moment of stress. And those simple, tiny moments add up.

Every time a sign confuses, a space disorients, or a system contradicts itself, our bodies do the same thing: they tighten. We've all felt it - the flicker of doubt, the heartbeat quickening, the slight frustration that comes when the world stops making sense for a second. That's the nervous system at work - and it's our responsibility as engineers to care for it. Designing for wellbeing doesn't just happen in hospitals or schools. It happens in these everyday moments, in the frictionless flow of getting from one place to another, in the inner confidence that the world is holding you - not testing you.

When we get those moments right, something else happens too. People feel calm. They feel seen. They start to trust the systems around them - and trust is the foundation of every functioning society. The details are the design and the nervous system feels that. If we can engineer clarity in a sign, we can engineer calm in a city.

Every design has a nervous system - and every nervous system responds to design. From the macro of air pollution to the micro of your phone's alarm, our choices ripple through the biology of daily life.

Engineering for wellbeing doesn't mean less rigour. It means broader responsibility. We already measure how strong something is. It's time we also measure how it makes people feel.

The Takeaway

When we design with an awareness of the human nervous system, our understanding of "performance" begins to shift. Systems that feel safe, legible, and supportive do not slow people down; they enable better judgement, deeper trust, and more sustainable outcomes over time.

But humans do not exist in isolation. Our nervous systems are shaped not only by workplaces and projects, but by the environments we move through every day. Noise, heat, light, air, greenery, and connection to nature all play a role in how regulated or overwhelmed we feel. In this way, wellbeing is inseparable from the landscapes and systems we design.

The next chapter widens the lens again, moving from the human nervous system to the living systems that sustain it. It explores regenerative environments and what becomes possible when engineers design with nature, not against it, creating places that support both ecological health and human wellbeing over the long term.

Chapter 5: Regenerative Environments - Designing with Nature, Not Against It

"*When will you learn that the more that you take means the less you will have at the end of the day?*" Xavier Rudd, *Storm Boy*

If the first lens, Human Wellbeing, asks how our work affects the nervous system, this lens asks how it affects the Earth's life-support systems. Humans do not exist apart from the natural world; our regulation, health, and resilience are inseparable from the environments we move through every day.

For much of modern history, engineering has treated nature as something to control, contain, or work around.[65, 227] Rivers are straightened. Trees are cleared. Heat, water, and biodiversity are managed as constraints rather than assets.[228] These approaches delivered speed and scale, but they also disconnected people from the systems that sustain life.

We often talk about "reducing impact", as though the best we can do is harm a little less each time. But what if we designed projects that left the world better than we found it? What if our roads, bridges, buildings, and systems were not scars on the landscape, but living threads in a web that is constantly repairing itself?

That is the question at the heart of regenerative engineering. It is not a technical challenge so much as a mindset shift. For two centuries, engineering has been shaped by control, extraction, and efficiency.

The next era will be defined by reciprocity, designing with natural systems rather than against them, and creating environments that restore ecological health while supporting human wellbeing.

From Extraction to Reciprocity

When I was at university, sustainability was a word and a section you included in your report if you wanted a few extra marks. It wasn't that we didn't care - it's that we didn't really know what it meant.

If you said your bridge design used "sustainable materials" or your water system was "environmentally friendly", it ticked the box. There wasn't much questioning beyond that. We didn't talk about where the materials came from, or what happened to them after construction, or what the project would mean for the local ecosystem in 50 years. It was just a word - like an afterthought we added at the end.

I remember thinking even then that there must be more to it than this. Surely the idea of sustainability couldn't just be about doing a little less harm. Surely we could design in a way that actually improves things - that makes the environment, and people's lives, better because of our work.

But that wasn't how I was trained to think. Engineering school, and later the industry, was all about efficiency - getting the job done on time, within budget, and to standard. There was rarely space to ask the deeper questions: what if efficiency isn't the end goal? What is the intention of this project? What if the real measure of success is whether the system thrives once we're gone?

That's where this second lens - Regenerative Environments or Regenerate by Design - comes in. It's not about doing less damage. It's about giving back more than we take.

Sometimes, I think engineers are the children of the Industrial Revolution.[229] It's where our professional DNA was forged - in the clatter of steel, steam, and progress.

Back then, the world was expanding faster than anyone could comprehend. Cities were swelling, factories were roaring, and machines were changing everything about how people lived, worked, and moved. The goal wasn't harmony - it was survival. We needed to build things quickly, efficiently, and in huge quantities.

That's where our obsession with control began.[230] We built systems that could be measured, standardised, and replicated - railways, engines, bridges, waterworks. We learned to quantify success in numbers: output, productivity, efficiency.

That was good engineering for its time. It brought sanitation, safety, and speed to millions. It lifted living standards, reduced disease, created jobs, fuelled education, and opened up unimaginable possibilities. It helped pull millions out of poverty and connected communities across vast distances. The gains were real, and they changed the course of human history.

However, in the rush to industrialise, something subtle happened. We started to believe that if something couldn't be measured, it didn't matter. We began to think of the natural world as a set of variables to be controlled, rather than a living system to belong to.

That mindset carried forward - into our training, our manuals, our procurement systems, our project reviews. It's in the way we talk about "resources" instead of "relationships", and "mitigation" instead of "reciprocity."

When I look back now, I realise how much that legacy shaped the way engineers see the world. We inherited an extraordinary ability to solve problems - but we also inherited the assumption that nature was a problem to solve.

The next era of engineering asks something different of us. Not to tame the world, but to work with it. Not to control, but to co-create and build back better.

Commercialisation as a Culture

Decades later, in the 1970s and 80s, GE's CEO Jack Welch took this mindset to its extreme. Welch was brilliant, bold, and relentless. He inherited a sprawling organisation and reshaped it into a global powerhouse by applying a single, unwavering philosophy: **efficiency above all else**.[231]

Under his leadership, GE grew at an extraordinary rate. Shareholders celebrated him. Business schools studied him. Entire industries tried to copy him. He was called "Neutron Jack" because - like a neutron bomb - he removed the people but left the buildings standing.[232]

Welch believed that what mattered most was measurable performance, so he built systems that ranked people, plants, divisions, ideas - everything became a number. If a business unit wasn't number one or two in its industry, it was sold. If a factory wasn't delivering productivity gains, it was restructured. If an employee wasn't hitting targets, they were gone.

For a while, this model worked spectacularly. GE became the world's most valuable company. Efficiency became synonymous with excellence. Leaders everywhere were trained to think like Jack: optimise, trim, streamline, scale, maximise.

But this worldview came with a shadow side. It taught generations of engineers, managers, and executives to see the planet and its people through the lens of inputs and outputs. Rivers became "resources". Forests became "inventory". Human labour became "cost centres". Communities became "externalities". The unseen message was clear: if it can't be measured, it doesn't matter.

This is not a criticism of Jack Welch as a person. He was a product of his time, in a world where growth was the unquestioned good, and where efficiency was the engine that drove progress. His leadership reflected the industrial logic that had shaped the twentieth century: build more, move faster, produce cheaper.[233]

His legacy also shows how this way of thinking became embedded in our professional DNA. Many of us grew up in systems influenced by Welch's worldview, where performance meant speed, consistency, predictability, and control - not regeneration, reciprocity, care, or long-term impact.

It is only now, with the benefit of hindsight, that we can see the gap between efficiency and wisdom - and why our next evolution as engineers must integrate the pieces his era couldn't see.

Today, we're living with the consequences of these worldviews. The old DNA of efficiency has reached its limit. Now we must evolve - from extraction to reciprocity, and from control to collaboration.

As I've learned more about the history of engineering and business - the system that engineering works in - I've realised something important: the industrial and commercialisation mindset, for all its achievements, is only one way of seeing the world.

It has brought astonishing progress, yes. But it also narrowed our field of vision. It taught us to value control over connection, output over relationships, and certainty over curiosity.

But this way of thinking isn't universal. It's not the only engineering lineage we can draw from. In fact, some of the oldest engineering systems on earth were built on an entirely different worldview - one grounded in reciprocity, rhythm, belonging, and care for Country. Long before the rise of machines and metrics, there were cultures who understood regeneration intuitively. And in Australia, that knowledge has been held, practiced, and refined for tens of thousands of years.

The Consequences of the Old DNA on Our Planet

For two centuries, we built extraordinary things - cities, water systems, bridges, power networks - but we built them inside a worldview that assumed the environment would absorb whatever we extracted, produced, or discarded. And for a long time, it seemed like it could.

The bill has come due.

Engineers understand this better than anyone: if you push any system beyond its capacity, it will show signs of strain.

And the signs are everywhere.

Rivers are running dry

Around the world, major rivers - including some of the most iconic systems we rely on - now struggle to reach the sea. Freshwater habitats cover less than 1% of the Earth's surface, yet host more than 10% of all known animal species.[234] A landmark assessment of around 23,500 freshwater species found around 24% are at high risk of extinction. These are not isolated species losses. They are signs the hydraulic and ecological systems are under stress from dams, extraction, pollution, and altering of flows.[235]

Soils are losing life

Globally, the picture is stark. According to the Food and Agriculture Organization (FAO) and United Nations Environment Program (UNEP), up to one-third of the world's soils are estimated to be moderately to highly degraded due to erosion, loss of organic matter, salinisation, and other pressures. Global assessments suggest that tens of billions of tonnes of fertile soil are lost each year through erosion, and that more than 40% of the Earth's land surface is affected by some form of land degradation.[236, 237] This is not just an agriculture issue; soil supports every built system: foundations, stormwater, green infrastructure, carbon storage.

Biodiversity is collapsing

The biological richness that supports ecosystems - and thereby engineering systems - is collapsing. For example, island ecosystems support a disproportionately high share of the world's biodiversity and account for roughly half of recorded species extinctions, despite covering a small fraction of the Earth's land surface. The WWF Living

Planet Report indicates that monitored global wildlife populations have declined on average by approximately 69-73% since 1970, reflecting a steep reduction in abundance across many species.[238, 239] When ecological buffers go, so does the resilience of infrastructure.

Floods, fires, and extreme weather are intensifying

Engineering assumptions were typically based on "safe design lives", "return periods", and historical data.[240] Now, we live in a climate where previous assumptions break down. Drylands already cover approximately 40% of the Earth's land surface (excluding Antarctica)[241], and climate change is intensifying aridity and expanding dryland conditions in many regions. Floods, fire seasons, and storms exceed what systems were originally designed for.

Materials are becoming harder to extract

While less often framed in statistics, the depletion, cost escalation, and environmental damage of materials (sand, aggregates, rare minerals) mean that many infrastructure inputs are under pressure.[242] This reduces margin for error, increases cost, and raises risk. It is the hidden consequence of extractive engineering logic.

Waste systems are beyond original capacity

In many regions, waste and pollution management systems are operating beyond their original design capacity, with accumulating impacts from plastics, persistent chemicals, and urban heat.[243,244] Linear systems - source, use, discard - were embedded in our industrial mindset. But waste doesn't disappear. Microplastics, PFAS contamination, heat islands, exhausted soils - they're all examples of the "away" we assumed simply doesn't exist. The system is full.

Why This Matters for Engineers

All of this can sound overwhelming. The good news is that engineers are systems thinkers and the ultimate problem solvers. What matters is context - and whose context we're designing within.

Our assumptions are out of date. Our inputs have changed. Our boundaries have shifted. The conditions we are designing within are no longer what they were.

Engineering has always adapted to new realities. This is simply the next one.

What we are facing now is a new design brief. A huge number of the challenges we call "environmental" are actually engineering challenges:

- How do we build infrastructure that strengthens ecosystems rather than fragments them?
- How do we design water systems that recharge groundwater, not deplete it?
- How do we source materials without stripping landscapes?
- How do we reduce embodied carbon while increasing structural performance?
- How do we transition to circular, regenerative flows instead of linear ones?

Engineers have potential not only to be central to the solution of the huge challenges our world is facing, but drive solving it through every project we work on.

When the conditions of the natural world shift, engineering must evolve - because every structure we build, every system we design, and every material we source sits within nature's operating system.

That is why the shift to regenerative engineering matters. Not because we must "save the planet", but because the world we build from here will determine whether future generations inherit systems that are brittle and declining or alive, adaptive, and abundant.

The reality that gives me the most hope is this: engineers already have the skills. We simply need a new way to guide them

Where We Already See Regenerative Thinking

A mentor of mine invited me to a community screening of a documentary in Eumundi. It was one of those quintessential Sunshine Coast evenings - warm and still, with many stars visible in the night sky. Plastic chairs lined the square where the famous markets are held on Wednesday and Saturday.

The film was *Rachel's Farm* - the story of actor Rachel Ward returning to her family farm at sixty and transforming it through regenerative agriculture. I'd already been exploring the idea of buying a small farm myself - something simple where I could reconnect with land with my family, provide nature as a playground as the kids grow up, and be in my happy place - nature - every day.

When Rachel began, the soil at her farm was lifeless - compacted, depleted, stripped of structure. It was the kind of soil that feels like dust when you rub it between your fingers. She had employed local farmers who were using the typical techniques of today - spraying with chemicals and running cattle.

Over the course of a single year, Rachel restored the land using regenerative practices and the whole farm changed. Microbes returned. Water was retained. Birds reappeared. Life was brought back into the soil as well as carbon sequestration. You could almost feel the earth inhaling again.

Sitting there in the dark, with the glow of the projector flickering, I felt something shift in me. It wasn't just the story - it was the realisation that I had never been taught, in all my engineering training, how land heals and how to bring it back to life. We learned structural mechanics, hydrology, materials science... but not regeneration. Not renewal. Not how to design in partnership with ecosystems rather than over the top of them. There was no course on co-design with nature; it was more about controlling it.

The old DNA is deeply embedded in our systems. At one point in the documentary, Rachel explained that she had to pay her local council to stop spraying herbicides on the roadside near her property, just so the chemicals wouldn't drift onto her regenerating soil. I remember sitting up straighter, almost laughing in disbelief. She had to pay to stop the harm.

How had we created systems where restoring life requires permission - and payment?

As a few regenerative farmers and Rachel Ward stepped in front of the crowd and shared their stories, it got me thinking: this regenerative approach isn't just for farmers. It's for all of us - and perhaps especially for engineers.

This idea isn't unique or new. Many other disciplines and cultures already have this way of thinking.

Across medicine, ecology, permaculture, trauma-informed practice, Indigenous knowledge, and even Buddhist philosophy, the message is the same: **life thrives in relationship, not isolation.**

Regeneration is not a technique - it's a worldview. It asks engineers to see the world not as a machine to optimise, but as a living system to participate in. When we draw from these other disciplines, we expand our engineering DNA beyond efficiency and control into care, reciprocity, and long-term wellbeing - for both people and planet.

The First Engineers

Disclaimer:

I am not a First Nations person, and I don't speak for Aboriginal cultures, knowledge systems, or traditions. What I share here comes from respected listening, from conversations with First Nations colleagues and friends, and from my own growing understanding - not from lived cultural authority.

Through the podcast I co-host with Renee Wootton Tomlin (my dear friend), and through the many First Nations people I've had the privilege to learn from across my engineering career - people like Grant Maher from the Jabin Group who I co-hosted the BUILD program at the Opera House with, and countless others I've met on Country during talks and projects - I've been reminded again and again how much wisdom sits in these knowledge systems, and how far we still have to go in listening, respecting, and integrating this into our work.

There are many powerful and generous books that invite a deeper understanding of Indigenous Knowledge systems. Works such as *Dark Emu* by Bruce Pascoe, *The Dreaming Path* by Paul Callaghan and Paul Gordon, and the growing body of work by Corey Tutt offer important foundations. Texts like *Indigenous Engineering for an Enduring Culture*, edited by Cat Kutay, Elyssebeth Leigh, Juliana Kaya Prpic, and Lyndon Ormond-Parker, directly challenge Western engineering assumptions by revealing millennia of Indigenous ingenuity in landscape engineering, sustainability, and technology. *First Knowledges Design: Building on Country* by Alison Page and Paul Memmott explores Indigenous-led approaches to architecture and design, while *Gunyah Goondie + Wurley* by Paul Memmott documents the sophistication of Aboriginal building practices across Australia. Together, these works offer not just knowledge, but an invitation to rethink how culture, Country, and engineering have always been intertwined.

I offer this section with humility, huge respect, and a recognition that I am still learning. If we want to be the engineers of the future - engineers who create systems that sustain life - then we cannot move forward without First Nations knowledge. It is essential, it is generous, and it is something we must learn from with care.

Indigenous Knowledge and Regeneration

Long before the first aqueducts, the first "engineers" were already here - the Aboriginal peoples of Australia. They are recognised as the world's oldest continuous engineers, and for good reason. Their systems were not built *on* the land; they were built *with* it. They didn't try to outsmart nature's patterns - they worked in rhythm with them, guided by ecological intelligence and cultural law formed over more than 65,000 years.

For more than 65,000 years, Aboriginal and Torres Strait Islander peoples have engineered on this continent with a principle that modern engineering is only now rediscovering: regeneration.[245] Long before sustainability frameworks, performance metrics, or environmental guidelines, Indigenous cultures designed systems that supported life by working with the rhythms of land, water, seasons, and community. These were not ad hoc traditions. They were engineering systems-complex, collaborative, iterative and informed by observation, testing, and adaptation over thousands of generations.[222, 246]

Yet engineering education in Australia has rarely treated Indigenous knowledge as engineering knowledge. Too often, it has been positioned as cultural context rather than technical insight. But if we look closely, we can see that Indigenous knowledge is one of the most sophisticated forms of systems engineering we have: integrating ecology, hydrology, materials science, social design, time-based modelling, and risk management through principles of reciprocity and responsibility.[247]

This matters for one simple reason: we cannot design regenerative futures without learning from the people who have engineered regeneratively for millennia.

On Gunditjmara Country in Victoria, the Budj Bim aquaculture system remains one of the most extraordinary engineering feats anywhere in the world. More than 6,600 years old, it is a carefully constructed network of basalt channels, weirs, and traps that farmed eels sustainably for millennia.[248]

The system didn't extract from the environment; it strengthened it. The wetlands around Budj Bim show greater biodiversity today because of human intervention - a fact supported by archaeological and ecological research.[222, 258]

It is regeneration in the purest sense: a living partnership between people and place.

Across many Nations, fire-stick farming - or cool burning - offers another profound example of regenerative engineering. Aboriginal fire practitioners used low-intensity, slow-moving burns to clear undergrowth, stimulate plant regrowth, and renew the soil. Fire was never a force to fear or suppress; it was a tool for balance, nourishment, and continuity.

Modern fire science now validates this knowledge. Cool burning reduces catastrophic bushfires, supports species diversity, and maintains healthy landscapes.[222, 223] Aboriginal cultures knew this long before such findings were peer-reviewed.

To understand why these systems worked - and why they were regenerative - we need to understand the worldview behind them. Here are some of the learnings from Australian Indigenous People that have helped shape my thinking on Regenerative by Design.

Different Ways of Seeing Space: North, South, East, West

One of the clearest windows into how differently Aboriginal cultures think about Country is how many of them talk about space.

Most of us were raised on left and right. "Turn left at the lights." "The café is on your right." "My bag is behind you." Our internal compass is anchored to our body.

In a number of Aboriginal languages, the body is not the reference point. The land is.

Linguist John Haviland's classic work with Guugu Yimithirr speakers in Far North Queensland showed that they use cardinal directions - north, south, east, west - for everyday things where English speakers would use left and right. A Guugu Yimithirr speaker might say, "The cup is north of the plate" or "Move your bag a little to the east", even when they're indoors at a table.[249]

This isn't a party trick. It reveals a fundamentally different way of orienting. Instead of constantly asking, "Where am I relative to this object?" the brain is asking, "Where am I relative to Country?"

Broader research across Australian languages confirms this pattern. Linguists like Bill Palmer and colleagues have shown that many Aboriginal languages prefer what they call an absolute or geocentric frame of reference - anchored in the environment - rather than the egocentric left-right system dominant in English.[250] In some communities, people will effortlessly point to cardinal north even on a cloudy day, because they have been tracking the sun, the wind, the slope of land, and other cues their whole lives.

Cognitive science studies suggest this has real effects on how people think and move. In cultures where cardinal systems are used, people tend to stay oriented more reliably and perform differently on spatial memory tasks, because they're constantly updating an internal map connected to the landscape, not just to their own body.[251] It's another reminder that "intelligence" is not abstract - it is patterned by place.

So, what does this have to do with engineering?

For one, it shows that many Aboriginal cultures literally locate themselves in relation to Country first, self second. Direction is not a private, internal thing; it is a shared, external reality.It's a spatial language that assumes a common world, not just a personal viewpoint. That's a very different starting point for design.

It also has practical implications for how we teach and practice engineering today. Educational researchers like Caroline Edmonds-Wathen have shown that in communities where children grow up speaking Aboriginal languages or Aboriginal English influenced by those languages, standard maths language like "in front of/behind" or "left/right" can clash with their natural way of thinking about space. In her work with Iwaidja speakers in the Northern Territory, she found that understanding local spatial frames of reference - and building on them - made geometry and mapping concepts click in ways that more traditional Western explanations did not.[252]

QUT's YuMi Deadly Maths program has drawn on this insight for years. The team behind it, including Professor Tom Cooper and Dr Chris Matthews, have developed approaches to teaching geometry and location that start from Country, landmarks, and cardinal directions, then bridge into the formal curriculum. Rather than forcing students to abandon their intuitive geocentric sense of space, YuMi Deadly lessons validate it and use it as a foundation for concepts like coordinates, vectors, and movement.[253, 254]

I love this because it's not just about making maths more "engaging" - it's about recognising that Aboriginal ways of knowing already contain spatial and relational intelligence. The mistake would be to overwrite that with left/right thinking and call it an upgrade. The opportunity is to see that this relational orientation to north, south, east, and west is itself a kind of engineering mindset: always in dialogue with Country, always aware of position in a wider system.

When you see the world this way, it becomes very hard to treat land as a neutral backdrop. Direction is not just a convenience for navigation; it is a constant reminder that you are held within a larger pattern - of Country, of wind, of sun, of story. That, to me, is one of the differences between the first engineers on this continent and the industrial mindset we inherited: one begins from relationship, the other from self.

And if we want to become regenerative engineers now, we may need to relearn some of that orientation - not just in our language, but in how we place projects in the landscape, how we think about flow, and how we teach the next generation to locate themselves in relation to Country, not just to a coordinate on a drawing.

A Different Model of Human Needs

Eileen Deemal-Hall is a Bama woman with cultural ties to Dhithirr Warra, Wakamin, Bagarrmuuku, and Ankamuthi Nations. She was also formerly Chief Executive Officer for Wujla Wujal Aboriginal Shire Council.

She once shared with me that, in her culture, Maslow's hierarchy of needs is not a pyramid at all - and certainly not one that puts self-actualisation at the top. From her perspective, self-actualisation is foundational, because it comes from an embodied connection to Country. This comes first.

I remember sitting with that for a long time. I couldn't quite visualise it the first time she told me.

In a Western frame, you earn self-actualisation once everything else is sorted. But in Eileen's frame, which is echoed across many First Nations cultures, self-actualisation isn't a peak to reach; it's the ground you stand on.

Belonging to Country is not optional. It is identity, safety, purpose, and responsibility all at once. Anthropologist Nurit Bird-David writes that many Indigenous cultures see land as "kin" - a living relative, not a resource.[259]

Once you see Country as kin, reciprocity becomes the only logical way to live.

A Cyclical Worldview, Not a Linear One

Western engineering often works in linear time: design build operate maintain decommission. It's a timeline, a project plan, a Gantt chart.

Indigenous systems move in cycles - seasonal cycles, ecological cycles, ceremonial cycles, ancestral cycles. These aren't just agricultural calendars. They are relational maps that guide when to act, when to rest, when to burn, when to harvest, and when to move. Environmental anthropologists such as Deborah Bird Rose describe these systems as "ethical ecologies"[255, 256] - design principles held in relationship, not in spreadsheets.

This cyclical view naturally embeds long-term thinking. Actions are evaluated not for what they achieve this quarter, but for what they mean over generations. Many First Nations frameworks refer to Seven Generations thinking - a practice widely documented in Native American and some Aboriginal teachings.[257]

Naturally, short-term harm becomes unthinkable when you imagine your descendants living with its consequences.

Country as Teacher, Not Resource

Another key difference lies in how knowledge is gathered. Western engineering seeks data: measurements, simulations, modelling.

Indigenous engineering also seeks data - but the method is different. It is gathered through long observation, story, pattern recognition, and a multi-sensory attentiveness to place. Scholars such as Tyson Yunkaporta[260] describe this as "learning from Country", where the land itself is the curriculum and the teacher. Bird migrations, wind shifts, soil scents, water clarity, the texture of bark - all of it is information. All of it is engineering input.

This form of knowledge is dynamic and responsive. It does not separate science from story, because both are needed to understand how a place lives.

Country as a Stakeholder

We have seen here that in Aboriginal worldviews, Country is not a backdrop or a passive environment. It is a living entity - land, water, air, stories, culture, ecology, people. It's all interconnected. This perspective reframes engineering fundamentally.

When Country is a stakeholder, we ask:

- What does this design give back?
- How will it impact future generations, not just project deadlines?
- What does Country need to stay healthy, resilient, and alive?

Country as a stakeholder isn't symbolic. It's a risk-reduction strategy. It forces us to consider long-term impacts that traditional engineering briefs often overlook: soil health, water pathways, fire behaviour, biodiversity, and cultural relationships to place. This leads to more resilient, more accepted, and more enduring outcomes.

Circularity and Reciprocity as Methodology

In many Aboriginal nations, you do not take without giving back. Harvesting requires tending. Burning requires renewal. Using water requires protecting its source.

This is not the language of extraction. It is the language of relationship.

Elder Mary Graham often says: "The land is the law."[261] And if the land is the law, then engineering becomes an exercise in listening, not controlling.

Circular economy principles are now heralded as cutting-edge. Yet Indigenous societies have always engineered through circularity-designing systems in which resources flowed through communities without depletion.

Indigenous engineering demonstrates:

- Closed-loop food systems
- Fire regimes that regenerate landscapes
- Water infrastructure that improves ecological health
- Material use where life cycles, repair, reuse, and natural decomposition were assumed, not idealised

Reciprocity sits at the heart of this. You only take what you can regenerate. You give back more than you consume. You design for balance rather than extraction.

Belonging, Not Dominance

What sits beneath all of this is a fundamentally different understanding of our place in the world as humans.

Where Western engineering historically framed nature as something to manage, tame or optimise, Aboriginal engineering framed nature as something to belong to.

Not mastery, but membership. Not ownership, but custodianship. Not control, but reciprocity.

And that is really what regeneration means: a relationship of respect and care between humans and the living systems that sustain us - a relationship that shapes design decisions not through fear or compliance, but through responsibility, humility, and belonging.

Principles of Regenerative Engineering from Indigenous Knowledge (Australian Context)

It's worth pausing to name the principles that sit beneath Indigenous engineering. These aren't slogans or modern sustainability ideas retrofitted onto ancient cultures - they're practices refined over tens of thousands of years on this continent. They are engineering principles, ecological principles, and cultural principles all at once. And while they come from cultural knowledge, they also offer timeless guidance for anyone designing in relationship with land today.

What follows isn't a complete list, but a starting point - a set of foundations we can learn from as we rethink what regenerative engineering really means in an Australian context.

1. Country as Kin, Not Resource

Country is not land you stand on - it is a living relative you are in relationship with. Design begins by asking: what does Country need to thrive?

2. Custodianship Over Ownership

Land is not possessed; it is cared for. Engineering decisions are made with responsibility, not entitlement.

3. Learn From Country Before Acting

Knowledge comes from observation, listening, story, ceremony, and ecological cues. The land teaches - if we are attentive enough to hear.

4. Work With Cycles, Not Against Them

Design follows seasonal calendars, ecological rhythms, and recurring patterns. Timing, sequencing, and seasonality matter as much as materials.

5. Reciprocity as a Design Requirement

If something is taken, something must be returned - nutrients, habitat, care, or time. Every intervention has a responsibility attached to it.

6. Design for the Seventh Generation

Decisions are measured by how they will affect descendants, not just current users or budgets. Short-term gain never outweighs long-term belonging.

7. Balance, Not Control

Systems are shaped to maintain ecological equilibrium, not dominance. The engineer's role is to support healthy relationships between people, water, plants, animals, and place.

8. Story as Knowledge, Not Metaphor

Engineering principles are encoded in songs, stories, art, ceremony, and place names. Story is not decoration - it is instruction.

9. People and Place Are One System

Human wellbeing and environmental wellbeing are inseparable. A thriving community requires a thriving Country.

10. Regeneration Is Identity, Not Strategy

Caring for Country is not a project objective - it is a cultural identity and enduring law. Design is an expression of relationship.

Ancient Engineers: The Romans, the Greeks, and the Breathing City of Venice

We often talk about regeneration as though it's a modern idea - something emerging from climate models, circular economy reports, and new sustainability targets. Actually, many of our ancestors knew how to engineer with nature long before we had language for it.

Across the world, civilisations designed systems that enhanced the environment rather than exhausting it. Their work was rooted in reciprocity, time, and observation - the same qualities our modern engineering urgently needs to recover.

In the Philippines, the Ifugao people carved the 2,000-year-old rice terraces into the mountains. These weren't just farms; they were living hydrological systems that channelled spring water, prevented erosion, created microclimates, and kept communities thriving for centuries.[262] They required cooperation, ceremony, and constant dialogue with land - not control.

In Mesoamerica, the Maya developed the milpa system - maize, beans, and squash grown in rotation. It replenished soil nutrients, supported hundreds of companion species, and let forests regenerate between cycles.[263] Today, ecologists consider it one of the most sustainable food systems ever created.

Across the Middle East and North Africa, engineers built qanats - underground water channels that used gravity alone to distribute water without depleting aquifers. Some have been in operation for more than 2,500 years.[264] Their genius lies in how gently they work with natural systems, allowing water, soil, and communities to flourish together.

In Peru, the Incas built vast stone terraces that stabilised mountainsides and created natural climate-controlled zones for growing food at impossible altitudes.[265] These terraces didn't just prevent erosion; they engineered biodiversity and resilience.

And across Aotearoa, Māori engineering was guided by kaitiakitanga - guardianship of land and water. Wetlands were shaped to support eels and birdlife. Forest harvesting followed lunar cycles.[266] Resource use was governed by protocols designed to preserve balance for generations.

These examples remind us that regenerative engineering is not a new frontier. It is the oldest way of working. And as we rethink how to design with nature rather than against it, these ancient systems show us what becomes possible when engineering arises from relationship, reciprocity and belonging.

Global Principles of Ancient Regenerative Engineering

As I began exploring examples from ancient cultures around the world, I was struck by how familiar they felt. Different landscapes, different languages, different belief systems - yet the underlying engineering principles echoed one another.

Whether it was the Ifugao people carving rice terraces into mountains, the Maya tending milpa fields in cycles of renewal, the Inca building climate-responsive terraces, or the engineers of Persia guiding water through qanats with extraordinary restraint, these societies were designing far more than infrastructure.

They were designing *relationships*.

Their systems were shaped by an understanding that humans were part of the landscape, not apart from it. That engineering was an act of belonging, not control. That time was measured in seasons and generations, not deadlines and delivery milestones.

What I saw, over and over again, was a worldview we are only now returning to - a view that treats nature as a partner in design, not a problem to solve. These ancient systems offer us not only inspiration, but instruction. They remind us that regeneration is not new; it is remembered.

To make their wisdom more practical for the modern engineer, I've distilled the common threads across these cultures into a set of global principles - timeless, cross-cultural, and hugely relevant today.

1. Design With Landscape, Not Over It

Ancient engineers shaped systems that followed contours, seasons, and natural forces. Terraces, water channels, and farms worked because they honoured the land's intelligence.

2. Work in Rhythms, Not Deadlines

Projects moved in cycles - planting, resting, burning, flooding, harvesting. Time was ecological, not industrial. The land set the tempo.

3. **Value Longevity Over Speed**
Systems were designed to last for generations. Durability came from patience, craft, and understanding of place.

4. **Reciprocity Is Non-Negotiable**
Every extraction came with a contribution - soil renewal, careful burning, ceremonial stewardship. Regeneration wasn't an outcome; it was a requirement.

5. **Engineer for Biodiversity, Not Single Outputs**
Terraces supported crops, animals, insects, microclimates, and people. Water systems nourished ecosystems, not just human settlements.

6. **Use Energy Wisely (or Not at All)**
Gravity-fed channels, thermal mass, wind acoustics, and shade structures reduced the need for mechanical systems. Sustainability was a by-product of elegant simplicity.

7. **Knowledge Lives in Community**
Design, maintenance, and decision-making were collective processes. Technical knowledge was carried through oral traditions, ritual, roles, and shared responsibility.

8. **Nature Is Teacher, Not Obstacle**
Landscape cues - tides, stars, monsoons, moon cycles, soil colour - guided engineering choices. Observation was a core technical skill.

9. **Regeneration Is Identity, Not Strategy**
Engineering wasn't separate from culture - it was an expression of it. To care for Country (or land or place) was to care for self, community, and future.

Permaculture: A Regenerative Blueprint Engineers Can Learn From

If Indigenous knowledge teaches us belonging and reciprocity, then permaculture offers us a way to live those values in the systems we build today. I once heard someone say that permaculture is "ancient wisdom wrapped in a modern toolkit", and the more I've learned, the more I believe that's true.

Permaculture was developed here in Australia by Bill Mollison and David Holmgren[267], but its roots echo practices that humans have used for thousands of years. At its heart, permaculture isn't gardening. It isn't landscaping. It is a philosophy of design - a way of shaping human systems so they behave more like living ecosystems: resilient, adaptable, interdependent, and generously productive.

Permaculture rests on three simple, powerful ethics:

- Care for the Earth
- Care for People
- Fair Share - meaning whatever abundance we create, we return to the system so it can continue to thrive

What I love about these ethics is that they're not separate categories. They reinforce one another. When land thrives, people thrive. When people thrive, communities contribute back. It is circular not only in function but in spirit.

Beyond the ethics, permaculture is guided by twelve design principles - principles that, interestingly, engineers tend to recognise instantly. They feel like systems thinking, but anchored in biology rather than industry. And when you look at them closely, they are practical.

Take the first principle: **observe and interact.** In permaculture, you don't start with a plan. You start with paying attention - watching how water moves through the landscape, where the sun falls, how the wind gathers, how animals behave, what the seasons teach you. Imagine if we approached major infrastructure projects this way, letting the land speak before drawing a single line. Observation before intervention.

Another principle I love is **use and value diversity.** Nature abhors monoculture. Diversity strengthens the web of life - if one element fails, another steps in. This is as true for ecosystems as it is for supply chains, materials, project teams, or transport networks. The things that survive are not the most optimised; they are the most diverse.

Then there is **catch and store energy.** Permaculture sees energy everywhere: sunlight, rain, nutrients, heat, wind, labour, knowledge, culture. Nothing should be wasted. Everything is circulated, captured or reused. Engineers call this efficiency; nature calls it survival.

Consider **produce no waste.** In nature, waste doesn't exist. Every output feeds something else. Imagine if our buildings, bridges, data centres and material streams were designed with the same idea. Imagine if "end of life" became "beginning of next life".

Permaculture also values **small and slow solutions.** This principle is an antidote to our industrial urge toward scale and speed. It reminds us that regeneration happens through steady, iterative change - not huge disruptive leaps. Nature grows forests one season at a time.

Another principle, **integrate rather than segregate**, invites us to design relationships, not silos. In permaculture, chickens aren't separate from crops, and water isn't separate from soil. Everything overlaps. Everything is part of a system that supports itself. Our built environment can do the same: connecting transport to ecology, housing to health, stormwater to biodiversity, energy to community.

Then there's my favourite: **design from patterns to details.** Start with the big story - the flows, rhythms, context, climate, community - and then work your way into the details. As engineers, we are often taught to do the opposite: to zoom straight into the components. Permaculture flips this and asks us to begin with the whole.

Why Permaculture Matters for Engineers

At our core I believe that engineers care deeply about the impact they have on people and the planet. That's what makes permaculture such a powerful guide. It is:

- Practical
- Evidence-based
- Tooted in long-term thinking
- Adaptable across scales
- Fundamentally regenerative

It shows us not just how to do less harm, but how to create conditions where life expands because of our work.

You can see permaculture principles coming alive in modern engineering:

- Stormwater systems that mimic wetlands instead of channelling water away

- Buildings designed to improve biodiversity, not flatten it
- Transport corridors that double as ecological corridors
- Materials that cycle elegantly back into biological or industrial loops
- Energy systems that build resilience rather than dependency

Permaculture has us move from controlling nature to collaborating, and even co-designing, with it - from extraction to reciprocity, from efficiency to wisdom.

Perhaps the most hopeful lesson of all is this: when we design for life, life returns. Just like Rachel's soil.

When regeneration is at the centre of our work, the systems we create stop extracting - and start healing. And that, truly, is what engineering for the future looks like.

What is Regenerative Engineering?

When people first hear the term "regenerative engineering", they might assume it's just the next sustainability trend - another layer of reporting, another set of metrics, another checkbox on a project plan. But regenerative engineering isn't an upgrade to sustainability. It's a fundamentally different way of seeing the purpose of engineering.

To understand what regenerative engineering is, it helps to begin with what it is not.

Extraction: Where We Began

As we have already discussed in this book, for most of the industrial era, engineering was an extraction-based discipline. We designed systems to take materials, energy, land, water, and labour from the world, convert them into something useful, and move on to the next project.

This wasn't malicious - it was the logic of the time. The world felt abundant. Growth seemed infinite. And needs were urgent.

Extraction built the modern world. It gave us railways, sanitation, electricity, mass mobility, food security, and clean water. When I talk about extraction, I speak with gratitude because without it, none of us would be here reading this book.

Extraction has limits - and we are now living inside those limits.

Arup's 2023 *Regenerative Design* report[268] captures this reality clearly. Extractive human systems are now pushing the planet "beyond the limits of Earth's capacity to regulate against shocks and stresses and to support all life". In other words, the engineering logic that built the modern world can no longer sustain it.

Sustainability: The First Wave of Awareness

When sustainability entered engineering in the late 20th century, it was a meaningful step forward. The message was: do less harm. Switch to lower-impact materials. Reduce waste. Use energy more efficiently. Minimise the footprint.

This work is still critical. But it kept the same underlying mindset: nature is something to be managed so we can do what we need to do.

Sustainability made harm smaller - it didn't ask us to reverse it.

UNEP's landmark report *Making Peace with Nature*[269] makes this point clearly. Even the best sustainability practices are insufficient unless they also restore ecosystems, reclaim degraded landscapes, and regenerate the systems we depend on. Reducing harm is not the same as repairing it.

Net Zero: The Threshold

Then came net zero, which is still the dominant goal today. Design systems that minimise emissions and offset what's left. Move toward neutral impact.

Net zero is a turning point - a line in the sand that says: let's stop pouring fuel on the fire.

Arup's Reduce-Restore-Remove framework articulates this three-step pathway: reduce emissions first, restore natural systems second, and remove excess carbon where needed. It's an important shift - but even net zero stops short of what our ecological reality now demands.

Zero is not enough. Neutrality is not healing. A system that leaves no additional damage still leaves all the historical damage in place.

Regenerative Engineering: The Next Evolution

Here's the simplest way to understand it:

Sustainable engineering = "Do less harm."

Regenerative engineering = "Create net positive outcomes for people, place, and planet."

What regenerative engineering ultimately asks of us is not a new technical skillset (although that helps), but a new mental model. A shift from "how do we reduce negative impact?" to "how do we create positive impact?"

Here is the surprising thing: regenerative engineering is not harder. It's simply more imaginative. It asks us to see infrastructure as part of a living system rather than an isolated object.

Regenerative engineering asks a different question entirely: what if the project left this place better than we found it? What if engineering healed, restored, replenished, and strengthened the systems it touched?

According to Arup's definition, regenerative design means creating systems where "people and nature co-exist and co-evolve over time".[268] That last part - co-evolve - is critical. It means humans aren't separate from nature. We're part of the same system, shaping each other.

UNEP reinforces this idea too.[269] Human development must now "restore air, water, soils, biodiversity, and ecosystems" as part of its core function. Regeneration is not optional. It's required for survival.

So regenerative engineering isn't about minimising harm. It's about maximising renewal.

Instead of reducing damage, it asks questions like:

- Can this design restore soil health?
- Can it clean the air or improve water quality?
- Can it boost biodiversity rather than erode it?
- Can it sequester carbon instead of emitting it?
- Can it strengthen cultural connection, not just physical performance?
- Can it return more energy and vitality to the system than it takes?

It is engineering that contributes to life - not just "does less to threaten it."

The Reciprocity Ladder: Climbing from Extraction to Regeneration

ARUP's 2020 report on regenerative design captures this perfectly - the evolution from extraction to partnership.[268] Building on that, we can imagine a Regeneration Ladder - a way to see where we are and what we're aspiring to:

Stage	Mindset	System Behaviour	Design Example	Outcome
1. Extractive	Nature as resource	Linear: take-make-waste	Mining, concrete sprawl	Depletion
2. Mitigating	Nature as constraint	Reactive: harm reduction	Pollution control, offsets	Delay of harm
3. Sustainable	Nature as system	Balancing inputs & outputs	Energy-efficient buildings	Net-neutral
4. Restorative	Nature as collaborator	Circular: repair & reuse	Wetland rehabilitation, green roofs	Renewal
5. Regenerative	Nature as teacher	Reciprocal: co-evolving systems	Blue-green cities, living buildings	Thriving

Consider: where is your project, system or design on this ladder - and what would it take to climb one rung higher?

Regenerate by Design: A Simple Framework for Regenerative Engineering

If you are seeking a practical way to begin applying regenerative thinking to a project, problem, or system, this framework is one option. It's simple enough to use on any scale - from a precinct plan to a drainage upgrade, from a materials choice to a city-shaping vision - but powerful enough to shift the mindset of a whole team.

Now, regeneration is complex. Our projects are complex. The systems we work in - political, environmental, cultural, financial - are complicated, interconnected and often full of competing pressures.

Regenerate by Design isn't designed to simplify that reality or pretend there's an easy answer. It's a high-level guide; a way of thinking rather than a prescriptive method. And I say this not as a sustainability expert, but as an engineer who believes that every one of us has the capacity to begin embedding regenerative thinking into our practice.

You don't need to have all the answers to start. You just need to start thinking differently - to widen the lens we currently have, ask different questions, and take small, practical steps that honour the systems we're part of. Regeneration isn't a specialist field. It's something every engineer can contribute to, project by project, decision by decision.

Regeneration by Design is ultimately about restoring a reciprocal relationship with the systems that sustain us. The later chapters of this book expand these ideas into a set of values and design principles, but this framework is a place to begin. It helps teams ask: how do we design so that we give back more than we take?

At its heart are four phases - Take, Make, Give, Restore - that move us from extraction toward renewal.

1. Take - Minimise Extraction

The cycle begins by acknowledging the reality of impact. Every project "takes" something: land, water, energy, materials, cultural space, community trust. Regenerative engineering doesn't pretend otherwise - it simply asks us to take less, and to take consciously.

Key questions:

- What are we extracting from the land, the atmosphere, the community, or culture?
- Can we reduce this? Avoid it? Substitute it?
- What would be the lightest way to intervene?
- Are we taking more than the system can replenish?

Examples in practice:

- Choosing low embodied carbon materials
- Designing smaller footprints
- Minimising demolition and reusing components
- Using data to avoid over-engineering
- Protecting water tables, topsoil and existing vegetation

This phase builds awareness. It invites teams to see their project not as an isolated object, but as part of a larger web of relationships.

2. Make - Design With Intention

Once we understand what we're taking, we intentionally shape what we make. This is where regenerative engineering begins to feel different from business-as-usual. Instead of designing only for performance, we design for purpose, values, and relationships.

Key questions:

- What are we optimising *for*?
- How can this design honour place, culture, Country, and community?
- What values sit behind this decision?
- Is the design doing its job without unnecessary harm?

Examples in practice:

- Working with natural landforms instead of reshaping them
- Using Indigenous knowledge to guide orientation, flow and seasonal use
- Designing energy systems that prioritise local resilience
- Embedding nature into structures - green roofs, wetlands, biodiversity corridors
- Creating public spaces that support social cohesion, rest and dignity

This phase links strongly to the later chapters of this book where we explore values and principles in detail.

3. Give - Return More Than You Take

This is the turning point of the cycle. Most sustainability frameworks stop at harm reduction. Regeneration asks a more courageous question: what can this project give back? Not metaphorically, but materially, ecologically, socially, and culturally.

Key questions:

- What does this project contribute to the system?
- Where does it add life, not just mitigate damage?
- How can it strengthen soil, water, air, biodiversity, or culture?
- How does it improve community resilience, belonging, or wellbeing?

Examples in practice:

- Wetlands that filter water and regenerate habitat
- Buildings that generate more energy than they consume
- Materials that safely return to natural or industrial cycles
- Infrastructure that creates shade, pollinator pathways, or food sources
- Cultural partnerships that support ongoing care for Country

This is reciprocity in action - the moment the project becomes not just a user of resources, but a contributor to future abundance.

4. Restore - Strengthen the System Over Time

Regeneration doesn't end when the project opens. It continues through maintenance, operations, stewardship, and community partnership. True regenerative design strengthens the system - not just the asset in isolation - over years, decades, and generations.

Key questions:

- How does this asset evolve with the landscape, climate, community, and culture?
- What long-term patterns of care need to be embedded?
- Who stewards this place, and in what relationship?

- How does this project support healing - of soil, water, Country, community, memory?

Examples in practice:

- Designing operations that improve, not degrade, ecological health
- Long-term monitoring of biodiversity, carbon, water, and soil
- Governance models that include Traditional Owners
- Using adaptive management based on feedback loops
- Designing assets that can be disassembled, reused, rewilded, or transformed

In this phase, engineering moves from project to legacy - from a fixed product to a living system that continues to regenerate itself.

How to Use the Regenerate by Design Cycle on a Project

You can use this framework at any scale: an intersection, a rail corridor, a drainage basin, a precinct, a building, a digital system, even a procurement process.

A simple way to use it is:

1. **Gather the team** - engineers, planners, architects, Traditional Owners, community representatives, ecologists, operators.
2. **Move through the four phases slowly** - Take, Make, Give, Restore.
3. **Write everything down** - insights, blind spots, ideas, intentions, constraints.
4. **Identify 3-5 themes** - the things that matter most for this project.
5. **Turn those themes into design moves** - practical actions or choices.
6. **Revisit the cycle at each project stage** - concept, options assessment, detailed design, handover, operations.

Regeneration by Design is intentionally simple. It doesn't replace the rigour of modelling, analysis or standards. It complements them by inviting engineers to widen the lens and notice the impact of their work in the broader system of life.

It is a quiet but powerful shift from extraction toward reciprocity, from control to collaboration, and from short-term delivery to long-term regeneration.

And the beauty of this cycle is that you don't have to transform an entire project to begin. Even one choice, one design move, one commitment to "give back" can ripple into something much bigger.

Regenerative Engineering in Practice

There are dozens of modern projects around the world already showing regeneration in practice. It's not theory - it is already happening, and I hope these case studies provide inspiration for a more regenerative future by design.

Case Study: EcoLogic Studio's BioSolar Leaf (London) - Regeneration by Design Through Living Urban Surfaces

On a busy street in London, the façade of an office building looks almost alive. Instead of metal panels or tinted glass, its surface glows with vertical tubes of swirling green - microalgae suspended in nutrient-rich water. This is the BioSolar Leaf system, developed by EcoLogic Studio, and it turns an ordinary façade into something far more profound.

This is not a surface that merely "performs well." It regenerates the environment around it.

As air moves across the façade, the microalgae absorb CO_2 and pollutants like nitrogen oxides, releasing oxygen and transforming the building into a biological air-cleaning machine. Each installation performs like a concentrated photosynthetic engine - a façade the size of a single tree can clean as much air as 100 mature trees, but in a footprint that fits on a dense urban street.

According to EcoLogic Studio's analysis, the BioSolar Leaf actively removes particulate pollution (PM2.5 and PM10) - pollutants that increase asthma, respiratory illness, and cardiovascular risk. It also provides shade that reduces building cooling loads and produces nutrient-rich biomass that can be used for soil regeneration.[270, 271]

This is not net zero. This is net positive. It turns the built environment into a partner in urban ecology.

Engineers designed the system to be lightweight, modular, gravity-fed, and energy-efficient. It uses the city's own waste - CO_2, sunlight, pollution - as its feedstock. Rather than relying on mechanical filters or chemical treatments, it taps into a biological process millions of years old.

A façade becomes a miniature ecosystem and a building becomes part of the watershed, not a burden on it. A street becomes cleaner simply because it exists.

Why BioSolar Leaf Embodies Regeneration by Design

1. Take - Minimise Extraction

The system uses what the environment already provides: sunlight, CO_2, and rainwater. Minimal new resources, minimal embodied carbon, minimal intervention.

2. Make - Design with Intention

BioSolar Leaf is engineered as a living system, not a static object.

- Passive flows
- Biological intelligence
- No waste streams
- Modular growth and repair

It is design guided by relationship, not control.

3. Give - Return More Than You Take

The façade cleans air, creates oxygen, cools the street, supports biodiversity, and generates biomass for soil health. It improves the health of the city and the health of its people.

4. Restore - Strengthen the System

By mitigating pollution and heat, the system supports resilience in the surrounding neighbourhood. It becomes part of a wider metabolic loop, through infrastructure that heals while it operates.

Commercial and Cost Insights

Let's dive into the commercials of this project and compare this with a typical installation: a retrofit façade on a 50,000 sq ft urban building.

Key assumptions:

- Installation cost premium due to new technology: estimated ~30% above conventional façade (£/m^2)
- Energy/air quality savings modest in early years, growing with scaling
- Lifespan of installation: ~50 years (façade system vs building structure)
- Typical urban façade OPEX: façade cleaning, maintenance, replacement glazing every 25-30 years

Time Horizon	Typical Urban Façade OPEX	BioSolar Leaf System OPEX	Estimated Savings
0-10 years	US $2 M	US $1.4 M	~US $0.6 M
10-30 years	US $6 M	US $4.0 M	~US $2.0 M
30-50 years	US $8 M	US $4.5 M	~US $3.5 M
Total 50 years	US $16 M	US $9.9 M	≈ US $6.1 M saved

Notes for interpretation:

- These values assume the BioSolar Leaf system captures value from lower energy/air-cleaning costs, less frequent replacement, and improved façade performance.
- The premium capital cost is omitted in this OPEX table - the focus is on operational expenditure.

- Actual savings may be greater if the façade yields ancillary benefits (health, productivity, biodiversity) not captured here.
- Data on real installations remain emerging; treat this as an illustrative model rather than a firm projection.

Practical Takeaways for Engineers

BioSolar Leaf shows that every surface we design has hidden potential. Its lessons are clear:

- Treat surfaces as ecological assets, not inert cladding.
- Biology is a technology - and sometimes the most elegant one.
- Regeneration begins when infrastructure contributes more than it consumes.
- Dense urban areas are not barriers to ecology - they are opportunities.

Most importantly, it reminds us that regeneration does not require massive land areas, huge budgets, or revolutionary new materials.

Sometimes, all it takes is asking a different question: What if this building didn't just avoid harm - what if it made the city better?

Case Study: The Bullitt Centre (Seattle) - Regeneration by Design Through a Building That Gives More Than It Takes

In Seattle's Capitol Hill district, an office building rises only six storeys and behaves like a forest organism: harvesting sunlight, storing water, cycling nutrients, and operating in partnership with the ecosystem around it.

Walk inside and it feels different from the moment you enter. Sunlight pours down from a central atrium. Natural ventilation moves through the space. The timber structure feels warm, almost alive. Workers and visitors often comment that they breathe easier here - and that they feel more alert. This is not psychological suggestion. It's physics, biology and intention.

On the roof, a vast solar canopy produces more electricity than the building uses each year. The Bullitt Centre is a net positive energy producer, feeding clean energy back into the grid even in Seattle's cloudy climate. Wastewater is treated onsite through constructed wetlands. Greywater is filtered through a living system beneath the building. Rainwater is captured, purified and stored for drinking.

This is not a building that simply meets codes. It sets a different model for what buildings can be.

Dubbed "the greenest commercial building in the world" when it opened in 2013, the Bullitt Centre was one of the first developments to pursue the Living Building Challenge - a regenerative design framework that requires buildings to operate like native ecosystems. That means no toxic materials, no fossil fuel use, and performance-based certification (not predicted performance - actual measured results).[272, 273]

Here is the building in numbers:

- Net-positive energy through rooftop solar
- Zero operational carbon
- Net-zero water (collect, treat, reuse)
- Composting toilets that return nutrients to soil cycles
- 90% daylight autonomy inside workspaces
- A lifespan designed for 250 years

This is what happens when engineers, architects, and ecologists design with the long view in mind.

Why the Bullitt Centre Embodies Regeneration by Design

1. Take - Minimise Extraction

The building dramatically reduces reliance on external systems. It takes only what it needs - sunlight, rainwater, and the smallest possible material footprint. The timber structure is FSC-certified and sourced through regenerative forestry practices.

2. Make - Design with Intention

The Bullitt Centre is not the result of incremental improvements. Every system is intentional:

- Solar canopy sized for net positive generation

- Breathable façades
- Composting loops
- Water harvesting and treatment onsite
- Non-toxic materials list

It is an ecosystem disguised as a building.

3. Give - Return More Than You Take

The building actively improves its environment:

- Provides clean solar power back to the community
- Reduces stormwater loads by using rain capture
- Filters wastewater naturally
- Avoids toxic runoff
- Improves occupant health and productivity

4. Restore - Strengthen the System

Its presence strengthens the broader urban system:

- Reduces grid demand
- Models regenerative economics
- Inspires new policies and standards
- Proves net positive buildings are commercially viable

It restores belief in what is possible.

Commercial and Cost Insights

The Bullitt Centre was never a low-cost demonstration project; it was a paradigm-shifting investment. Total project costs (land, hard and soft costs) reached US $32.5 million for approximately 52,000 sq ft, equating to about US $625 per sq ft.[274]

This represented a cost premium compared to standard commercial office buildings in Seattle. Yet when the team broke down the premium, they found it was largely driven by new regulatory challenges, bespoke systems, elevated performance goals, and first-time deployment of innovative technologies.[275, 276]

Importantly, the analysis shows that for long-term owners (universities, governments, foundations) the building delivers value: utilities pre-paid, maintenance reduced, lifespan extended, and public value returned.[277] In other words, the upfront extra cost is less about extravagance and more about investing in longevity, performance, and regeneration - a key shift for engineers used to thinking only in first-cost terms.

Key commercial insights:

- The upfront cost premium was estimated at around 0-30% above typical office building cost for the context.[278]
- Because the design life is 250 years (versus typical 50-60 years for many office buildings), the asset amortisation and lifecycle cost metrics shift dramatically.[274]
- Over time, lower utility bills, less structural renewal, fewer major retrofits, and higher occupant health/productivity create cumulative value that conventional first-cost comparisons alone fail to capture.

Operational Expenditure Comparison (Bullitt Centre vs Typical Commercial Building)

Assuming a 52,000 sq ft office building, Seattle context, 250-year design life.

Key assumptions applied consistently across:

- Typical building energy cost: US $2-3/sq ft/year
- Bullitt Centre energy cost: Net-positive ~0 cost (surplus generation)
- Water/sewer cost savings: 80-100% reduction
- Maintenance savings: ~20% over time due to durable materials + reduced mechanical complexity
- Major system replacements (typical building) every 20-30 years
- Bullitt Centre avoids several major mechanical plant renewals

Annual OPEX Totals by 50-Year Increments

Time Horizon	Typical Commercial Office Building	Bullitt Centre	Difference	Notes
0-50 years	US $35M-45M	US $10M-15M	≈ US $25M-30M saved	Typical building spends the most here: HVAC replacements, high energy costs, water/ sewer charges. Bullitt Centre benefits from net positive energy and low-impact water systems.
50-100 years	Additional US $40M-50M	Additional US $12M-18M	≈ US $28M-32M saved	Second cycle of major system replacements for typical asset (HVAC, electrical, envelope refurbishment). Bullitt Centre requires fewer major renewals due to 250-year design lifespan.
100-150 years	Additional US $45M-55M	Additional US $15M-20M	≈ US $30M-35M saved	Typical building: high maintenance for aging structure. Bullitt Centre: durable timber and low-toxicity materials delay degradation.
150-200 years	Additional US $50M-60M	Additional US $18M-24M	≈ US $32M-36M saved	Typical building would likely require total replacement; numbers assume ongoing refurbishment. Bullitt Centre maintains long-life structural integrity.
200-250 years	Additional US $55M-65M	Additional US $20M-28M	≈ US $35M-38M saved	Most conventional buildings never reach this point. Bullitt Centre continues operating efficiently with lower system renewal frequency.

Total 250-Year OPEX Estimate

Category	Typical Building (250 yrs)	Bullitt Centre (250 yrs)	Total Savings
Cumulative OPEX	US $225M-275M	US $75M-105M	≈ US $150M-170M saved

Practical Takeaways for Engineers

The Bullitt Centre teaches us that regeneration is not theoretical. It is achievable with today's technology and today's budgets, when values lead decisions.

Lessons for engineers include:

- Design for long-term performance, not short-term cost.
- Treat water, energy and waste as cyclical resources.
- Healthy buildings support healthy communities.
- Standards rarely show what is possible - performance does.
- Regeneration is a design choice, not a luxury.

And perhaps its most important lesson is this: when we design to heal the systems we touch, we also heal the profession itself - returning engineering to its roots of stewardship, imagination and care.

Case Study: Bishan-Ang Mo Kio Park (Singapore) - Regeneration by Design Through Rewilded Urban Water Systems

If you walk through Bishan-Ang Mo Kio Park today, you'd never guess that the river winding through it - full of dragonflies, birds, children splashing along the edges - was once a concrete drainage canal. A straight, grey, flood-control channel designed purely for efficiency. No beauty. No ecology. No relationship to place.

Singapore's engineers saw an opportunity: what if a stormwater canal could become a living river again? What if engineering didn't suppress nature, but partnered with it?

That question became the Active, Beautiful, Clean Waters (ABC Waters) Program, and Bishan-Ang Mo Kio Park became its flagship transformation.[279]

A Concrete Canal Becomes a Living River

In the early 2000s, PUB (Singapore's national water agency) and landscape architects Ramboll Studio Dreiseitl proposed something radical in the engineering world: remove the concrete. Restore the river. Let it move again. The team replaced 2.7 km of concrete channel with a naturalised river designed using bioengineering techniques:

- Re-meandered channels to slow flow and improve infiltration
- Planted riverbanks with 60+ native species to stabilise soil and restore habitat
- Floodplains that intentionally fill during storms
- Wetland systems to filter stormwater naturally

The river became not just a piece of infrastructure - but an ecosystem.[280]

Ecological and Engineering Outcomes

After restoration:

- Biodiversity increased by over 30%, including fish, butterflies, birds, and amphibians.
- Stormwater resilience improved significantly due to natural floodplains.
- Urban cooling increased - temperature drops of up to 2-4°C in surrounding areas during peak heat.
- Water quality improved, with wetlands filtering nutrients and pollutants.
- The park now attracts millions of visitors per year, strengthening community wellbeing.

A flood-control asset became a thriving public realm - cool, green, resilient, and loved.

Why Bishan-Ang Mo Kio Park Embodies Regeneration by Design

1. Take - Minimise Extraction

Instead of pouring more concrete, engineers removed it. Instead of building bigger drainage pipes, they built ecological capacity. Materials were reused, soil was restored, and the river's own behaviour became part of the design system.

2. Make - Design with Intention

This project required a huge amount of collaboration between hydrologists, ecologists, and landscape architects. The intention was not efficiency - it was relationship:

- To how water naturally wants to move
- To how soil stabilises through roots
- To how floodplains breathe with the seasons
- To how community gathers around water

It is engineering guided by ecology rather than opposition to it.

3. Give - Return More Than You Take

The park now gives back continuously:

- Habitat for wildlife
- Cooling for the suburb
- Clean water
- Recreation and mental health benefits
- Climate resilience

4. Restore - Strengthen the System

The restored river now supports a larger ecological network. It improves the resilience of downstream waterways, reduces pressure on stormwater systems, and supports microclimates across the district. Water infrastructure becomes a keystone for urban health.

Commercial and Cost Insights

Over a 60-year horizon, the naturalised river and park version of the project is estimated to save around US $15 million compared to a traditional concrete canal in the same location. These savings stem not just from green aesthetics, but from genuinely lower operational demands, lower flood risk, and higher ecosystem function.

For engineers, this emphasises how investing in ecological infrastructure isn't just a "nice-to-have" - it changes the long-term cost-profile of the asset.

Key baseline data:

- Project capital cost: ≈ S$ 76 million (~US$ 56 million) for the river restoration and park transformation.[281, 282]
- Conventional alternative (re-lining concrete channel) reportedly cost around 15% more than the ecological option.[282]
- Operational savings assumed from reduced flood management cost, lower channel maintenance, increased ecosystem service value (cooling, biodiversity, recreation).

Operational Expenditure Comparison (Estimated)

Time Horizon	Typical Concrete Canal / Drainage Channel	Naturalised River & Park (Bishan-AMK)	Estimated Savings
0-20 years	US $10 M (maintenance, flood risk costs)	US $6 M (lower maintenance /flood risk)	~US $4 M
20-40 years	US $12 M	US $7 M	~US $5 M
40-60 years	US $14 M	US $8 M	~US $6 M
Total 0-60 yrs	US $36 M	US $21 M	≈ US $15 M saved

How to Interpret This Table

- The "Typical Concrete Canal" column models a conventional hard-engineered drainage channel (straight, concrete-lined, high maintenance) situated in a tropical urban environment.
- The "Naturalised River & Park" column models the Bishan-AMK project: an ecological infrastructure system that integrates public park, naturalised river, floodplain, and ecosystem services.
- Savings derive from: lower channel maintenance, reduced flood-control costs, and ecosystem services (cooling, biodiversity) which are not usually quantified in standard infrastructure OPEX.

- The savings estimate is conservative; the actual value may be higher when including health, recreation, property value uplift, and ecosystem resilience.

Practical Takeaways for Engineers

Bishan-Ang Mo Kio Park offers some great key lessons:

- Nature is an engineer - sometimes better than we are.
- Ecology can replace concrete when systems are designed intentionally.
- Flood resilience improves when water has space, not when it is confined.
- Infrastructure can be beautiful *and* functional - the two are not opposites.
- Regeneration doesn't cost more in the long run - natural systems reduce operational burdens dramatically.
- Cities can hold wildness - and people thrive when they do.

This park has become a global benchmark for regenerative engineering. It demonstrates what happens when engineers stop asking, "How do we control this river?" and begin asking, "How do we help this river live?"

Case Study: Øresund Motorway Rewilding (Denmark) - Regeneration by Design Through Living Transport Corridors

If you drive along the Øresund Motorway today - the major corridor connecting Denmark to Sweden - you'll see wildflowers, native grasses, bees, butterflies, and birds moving along the edges of the highway. It feels almost like the road is threading through a meadow rather than cutting one apart. But this wasn't always the case.

For decades, the motorway was like most highways in the world: linear, hot, noisy, ecologically barren, and functioning as a barrier through the landscape.

But Danish engineers asked a different question: what if a motorway could function like an ecological corridor instead of an ecological scar?

This thinking led to a rewilding program that transformed the verges and medians into biodiverse, connected ecosystems supporting pollinators, small mammals, and plant communities that once struggled to survive in fragmented landscapes.

A Motorway Verge Becomes an Ecological Corridor

Beginning in the 2000s, Vejdirektoratet (the Danish Road Directorate) and ecological design partners launched a long-term rewilding strategy across major transport routes, with Øresund becoming one of the flagship corridors.

The strategy included:

- Converting mown grass verges into wildflower meadows
- Restoring native plant communities adapted to coastal conditions
- Introducing pollinator-friendly species to support bees and butterflies
- Creating structural diversity - logs, rocks, grasses - for habitat richness
- Reducing mowing frequency from 12+ times per year to 1-2 targeted cuts
- Installing wildlife crossings and continuous habitat bands
- Managing stormwater as a landscape asset rather than a drainage liability

The motorway remains a motorway - but the land around it has become alive again.

Ecological and Engineering Outcomes

After rewilding the corridor:

- Pollinator numbers increased significantly, with bee and butterfly diversity climbing across key monitoring sites.
- Vegetation species richness increased, returning many native meadow plants to the area.
- Heat and reflected glare decreased, due to greater ground cover and shading from vegetation.

- Maintenance costs dropped, as mowing and chemical use plummeted.
- Stormwater infiltration improved, reducing runoff pressures on drainage infrastructure.
- Wildlife movement increased, with small mammals and insects crossing the corridor more safely.

A transport asset became ecological infrastructure - still serving its mobility purpose, but now supporting the landscape rather than degrading it.

Why Øresund Motorway Rewilding Embodies Regeneration by Design

1. Take - Minimise Extraction

The project avoided reconstructing major elements of the motorway. Instead of adding new materials, engineers worked with what was already there - soil, slopes, verges, microclimates - and returned them to ecological function.

2. Make - Design with Intention

Rewilding was not a "plant and pray" exercise. It required ecological surveys, hydrology assessments, soil testing, and long-term monitoring. The intention was to restore landscape processes:

- Succession
- Seasonal productivity
- Pollination
- Infiltration
- Habitat connectivity

This is design based on ecological intelligence, not aesthetic greening.

3. Give - Return More Than You Take

The motorway now gives back through:

- Habitat for insects, birds and mammals
- Improved air filtration from plants

- Natural cooling of asphalt surfaces
- Reduced noise reflection due to soft vegetation
- Stormwater quality improvements
- Carbon sequestration in soils and vegetation

The corridor contributes positively to the region's ecological health.

4. Restore - Strengthen the System

The rewilded verge acts as part of a regional ecological network, linking coastal habitats, wetlands, and woodlands. It increases landscape resilience, especially for pollinators whose decline threatens global food systems.

Commercial and Cost Insights

The most interesting commercial and insights come from comparing long-term maintenance:

- Traditional motorway verges require frequent mowing, weed control, pest management, and uniform landscaping.
- Rewilded verges require selective mowing, ecological monitoring, and targeted intervention.

Across European rewilding projects of similar scale, the data shows:

- 30-50% reduction in annual mowing costs
- Significant reductions in herbicide use
- Lower stormwater treatment costs **due to natural filtration**
- Long-term savings from reduced erosion and slope failure

Using these benchmarks, the Øresund corridor demonstrates a shift in OPEX behaviour, with a lower operational cost paired with higher ecosystem performance.

Operational Expenditure Comparison (Estimated)

Time Horizon	Typical Motorway Verge Maintenance	Rewilded Ecological Verge	Estimated Savings
0-20 years	US $18 M	US $12 M	~US $6 M
20-40 years	US $20 M	US $13 M	~US $7 M
40-60 years	US $22 M	US $14 M	~US $8 M
Total 0-60 yrs	US $60 M	US $39 M	≈ US $21 M saved

How to Interpret This Table

- Based on typical mowing costs, chemical use, stormwater treatment, slope maintenance, and erosion control.
- Ecological verges reduce high-frequency interventions and rely more on slow, stable natural systems.
- The savings only reflect maintenance - not counted are the ecological, climatic, air-quality, and community value gains, which would push savings far higher.

Practical Takeaways for Engineers

The Øresund Motorway corridor shows that transport infrastructure doesn't need to be environmentally destructive. Key lessons include:

- Linear infrastructure can become ecological infrastructure.
- Low-maintenance ecosystems often outperform engineered landscaping.
- Restoring habitat is compatible with safety, drainage and mobility requirements.
- Biodiversity corridors can be retrofitted, not just newly built.
- Regeneration often saves money in the long run.

But perhaps the most important lesson is this: infrastructure doesn't have to dominate the landscape - it can belong to it. A motorway verge can be a pollinator corridor. A drainage ditch can be a habitat. A transport route can be a living system.

This is regeneration made practical, applicable, and scalable.

Case Study: Greening the Pipeline (Melbourne) - Regeneration by Design Through Turning Old Assets Into Living Systems

If you walk through Truganina today in Melbourne's west, you might find it hard to imagine that the vibrant linear park you're strolling through was once a decommissioned sewer pipeline corridor. A long, fenced-off strip of land, barren and underused, sitting between suburbs. A forgotten piece of infrastructure.[283]

But Melbourne Water, Wyndham City Council and community partners asked some important questions: what if obsolete infrastructure could become ecological and social infrastructure? What if a disused pipe corridor could be transformed into a living landscape that cools, filters, connects, and restores?

That question became Greening the Pipeline. The first completed section, the 1.3 km Federation Trail pilot,[284] became a compelling example of regeneration by design in Australian suburbs.

A Pipeline Easement Becomes a Linear Wetland Park

The Western Trunk Sewer, built in the 1960s, once carried wastewater across Melbourne. When it was decommissioned, it left a long, empty corridor that cut through growing suburbs - a physical barrier that did nothing for community, ecology, or climate resilience.

Instead of simply mowing the grass and maintaining a buffer, the project team proposed a new idea:

- Transform the corridor into a linear park
- Reintroduce wetlands to filter stormwater naturally
- Plant native vegetation to increase biodiversity
- Provide shade, cooling and habitat along the trail
- Create active transport links for walking and cycling

What was once a forgotten strip of land became a green spine running through a neighbourhood.

Ecological, Social and Engineering Outcomes

After regeneration:

- Biodiversity increased, with birds, pollinators, frogs, and native plant species returning to the corridor.
- Urban cooling improved, reducing peak summer temperatures in adjacent streets and homes.
- Stormwater quality improved, with wetlands filtering nutrients and sediments before entering waterways.
- Community wellbeing increased, shown through consistent public use of seating, paths, play spaces and shade.
- Active transport use increased, with more people choosing to walk and cycle along the safe, shaded corridor.

Why Greening the Pipeline Embodies Regeneration by Design

1. Take - Minimise Extraction

The project reused an existing corridor instead of clearing new land. Materials were simple and local. Stormwater from the surrounding catchment became the water source for wetlands.

2. Make - Design with Intention

Engineers, ecologists, planners, and community members worked together to create a system based on natural processes:

- Wetlands as filtration
- Trees and vegetation as cooling systems
- Soil and microtopography as water managers
- Meandering paths as social connectors

3. Give - Return More Than You Take

The corridor now gives back through:

- Cleaner water entering the Werribee River catchment

- Increased habitat and pollinator activity
- Improved public health through cooling and recreation
- Shading that reduces nearby home energy use
- Safer walking and cycling routes

It makes the suburb better every day - simply by existing.

4. Restore - Strengthen the System

The project strengthens regional ecological networks by:

- Connecting fragmented habitats
- Supporting plant and animal movement
- Improving catchment resilience
- Reducing heat stress across the urban fabric

The old sewer line now contributes to the health of the whole district.

Commercial and Cost Insights

Greening the Pipeline demonstrates that regenerative infrastructure can be cost-saving over the long term.

Baseline Capital Cost

Federation Trail Pilot (1.3 km):
Estimated capital cost: ~A$ 5-6 million (pilot phase publicly reported ranges)

A conventional alternative - mowing, fencing, drainage management - would have cost far less upfront, but provided no ecological or social value and higher long-term OPEX.

Operational Expenditure Comparison (Estimated)

Time Horizon	Typical Decommissioned Pipeline Maintenance	Regenerated Linear Park (Wetlands + Vegetation)	Estimated Savings
0-20 years	A$ 6.5 M (mowing, weed control, drainage)	A$ 4.0 M (selective maintenance + wetland mgmt)	~A$ 2.5 M
20-40 years	A$ 7.0 M	A$ 4.2 M	~A$ 2.8 M
40-60 years	A$ 7.5 M	A$ 4.5 M	~A$ 3.0 M
Total 0-60 yrs	A$ 21 M	A$ 12.7 M	≈ A$ 8.3 M saved

How to Interpret This Table

- Typical pipeline maintenance is high OPEX because the land has no function except "not causing a problem."
- Regenerative landscapes cost less because nature self-maintains once established.
- Wetlands reduce stormwater management expenses.
- Shaded, vegetated corridors require fewer chemical or mechanical interventions.
- This estimate does *not* include the economic uplift from amenity, property value, health savings or reduced cooling loads - factors known to add millions more in long-term community benefit.

Practical Takeaways for Engineers

This project illustrates key lessons for modern practice:

- Redundant infrastructure can be reimagined, not removed.
- Stormwater is a resource, not a nuisance.
- Linear public spaces can double as ecological corridors.

- Nature-based solutions can reduce OPEX, not increase it.
- Community infrastructure can be ecological infrastructure - the two do not compete.

Regeneration isn't always about new builds. It's often about rediscovering the hidden potential of what's already there. An old sewer line can become a cooling spine. A forgotten easement can become habitat. A remnant corridor can become a place people love.

Case Study: The Gateway Moorings Project (Queensland) - Regeneration by Design Through Restoring Seagrass and Protecting Dugongs

If you stand on the edge of Moreton Bay on a still morning, the water looks like glass. But beneath that surface lies one of the most fragile and important ecosystems in Australia - seagrass meadows, home to turtles, fish nurseries, and one of Queensland's most iconic species: the dugong.

For decades, these meadows were silently suffering. Traditional boat moorings - the kind with chains that drag and scour the seafloor - slowly carved circular scars through the seagrass. Each mooring created a bare patch of sand. Hundreds of moorings created a patchwork of ecological loss.[285]

When engineers and marine ecologists examined the impact, they wondered: what if moorings didn't damage seagrass? What if they helped restore it? What if marine infrastructure could support the very ecosystems it sits within?

This idea shaped the Gateway Moorings Project - an initiative in Queensland that replaced damaging swing moorings with seagrass-friendly systems, restoring habitat for dugongs and improving the health of Moreton Bay.[286]

From Scouring Chains to Seagrass-Friendly Systems

Traditional swing moorings drag against the seafloor as boats move with tides and wind. The scouring action rips up seagrass, prevents regrowth, and creates clear "bald spots" up to 10-20 metres wide.

In Moreton Bay - a habitat for one of the largest dugong populations in the world - this impact was noticeable.

The regenerative alternative was simple but powerful:

- Replace scouring chains with eco-moorings using elastic tethers that lift off the seafloor
- Stabilise boats without dragging
- Allow seagrass to regrow naturally
- Monitor and measure ecological recovery

Whereas a traditional mooring destroys habitat, an eco-mooring protects it. And when you install dozens of them? You can restore an entire underwater meadow.

Ecological Outcomes

Within months of installation, early signs of recovery appeared. Within years, the change was clear:

- Seagrass coverage increased across former mooring scars
- Biodiversity improved, including juvenile fish, crustaceans and invertebrates
- Dugong feeding areas expanded, as restored meadows reconnected previously fragmented patches
- Water quality improved, with seagrass stabilising sediment and supporting nutrient cycles

This wasn't just mitigation; it was regeneration - a system giving back more than it takes.

Why the Gateway Moorings Project Embodies Regeneration by Design

1. Take - Minimise Extraction

The project removed destructive components entirely. Instead of adding more infrastructure or reinforcing the seabed, engineers took away the harm - the very definition of minimising extraction.

2. Make - Design with Intention

Eco-moorings were chosen because they work *with* the dynamics of the ocean:

- Elastic, flexible risers
- Minimal seabed contact
- Ability to withstand tidal and storm forces
- Designs tested with ecologists to ensure habitat compatibility

This is engineering shaped by the behaviour of water, wind, and marine ecosystems.

3. Give - Return More Than You Take

The project gives back through:

- Rebuilding food sources for dugongs and turtles
- Restoring carbon-rich seagrass beds
- Improving water clarity
- Supporting nurseries for commercially important fish species
- Reducing long-term environmental damage

4. Restore - Strengthen the System

By restoring seagrass, engineers restored the system's foundation:

- Stronger habitat
- Better sediment stability
- Improved ecosystem resilience
- Reduced cumulative human impact

A single intervention strengthened an interdependent network.

Commercial and Cost Insights

Eco-mooring replacement is not just environmentally beneficial - it can be financially prudent.

Baseline Capital Cost

Eco-moorings typically cost around AU$ 3,000-4,000 per mooring, including installation. Traditional swing moorings can be AU$ 1,500-2,000 - cheaper upfront but far more damaging.

Operational Expenditure Comparison (Estimated)

Time Horizon	Traditional Swing Moorings	Eco-Moorings (Gateway Project)	Estimated Savings
0-10 years	AU$ 0.9 M (high seabed impact, frequent chain replacement)	AU$ 0.6 M (low maintenance, no chain wear)	~AU$ 0.3 M
10-20 years	AU$ 1.0 M	AU$ 0.6 M	~AU$ 0.4 M
20-40 years	AU$ 2.2 M	AU$ 1.2 M	~AU$ 1.0 M
Total 0-40 yrs	AU$ 4.1 M	AU$ 2.4 M	≈ AU$ 1.7 M saved

How to Interpret This Table

- Traditional moorings incur high maintenance due to chain abrasion, corrosion and replacement.
- Eco-moorings significantly reduce maintenance because nothing drags on the seabed.
- The largest savings come from reduced environmental damage, which avoids restoration costs and supports economic activities such as fishing and tourism.

Practical Takeaways for Engineers

The Gateway Moorings Project shows that regeneration can be:

- Simple
- Low-cost
- Scalable
- Achievable with existing technology

And it reinforces three powerful lessons:

- Small design changes can have major ecological consequences.
- Restoring habitat often begins by redesigning the things that damage it.
- Regeneration isn't an "extra" - it can be core to the engineering brief.

Most importantly, it asks a question every engineer should ask themselves: if a tiny piece of marine infrastructure can regenerate habitat, what else could we redesign to give more than it takes?

Case Study: The Great Green Wall (Africa) - Regeneration by Design at Continental Scale

If you travel through the Sahel - the band of land stretching across Africa just below the Sahara - you notice something subtle but unmistakable. The land is shifting.

In some places, soil once cracked and bare is slowly turning green again. Small trees take root where nothing grew a decade ago. Villages that once struggled for water are beginning to stabilise their wells. This landscape has been shaped by centuries of wind, drought, and changing climate, but it is now being reshaped by one of the most ambitious regenerative projects on Earth: the Great Green Wall.

The Great Green Wall began as a simple idea: a living barrier of trees stretching 8,000 kilometres across Africa to slow desertification. But over time, engineers, ecologists, Indigenous knowledge holders, and communities reframed it into something more sophisticated - a mosaic of regenerative systems that restore soils, increase rainfall infiltration, strengthen food security, and rebuild ecological resilience.[287]

At its core, the project is not just about trees. It is about repairing the relationship between people and land. Across Senegal, Niger, Ethiopia, and Burkina Faso, regenerative practices now include water harvesting basins, sand dune stabilisation, farmer-managed natural regeneration, and the return of native species adapted to harsh climates.

Outcomes are emerging across the region:

- Over 18 million hectares of degraded land restored.
- Drought-resistant trees increasing soil moisture and reducing erosion.
- Communities reporting more stable agriculture and groundwater recharge.
- Landscape cooling effects observable through satellite assessments.[288, 289]

It is not perfect, nor is it linear. Some areas face conflict, drought or political instability. Yet the overall direction is clear: even severely degraded land can recover when human systems align with ecological systems.

Why the Great Green Wall Embodies Regeneration by Design

1. Take - Minimise Extraction

Rather than importing fertilisers, chemicals, or machinery, the project relies on local materials and ancestral agroecological knowledge.

2. Make - Design with Intention

Engineers design interventions around rainfall patterns, soil types, and traditional land stewardship practices that have lasted centuries.

3. Give - Return More Than You Take

Restored soils, improved microclimates, increased biodiversity, and strengthened food systems continually give back to communities.

4. Restore - Strengthen the System

Over time, restored land stabilises entire regional ecosystems - including rainfall cycles, wildlife routes, and human livelihoods.

Practical Takeaways for Engineers

- Regeneration is not a luxury; it is resilience.
- Land and climate systems can recover when engineering aligns with ecological wisdom.

- Projects can be both large-scale and local at the same time.
- Even in environments under extreme strain, restoration is possible.

Case Study: Floating Wetlands - Regeneration by Design on Urban Waterways

Along the edge of a city river - whether in Chicago, Sydney or Rotterdam - you sometimes see a curious sight: clusters of floating platforms covered in reeds, sedges and native plantings. Birds rest on the edges; small fish dart beneath them. These are floating wetlands, an emerging form of ecological infrastructure designed to bring life back into highly engineered waterways.[290]

Urban rivers are often straightened, deepened, and sealed with concrete. The ecological edges - where most aquatic life thrives - disappear. Floating wetlands recreate those missing edges. They mimic the structure of natural wetlands, providing shade, habitat, oxygenation, and filtration on surfaces that were once dead zones.

Engineers design the buoyant structures so that plant roots hang underwater, creating a living mat that draws nutrients, breaks down pollutants, and stabilises the water column. In Chicago, these systems have reduced nitrogen and phosphorus loads, helping restore sections of the river once deemed ecologically "dead." In Sydney, the Living Seawalls project - a cousin to floating wetlands - has increased marine biodiversity by up to 36% on otherwise sterile harbour walls.[291]

What makes floating wetlands remarkable is that they fit into places where traditional ecological restoration seems impossible. They bring pockets of life into canals, marinas, industrial waterfronts, and flood channels.[292]

They are a reminder that regeneration is not only for wild landscapes - it can happen in the tightest corners of a dense city.

Why Floating Wetlands Embody Regeneration by Design

1. Take - Minimise Extraction

Systems are built from long-lasting, often recycled materials and rely on plants rather than chemical treatment.

2. **Make - Design with Intention**
Engineers design structures that move with tides, flows, and wind - an engineering approach shaped by nature's rhythms.

3. **Give - Return More Than You Take**
Wetlands clean water, create habitat, cool the surface, increase biodiversity, and support ecological processes long after installation.

4. **Restore - Strengthen the System**
They reconnect fragmented waterways, improve ecological function, and reduce heat and pollution at a systems level.

Practical Takeaways for Engineers

- Even highly constrained sites can host ecological regeneration.
- Edges - often ignored - can become powerful ecological assets.
- Nature-based solutions can outperform traditional mechanical or chemical options.

Case Study: Hive Sensor Networks (UK/ Netherlands) - Regeneration by Design Through Pollinator Intelligence

In fields and rooftops across parts of the UK and the Netherlands, there are beehives that do more than pollinate. They listen. These hives are equipped with small sensors that track temperature, sound frequencies, hive weight, air quality, and pollen movements. Together, they form a distributed environmental sensing network - one that uses bees as indicators of ecological health.

This idea emerged from a simple fact: bees notice things humans and machines often miss. They are sensitive to pollutants, changes in plant cycles, heat stress, and shifts in local ecology. Engineers realised that hive data could act as an early warning system for environmental decline - and as a map for ecological opportunity.[293]

The system works through a combination of IoT sensors, machine learning, and biological observation. Data is shared with councils, urban planners, and ecologists, helping cities design green corridors, identify pollution hotspots, and understand seasonal rhythms at a granular level.[294]

This merging of ecology and engineering creates a feedback loop. Instead of designing in isolation from the living world, engineers begin to co-design with it - guided by the patterns and behaviours of the bees themselves.

Why Hive Sensor Networks Embody Regeneration by Design

1. **Take - Minimise Extraction**

The system leverages the natural behaviour of bees rather than extracting resources from the environment.

2. **Make - Design with Intention**

Data is made meaningful by linking engineering decisions to ecological signals - a rare but powerful design approach.

3. **Give - Return More Than You Take**

Cities benefit from increased pollinator health, improved green networks, and better decision-making around land use and biodiversity.

4. **Restore - Strengthen the System**

The project helps regenerate urban pollinator populations, reinforcing food webs and ecological resilience.

Practical Takeaways for Engineers

- Ecological allies (like bees) can become powerful design partners.
- Environmental monitoring can be distributed, low-impact and nature-led.
- Regeneration begins when systems reconnect with local ecology.

Case Study: Māori-Led Landscape Restoration (Aotearoa New Zealand) - Regeneration by Design Through Kinship With Land

Across Aotearoa, there are landscapes undergoing a seemingly hidden form of repair - not through heavy machinery or large civil works, but through cultural leadership.

Many Māori iwi (tribes) are restoring rivers, forests, and wetlands using a worldview rooted in whakapapa (genealogy), which sees all living things - water, soil, plants, animals, humans - as kin. This worldview shapes design decisions in ways that feel radically different to conventional engineering, yet aligned with regeneration.

One example is the restoration of the Waikato River, where co-governance between Māori and the Crown has reframed the river not as a resource but as an ancestor.[295] Engineering choices - from riparian planting and sediment management to wastewater discharge and flood works - are now guided by the principle that the river's health is inseparable from the health of people. This shift has changed the rhythm of decision-making: long-view rather than short-term, relationship before efficiency, reciprocity instead of extraction.[296]

Other iwi-led projects, such as the restoration of Te Urewera and wetlands near Whakatāne, follow similar patterns: reforesting degraded land, reintroducing native species, healing waterways through traditional knowledge and contemporary ecological science. These are landscapes designed not just to function, but to thrive.[297]

Why Māori-Led Restoration Embodies Regeneration by Design

1. Take - Minimise Extraction

Interventions are minimal - planting, erosion control, species reintroduction. Rather than imposing structure, teams remove pressures that prevent natural systems from recovering.

2. Make - Design with Intention

Design is guided by cultural protocols, seasonal knowledge, and ecological rhythms. Decisions are slower, deeper and relational.

3. Give - Return More Than You Take

Projects increase biodiversity, stabilise rivers, rehydrate soils, and strengthen cultural identity and intergenerational knowledge.

4. Restore - Strengthen the System

Healthy river systems support fisheries, forests, climate regulation, and community wellbeing. Regeneration here is both ecological and cultural.

Practical Takeaways for Engineers

- Indigenous governance structures can strengthen environmental performance.
- Cultural worldviews can guide multi-century decision-making.
- Regeneration includes community, identity, and ecological function - not just land.

Case Study: Mycelium Materials (Global) - Regeneration by Design Through Living Construction Materials

In labs and fabrication studios around the world, a new building material is emerging - one that grows itself. Mycelium, the root network of fungi, can be cultivated into bricks, insulation panels, acoustic boards, and lightweight structural elements. These materials are grown using agricultural byproducts - straw, sawdust, husks - and take on the strength and shape of their moulds.[298]

From an engineering perspective, mycelium is compelling because it behaves more like a biological system than an industrial one. It can be grown locally with minimal energy, emits almost no carbon during production, and biodegrades at the end of life without leaving microplastics or toxins. Some properties - fire resistance, acoustic absorption, thermal performance - rival or outperform conventional products.[299] Projects like the MycoTree Pavilion in Seoul or the Hy-Fi Tower in New York have demonstrated how structurally efficient mycelium shells can be, and companies such as MycoWorks and Ecovative are now scaling production for commercial building use.[300]

This is not just a new material - it is a new philosophy of making. Instead of extracting rock or sand, engineers cultivate a living material that reinforces the ecosystems it comes from.

Why Mycelium Materials Embody Regeneration by Design

1. Take - Minimise Extraction

Production relies on agricultural waste, not mined or heavily processed raw materials.

2. Make - Design with Intention

The material grows into its final form, reducing manufacturing waste and enabling local, decentralised production.

3. Give - Return More Than You Take

Mycelium materials improve soil health when composted and avoid the long-term pollution associated with petrochemical products.

4. Restore - Strengthen the System

By reducing demand for high-carbon materials, mycelium contributes to broader systemic climate and biodiversity resilience.

Practical Takeaways for Engineers

- Living materials can outperform industrial ones in key applications.
- Designing with biology creates circular life cycles, not waste streams.
- Regeneration in materials is both feasible and scalable.

Case Study: Room for the River (The Netherlands) - Regeneration by Design Through Rebalancing Water and Land

Along the Rhine-Meuse-Scheldt delta, the Netherlands has long fought water through an intricate system of dikes, levees, and pumps. But rising sea levels and more extreme rainfall made one thing clear: the old model - control at all costs - could no longer guarantee safety. Engineers had to rethink their relationship with water.

The national Room for the River program reversed generations of hydraulic thinking. Instead of tightening walls and raising barriers, engineers gave the rivers more space. Floodplains were expanded. Embankments were moved back. Side channels were reconnected. Old industrial lands were reshaped into wetlands. Cities redesigned plazas and riverfronts to flood safely during peak events.[301, 302]

What emerged was a hybrid of engineering and ecology: places that manage flood risk while restoring habitats, cooling cities, and creating valued public spaces.[303]

Why Room for the River Embodies Regeneration by Design

1. Take - Minimise Extraction

Rather than adding more concrete, engineers removed structures that forced rivers into unnatural confinement.

2. Make - Design with Intention

Designs were informed by hydrology, ecology, and community needs, integrating flood safety with public space and habitat creation.

3. Give - Return More Than You Take

The river system regained ecological function: improved water quality, increased biodiversity, and new wetlands.

4. Restore - Strengthen the System

By giving water space, the system's resilience increased dramatically compared to hard barriers alone.

Practical Takeaways for Engineers

- Sometimes the safest engineering solution is ecological, not structural.
- Making space for natural processes improves long-term system stability.
- Flood infrastructure can support community, biodiversity and climate resilience simultaneously.

Case Study: Restored Oyster Reefs (Gulf of Mexico & Global) - Regeneration by Design Through Living Coastal Infrastructure

Along the coastlines of the Gulf of Mexico, engineers and marine ecologists have been rebuilding one of nature's most powerful - and underappreciated - forms of infrastructure: oyster reefs.

For decades, coastal erosion, overharvesting and pollution devastated oyster populations. Without reefs, shorelines lost their natural protection and water quality plummeted. But in recent years, restoration projects have brought back kilometres of living reef using recycled shells, biodegradable structures, and strategic seeding.

An oyster reef is not a wall. It is a living breakwater that grows stronger over time. It filters water, stabilises sediment, dampens waves, supports fisheries, and increases biodiversity - all while regenerating itself.[304]

In Louisiana's Biloxi Marsh, restored reefs have reduced shoreline erosion rates by up to 40%. In Alabama's Mobile Bay, reefs now filter millions of litres of water per day. And in Florida and Texas, "reef gardens" are restoring habitats in places where traditional engineering failed.[305]

Why Oyster Reefs Embody Regeneration by Design

1. Take - Minimise Extraction

Most materials are reclaimed shells or biodegradable substrates - no concrete seawalls required.

2. Make - Design with Intention

Structures are shaped to maximise water flow, habitat creation, and self-repair.

3. Give - Return More Than You Take

Reefs filter pollutants, support fisheries, reduce wave energy, and improve water clarity.

4. Restore - Strengthen the System

As reefs grow, they enhance the resilience of entire coastal ecosystems - a dynamic, self-sustaining defence system.

Practical Takeaways for Engineers

- Not all infrastructure has to be inert; some of the best systems are living.
- Restorative coastal design often outperforms hard seawalls in storms.
- Ecological function should be considered an engineering parameter.

Case Study: Wakhan Corridor Solar Microgrids (Afghanistan) - Regeneration by Design Through Energy and Community Resilience

The Wakhan Corridor is one of the most remote regions on Earth - a narrow, mountainous stretch between Afghanistan, Pakistan, China, and Tajikistan. Until recently, many Wakhi and Kyrgyz communities here lived with no reliable electricity. Diesel was scarce and expensive. Winters were long and severe.

Small-scale solar microgrids, developed in partnership with local communities, have changed this profoundly. Installed with local labour, maintained through local cooperatives, and designed for extreme climates, these microgrids replaced diesel dependence with clean, decentralised, community-owned energy.[306]

For landscapes already fragile from climate pressure, this shift has ripple effects: lower deforestation rates, better indoor air quality, safer winter living conditions, and more stable local economies.

These are not monumental engineering works - they are interventions that transform a system from extraction and scarcity to resilience and belonging.[307]

Why the Wakhan Microgrids Embody Regeneration by Design

1. Take - Minimise Extraction

Diesel reliance is reduced or eliminated, decreasing emissions and pressure on fragile mountain ecosystems.

2. Make - Design with Intention

Systems are sized to community needs, designed to be repairable, and built using participatory processes.

3. Give - Return More Than You Take

Clean energy improves health, learning environments, safety, and local livelihoods.

4. Restore - Strengthen the System

By reducing pressure on local forests and providing reliable power, microgrids support ecological and social resilience simultaneously.

Practical Takeaways for Engineers

- Regeneration can be small-scale and still transformative.
- Infrastructure becomes resilient when communities co-own and co-maintain it.
- Energy design should match place, culture and ecosystem - not simply the grid.

Case Study: Tokyo's Satoyama Revival - Regeneration by Design Through Rebalanced Human-Nature Edges

At the outer edges of Tokyo, where the megacity softens into forested hills and rice terraces, lies a landscape the Japanese call satoyama - the meeting place between village and mountain, where people have tended forests, fields, and waterways in reciprocity for centuries.[308]

For much of the 20th century, satoyama areas declined. Urban expansion, mechanised agriculture, and economic pressures left terraces abandoned, forests unmanaged and biodiversity shrinking. These once-vibrant ecological-cultural systems began falling out of balance.[309]

But in recent decades, satoyama has come back to life.

From Saitama to Kanagawa, local volunteers, ecologists, engineers, and community groups have restored irrigation channels, revived coppiced forests, repaired terraces, reintroduced native species, and rebuilt the subtle hydrological patterns that once sustained these landscapes. Schools bring children to learn seasonal rhythms. Universities use satoyama as living laboratories. Some municipalities have integrated satoyama into urban masterplans as green infrastructure, flood buffers, and biodiversity corridors.

What emerges is not a nostalgic return to the past, but a living example of regenerative engineering shaped through care, cultural memory, and ecological intelligence.

Why Satoyama Embodies Regeneration by Design

1. Take - Minimise Extraction

Restoration begins with minimal intervention: returning to traditional low-impact land management, reducing chemical inputs, and allowing natural systems to recover.

2. Make - Design with Intention

Engineers and ecologists restore ancient irrigation channels, terraces, and forest rhythms because they work. They regulate water, stabilise slopes, reduce fire risk, and support biodiversity.

3. Give - Return More Than You Take

Satoyama systems now provide flood mitigation, pollinator habitat, local food, clean water, and cultural education - benefits that extend far beyond the communities who maintain them.

4. Restore - Strengthen the System

Each restoration node increases resilience across the wider Tokyo watershed, providing ecological stepping stones that knit fragmented ecosystems back together.

Practical Takeaways for Engineers

- Ancient land systems often solved problems we now face - erosion, water scarcity, heat, biodiversity decline - with elegance and minimal technology.
- Green infrastructure is not "soft." Properly maintained, it stabilises slopes, manages hydrology, and reduces hazards.
- Regeneration is often slow, seasonal work - but its effects compound over time.

Satoyama's revival reminds us that regenerative engineering is not always futuristic. Sometimes, it begins by remembering what the land already knows.

Case Study: Copenhill (Copenhagen, Denmark) - Regeneration by Design Through Energy, Landscape and Play

When you first see Copenhill - a waste-to-energy plant in Copenhagen - you don't immediately think "power station." You see people skiing down its green slopes, climbers scaling a 280-metre wall, and runners making their way along rooftop trails planted with native grasses. Smoke legacy you gainnestacks rise from a building that feels more like a mountain park than industrial infrastructure.[310]

Copenhill (also known as Amager Bakke) was designed with an idea both ambitious and disarmingly simple: if a city must have an energy plant, why shouldn't it make the city better?

Inside, it converts non-recyclable waste into clean energy for around 150,000 homes. Outside, it offers one of the most inventive pieces of public realm in the world - a reclaimed industrial site transformed into a year-round recreation landscape.[311]

It is part factory, part park, part climate solution.

Why Copenhill Embodies Regeneration by Design

1. Take - Minimise Extraction

By using waste already produced by the city, Copenhill reduces reliance on fossil fuels and decreases landfill volume.

2. Make - Design with Intention

The building is engineered as both power plant and public space:

- A ski slope designed to absorb heat from the structure
- A rooftop trail system doubling as stormwater management
- A façade built for climbing and insulation
- A plant designed to be one of the world's cleanest WtE facilities

This is infrastructure that asks, "Why not do more than one thing well?"

3. Give - Return More Than You Take

Copenhill gives back in multiple forms:

- Clean energy
- Drastically reduced landfill emissions
- Habitat for native grasses and insects
- Recreational space in a dense city
- Educational tours for schools and visitors

It transforms a necessary piece of infrastructure into a civic asset.

4. Restore - Strengthen the System

The project improves urban resilience by combining energy security, climate mitigation, stormwater management, and accessible public open space - all on a single footprint that otherwise would have been closed off to the public.

Commercial and Cost Insights

Copenhill cost roughly €670 million to build - comparable to other high-performance waste-to-energy plants of similar capacity. The difference lies in how that funding was leveraged.

Additional public-benefit features (ski slope, trails, landscaping, climbing wall) added roughly €10-12 million, a small fraction of the total budget.[310]

Operational costs are offset by:

- Energy sales
- District heating revenue
- Reduced landfill costs
- Tourism and brand value
- Exceptionally high energy efficiency

Over its 30+ year design life, the plant is projected to save tens of millions in avoided landfill emissions and urban cooling costs, while providing year-round community amenities at negligible additional operational cost.

Practical Takeaways for Engineers

- Multi-benefit design often costs little more upfront, but yields exponentially more value over time.
- Regenerative engineering doesn't reject industry - it asks industry to participate in the wellbeing of place.
- Sometimes the most effective sustainable design choice is simply asking, "What else could this piece of infrastructure do for the city?"

Copenhill expands the imagination of what is possible - and shows that regeneration doesn't eliminate engineering rigor. It elevates it.

What These Case Studies Show Us

As we step back from these case studies - algae façades in London, rewilded rivers in Singapore, seagrass revived for dugongs in Queensland, desert parks that cool entire neighbourhoods, and forests restored at the edge of Tokyo - clarity starts to appear.

Regeneration isn't a dream for someday. It's already happening now, in pockets all over the world. Each project shows, in its own way, that engineering can be a force that gives back rather than takes away.

What stands out is not that these projects are perfect - none of them are - but that they ask different questions from the beginning. Instead of "How do we minimise our footprint?" they ask, "How can this place thrive because we built here?" Instead of treating nature as a constraint to be controlled, they treat it as a collaborator. And instead of assuming that resilience comes only from stronger materials, they look to ecological intelligence - the patterns and processes that have sustained life for millions of years.

The more closely you look, the more reassuring this becomes. For me personally, finding so many case studies from across the world that already exist has been incredibly reassuring that regenerative engineering is possible - and it's already happening now!

Regeneration isn't a single leap. It's a series of thoughtful choices - small, practical, cumulative. A shift in how we see water, soil, energy, species, communities, and climate. A willingness to learn from the places we build in, not simply build over them. And perhaps most importantly, a recognition that the systems we design continue shaping the world long after construction ends.

For centuries, we've built as though we were separate from nature - as though the planet were simply the backdrop to human progress. But every bridge, every building, every road is part of an ecosystem that breathes, moves, and responds. Regeneration isn't about "going back to nature." It's about remembering that we were never actually outside it.

When we design with that awareness - when we pause long enough to ask, "What would nature do here?" - something shifts. Our work begins to carry the enduring wisdom that the first engineers understood: that life thrives through relationship, rhythm, and renewal, not domination. That engineering can hold the humility to work with the natural world, not over it. And that the things we build can be shaped not only for efficiency, but for empathy; not only for compliance, but for reciprocity.

This is how we move from efficiency to empathy. From sustainability to regeneration. From short-term projects to long-term, living legacies.

The Opportunity for Engineers

If there's one message I hope emerges from this chapter, it's that regenerative engineering is not a radical departure from our profession - it's the natural evolution of it.

Engineers already sit at the intersection of systems, materials, people, and place. We already know how to work with constraints, complexity and risk. We already design the physical backbone of society. And because of that, we are uniquely positioned to lead this shift - not through ideology, but through skill, curiosity, and practical wisdom.

The opportunity ahead is extraordinary:

- To design infrastructure that restores ecosystems instead of fragmenting them
- To revive rivers, soils, forests and microclimates as part of the engineering brief
- To choose materials that nourish cycles instead of breaking them
- To create energy systems that build resilience rather than dependency

- To imagine cities where people and ecology thrive together

This doesn't demand that we become ecologists overnight. It simply invites us to bring a different worldview to the table - one that treats nature as an ally, not an obstacle.

Every engineer has the capacity to begin this shift. Even small moves - a different question at feasibility stage, a partnership with local ecologists, a willingness to restore rather than replace - reshape the long-term arc of a project. Every project, no matter how small, is part of a larger system, and every system we touch has the potential to heal or to harm.

The Takeaway

When engineers design with nature, the time horizon inevitably stretches. Living systems operate on cycles longer than project schedules, business cases, or political terms. They remind us that what we build today will shape conditions long after we are gone.

In the chapters ahead, we explore the values that guide this work, the principles that help us design not only for the present, but for the many generations who will inherit what we build. Because regeneration is ultimately about legacy: the story a project leaves behind, the way it changes a place, and how it continues to serve long after we are gone. Once we accept that our projects participate in living systems, the question of legacy becomes unavoidable. What will this place be like in 10 years, in 50, in 100? Who will benefit, and who will bear the consequences?

The next chapter turns directly to that long view, exploring generational legacy and what it means to design for people we will never meet.

Chapter 6:
Generational Legacy - Designing for People You'll Never Meet

Every engineering decision extends forward in time. Long after the drawings are archived and the project teams have moved on, what we build continues to shape lives, landscapes, and possibilities for people we will never meet.

This chapter is about generational legacy. It asks what responsibility engineers carry to the future, and how our definitions of success change when we design not just for delivery, but for inheritance. Legacy is not about permanence or monumentality. It is about whether what we leave behind can be cared for, adapted, and loved by those who come next.

Designing for future generations requires a different mindset. It means looking beyond design life and compliance, and asking harder questions about continuity, memory, and responsibility. This chapter explores how engineering decisions made today determine the burden or benefit we pass forward, and what becomes possible when we choose to design for people we will never meet.

Design Life Versus Design Health

I recently read a book called *Outlive*,[312] which was about extending your healthspan - not just how long you live, but how long you live well. The author asks a confronting question: how long do you want to live, and what are you willing to do today so that your future self gets to enjoy that life?

My first reaction was: "Obviously, I'd love to get to 100." My second reaction was: "Oh. That means I have to start designing my life differently now."

This chapter is about exactly that mindset - not for a human body, but for the things we build.

As engineers, we constantly talk about "design life". This bridge has a design life of 50 years. This tunnel, 100 years. This asset, 120 years. We say those numbers like they're sensible, like they're neutral.

But here's the question we almost never ask: what does 100 years actually mean? Who is using this in 100 years? What does their world look like? What will they need from what we've left behind?

When I was working on a major tunnel project in Australia, the asset had an intended life of 100 years. At the time I was involved, we were only 20 years into that design life. On paper, the tunnel was performing well. Structurally, it was fine. The engineering was "as expected", which was a great result.

However, when we actually sat down with the people who operate and maintain it, something else emerged.

Maintenance logs - crucial records of faults, wear, and incidents - were being kept in binders and scribbled notes on someone's desk. There was no consistent digital system. No shared memory. No easy way to see pattern or risk across the asset. And this was a tunnel designed to run for another eighty years.

We ended up mapping the entire tunnel in high definition and helping the team move to a shared digital platform so problems - cracks, leaks, corrosion, performance issues - could be captured, tracked, and compared over time. That project wasn't just about structural performance; it was about memory and continuity.

A piece of infrastructure can meet its technical requirements and still fail its legacy, because legacy isn't just: "does it stand?" Legacy is: "can it be cared for?"

We talk a lot in this profession about project delivery. We hardly talk about project inheritance.

We act like the end of construction is the end of responsibility, but it's not. All we've done is hand the next generation a living thing to maintain, operate, upgrade, defend, explain, love, and justify.

Here's the hard part: the people who inherit what we build are almost never in the room when we decide what to build.

That's the tension of legacy and it's why we need to change.

The "100-year design life" doesn't mean: "this will be thriving in 100 years." It usually means: "we've designed it so it shouldn't catastrophically fail within that window, based on the assumptions we made at the time."

That's a different claim.

It doesn't tell us whether the asset will still be loved, or useful, or safe in a changed climate, or culturally acceptable, or financially maintainable, or even repairable with the skills, tools, and materials future teams will actually have.

There's a growing recognition in medicine that lifespan and healthspan are not the same thing. Many people are living longer, but not necessarily living *well* longer. The years may stretch out, but the quality of those years doesn't always follow.

Infrastructure and engineering projects are the same.

Every asset has a lifespan - the number of years it can physically exist. But very few assets have a designed healthspan - the number of years it continues to serve people well, remain safe and relevant, support its environment, and create value rather than extract it.

We design for how long something stands. We almost never design for how well it lives.

That gap is where legacy is either built... or lost.

Paradise Dam in Queensland is a perfect example. The dam was built quickly and cheaply using a fast, efficient construction method. Technically, on day one, the engineers could prove performance. But within two decades, there were serious safety concerns. The dam has since required massive remediation and partial deconstruction at enormous cost. The "cheap" option is now a multi-billion dollar burden.

That's not legacy. That's debt. Now someone has to carry the consequences of a decision they didn't make.

We've normalised this across industry. We build what we can justify today, let someone else absorb the cost tomorrow, and call it "value for money." That logic is cracking.

When we design only for design life or lifespan, we leave future generations with aging, brittle, misaligned systems. When we design for healthspan and legacy, something very different becomes possible - assets that adapt, remain fit for purpose, honour changing climates and cultures, and don't burden the people who inherit them.

Entering the Era of Maintenance

For most of modern engineering, our focus was expansion: build new highways, new pipelines, new airports, new transmission lines, new towers. The story of engineering was growth.

But we're at a turning point.

We are now moving into an era where our job is not just to build, but to tend. Maintain. Restore. Adapt. Keep alive.

We are managing assets that have reached, or are exceeding, their original design life. We are inheriting concrete that's already tired. Water systems built for a climate that no longer exists. Transport networks planned for cities that have doubled in scale. Power systems designed for a fossil economy that is disappearing beneath our feet.

That requires a different identity. It's not only: can you design it? It's: can someone maintain it? Can someone upgrade it without tearing it down? Can someone live with it?

In medicine, we've moved from "Can we keep you alive through the surgery?" to "Can we help you live well for decades?" We've shifted from emergency rescue to lifelong care and are starting to see a shift to preventative medicine.

Engineering must go through the same shift.

We have to stop seeing infrastructure as a finished object and start seeing it as a living system - one that needs diagnosis, attention, nourishment, memory, and adaptation across generations.

That's legacy thinking.

The Old Story We Inherited - and Why It's Not Enough Anymore

We have inherited an engineering mindset that assumes two things:

1. Resources are infinite.
2. Someone else will deal with the consequences.

We now know neither is true.

The paradigm we're stepping into is different. It says:

- The land is not passive.
- The community is not an output variable.
- The person who inherits your design is not hypothetical.
- Country is not a site. Country is a living entity.

We have seen time and time again First Nations exemplifying legacy thinking. It is also engineering thinking, if we let it be.

Because "Care for Country" is, in practice, asset stewardship. It's long-term risk management. It's intergenerational planning. It is designing for people you'll never meet.

We see this in:

- **Budj Bim** on Gunditjmara Country: an Indigenous aquaculture system more than 6,000 years old, with stone channels and traps designed to farm eels without exhausting the wetlands.[248] This wasn't extraction - it was partnership. The system increased food security and biodiversity at the same time.
- **Fire-stick farming (cool burning):** For tens of thousands of years, Aboriginal fire practitioners have used low, controlled burns to renew ecosystems, prevent catastrophic bushfires, and support habitat regeneration. Fire wasn't treated as an enemy. It was treated as medicine for Country.[222]
- **Shannon Kilmartin-Lynch's work** at Monash: drawing on Indigenous knowledge of and incorporating that knowledge into concrete technologies to make structures stronger and longer-lasting, with less waste.[313]

These are not stories from the past. These are instructions for the future.

What legacy actually asks of us

Legacy is not about building something that lasts forever. That's not realistic. We cannot and should not attempt to make every culvert immortal.

Instead, legacy asks: will this still make sense, have purpose, and positively contribute to people and planet in 50 or 100 years? That's very different from: will it still be standing?

Sometimes the bravest legacy choice is not to build at all. Sometimes legacy is building lighter, so a place can heal. Sometimes legacy is designing something beautiful and adaptable so that in 80 years, instead of tearing it down, people will fight to protect it and repurpose it.

Look at the Eiffel Tower. It was never meant to be permanent. It was temporary infrastructure for an exhibition.[314] But people loved it. They claimed it. It evolved. It found new uses - as a radio mast, a scientific instrument, a cultural symbol. Beauty bought it time. Love gave it protection. Adaptability gave it purpose.

What about the High Line in New York? It was an abandoned elevated rail line. By an old efficiency mindset, it was "waste", but by a legacy mindset, it was potential. Today it's a public park, an identity anchor, a community connector, an ecological corridor. Not bad for "scrap steel."[208]

And look at the Sydney Opera House seawall. For years, it was just concrete on the harbour. Now, engineered reef panels have turned part of that wall into habitat marine life is returning to.[315] The Opera House is no longer just a building. It's a living edge.

That's legacy. That's what it looks like when we stop treating infrastructure as an object and start treating it as part of an ecology, a memory, and a story.

What Did You Do Once You Knew?

There's a line from a poem by Drew Dellinger that lingers with me every time I think about legacy: *"What did you do once you knew?"*[316]

It's a question that reaches across generations. Once we understand the impact of our decisions - on people, on place, on the ecosystems that sustain us - we can't unknow it. And legacy isn't built from certainty or perfection; it's built from what we choose to do next.

Engineers rarely work with the luxury of ideal conditions. We work in constraints, budgets, politics, timelines. But we also work with influence - the cumulative power of each design decision, each brief we question, each assumption we widen.

Legacy begins the moment we acknowledge that our work will outlive us. Once we know that, the only question left is: what will we do with that knowing?

Legacy Thinking

The First People Who Built for Futures They Would Never See

When we talk about "legacy thinking", it's easy to imagine it as something new - a modern response to climate change, aging infrastructure, or shifting social expectations.

But humans have been thinking in legacy for thousands of years, it was critical for their survival. Some of the most enduring engineering and cultural systems on the planet were created by people who never expected to see the long-term fruits of their work. They weren't designing for quarterly targets or political cycles. They were designing for descendants.

Ancient engineers, architects, navigators, and knowledge holders had a very different relationship with time. Their work was slow, deliberate, and woven into the rhythms of place. Structures were built not to impress, but to belong. Systems were shaped by ecological wisdom, not extraction. Knowledge was something you protected and passed down, not something you hoarded.

What's striking is how aligned this is with our modern aspirations for sustainability and regeneration. These ancient builders weren't perfect - no civilisation is - but they understood something we're only just rediscovering: that engineering is, at its heart, an intergenerational act.

These case studies are reminders of what becomes possible when we design with a time horizon longer than our own careers, our own lifetimes, or even our own civilisation.

Imhotep - Designing for Millennia (Ancient Egypt)

More than 4,500 years ago, Imhotep stood on the stone plateau at Saqqara imagining something the world had never seen: a structure built not for decades, but for eternity.

His Step Pyramid wasn't simply a tomb - it was a systems experiment. Imhotep aligned architecture with astronomy, material science with ritual significance, geometry with climatic knowledge. He wasn't designing for a king; he was designing for time itself.[317]

What makes Imhotep's legacy so extraordinary is not just the stone that remains, but the thinking behind it: build in a way that future generations can inherit meaning, not just mass. The pyramid still stands because he designed it with an eye that reached far beyond his own lifetime - a lesson engineers still grapple with today.

The Mauryan Engineers & Ashoka - Infrastructure as Moral Responsibility (India, 3rd Century BCE)

In the time of Emperor Ashoka, engineering was considered a form of ethical leadership. After turning away from conquest, Ashoka directed his empire's engineers to build not monuments, but care systems: thousands of wells for travellers, rest houses shaded by trees planted every half-kilometre, irrigation works to stabilise harvests, animal hospitals, and some of the earliest environmental protections ever recorded.[318]

These were not vanity projects. They were practical, regenerative interventions meant to support millions - long after Ashoka himself was gone. The Mauryan engineers embedded moral intent into physical systems. Their mindset? Infrastructure must carry compassion across generations.

The Polynesian Wayfinders - Knowledge as an Intergenerational Asset (Pacific Islands)

Out on the open ocean, with no land in sight for thousands of kilometres, the ancient Polynesian navigators held knowledge that lived not in textbooks, but in bodies. They read stars, swells, winds, birds, and the colour of the sky. They memorised patterns passed down through centuries, each generation adding its own observations.[319]

Their engineering was not mechanical - it was ecological and relational. They understood that knowledge, like an ecosystem, must be tended if it is to survive. Wayfinding was legacy thinking in its purest form: you do not travel for yourself; you travel so the next generation knows the way.

The Venetian Lagoon Engineers - Cities That Move With Water (Italy)

Centuries before "resilience" became a design buzzword, Venice was already practicing it. Built on shifting mudflats, the city's engineers embraced the tides rather than fighting them. They designed canals that flushed with the ocean, foundations that flexed with settlement, and a wastewater system that relied on the lagoon's natural rhythms instead of pipes.[320]

The genius of Venice is not its beauty, but its adaptability. For generations, engineers sustained a delicate balance between human life and a living estuary. Their mindset reminds us long-term legacy emerges when we treat nature as co-engineer, not obstacle.

The Leaders Who Show Us What Future-Focused Engineering Really Looks Like

Legacy thinking isn't a relic of the past. It's alive today - in activists, designers, economists, engineers, and First Nations leaders who are showing us what it looks like to build systems that last, systems that heal, and systems that future generations will thank us for.

These modern leaders aren't connected by profession or industry. Some work with cities, others with forests, rivers, global governance, economics, or cultural knowledge. What unites them is their orientation toward the horizon. They act from the belief that leadership is not about being remembered - it's about ensuring that what comes after us is stronger, fairer, and more resilient than what came before.

Their work challenges a mindset that engineering has carried for too long: that our job ends when the project is delivered. These leaders remind us that delivery is only the midpoint of legacy - and that our true responsibility is to the people and places who will inherit what we build.

As you read these stories, my hope is that they expand what leadership can mean for engineers. Legacy thinkers don't have all the answers. They simply take the long view, act with intention, and ask themselves the powerful question: "What will my work make possible long after I'm gone?"

Wangari Maathai - Regeneration at Human Scale (Kenya)

When Wangari Maathai planted her first trees with a small group of women in rural Kenya, she wasn't planning a global movement - she was responding to eroded soil, failing crops, and disappearing water. Yet from those early efforts grew the Green Belt Movement and more than 50 million trees, many of which still stand today.

Maathai understood legacy as something humble and local: start where you are, with what you have. Her work restored ecosystems, strengthened community leadership, and shifted national policy.[321] She showed the world that regeneration isn't a concept. It's a practice - one that grows one tree, one woman, one season at a time.

Jan Gehl - Cities Built for Life, Not Speed (Denmark)

When Danish urbanist Jan Gehl began studying how people actually used public space in the 1960s, cities everywhere were being redesigned around cars. Gehl stood quietly on a footpath with a notebook and started counting not vehicles, but people: how they walked, paused, gathered, and connected.

Over the next 50 years, he helped cities from Copenhagen to Melbourne rediscover themselves. Streets narrowed, sidewalks widened, public squares reopened, children reappeared in city centres, and cycling became a major mode of transport.[322] Gehl was not just redesigning streets; he was redesigning values - reminding us that a city's legacy is measured in quality of life, not traffic speed.

Kate Raworth - Economics With Boundaries and Belonging (UK/Global)

Kate Raworth challenged one of the most entrenched ideas of our time: endless economic growth. Her Doughnut Economics model invited cities and nations to balance the needs of people with the limits of the planet. Amsterdam, Copenhagen, Melbourne, and others are now using her framework to guide urban planning, procurement, and climate action.[105]

Raworth's leadership is legacy thinking because she changed the questions we ask. Instead of "How do we grow?" she asks, "How do we thrive within planetary boundaries?" That shift will influence generations long after spreadsheets have changed.

Christiana Figueres - Choosing Optimism as a Strategy (Costa Rica/Global)

When Christiana Figueres took over the UN climate negotiations after the failed Copenhagen talks, the global mood was despair. She rebuilt the process from the inside out, driven by what she called "stubborn optimism" - the belief that long-term change is only possible when people feel it is.[323]

Her leadership resulted in the Paris Agreement, one of the most significant collective climate commitments in history. Figueres' legacy thinking wasn't technical. It was relational: build trust, build connection, build courage - and the systems will follow.

Eileen Deemal-Hall - Engineering With Country, Not On It (Australia)

Eileen Deemal-Hall, a respected Aboriginal leader and CEO, teaches a worldview that reframes engineering entirely. In her culture, self-actualisation doesn't sit at the top of a pyramid; it begins with a deep connection to Country. Everything - design, decision-making, leadership - flows from that relationship.[324]

Her collaborations with engineers show what it looks like to design *with* Country's knowledge and cycles, not impose upon them. Whether working on community infrastructure, environmental restoration, or cultural design, she embodies a principle engineers everywhere are rediscovering: legacy is a relationship you honour every day.

Becoming a Legacy Thinker

Legacy is not an accident. It's a practice. A way of thinking. A discipline of imagining the future as if someone you love lives there.

None of these people worked with unlimited resources or perfect conditions. What they had was clarity of intention and the courage to care about consequences that would unfold long after they were gone.

That is the heart of engineering legacy.

Their stories aren't just history or inspiration; they are invitations. Invitations to again widen our lens, widen our field of vision. To remember that what we build continues living after we've left the project. To recognise that our decisions ripple forward into communities, ecosystems, cultures, and futures we will never personally see.

This next section offers a way to bring the mindset of those legacy thinkers into everyday engineering practice. Something practical and honest. Something you can use in the messy realities of projects, competing priorities, and deadlines. A way to weave long-term thinking into design conversations without needing a philosophy degree or a spare decade.

The Legacy Scorecard - Thinking for a Hundred Years

The people we explored in the previous section - ancient and modern - shared a common intuition: they understood that their choices would ripple forward into lives they would never meet.

Legacy thinking wasn't a slogan for them; it was a worldview. It shaped how they built, how they healed, how they taught, and how they led.

They acted with the awareness that time is not a line but a circle, and that we are always shaping a future that will eventually shape us back.

As engineers, we stand inside that same circle by simply being engineers and having the responsibility that we are creating our world now and for the future every day. We may not carve stone for future astronomers or plant forests whose shade we will never sit in, but our work carries generational consequences all the same.

That is why I created the Legacy by Design Scorecard - a simple way to slow the pace of delivery long enough to ask the question: "What will this choice mean in 50, 100, or 200 - and for whom?"

This scorecard isn't a checklist or a compliance exercise. It's to encourage a way of thinking. A pause. A moment to lift our gaze beyond the design life and into something more meaningful: the *health* of the systems we leave behind.

We talk a lot about design life or asset lifespan in engineering, but the language is incomplete. It tells us how long something might physically survive, not how long it will continue to matter, or be loved, or be useful, or be safe in a changing climate.

Earlier in this chapter, I used an analogy from medicine: lifespan is how long a human body stays alive. Healthspan is how long a person stays well.

In engineering language, those map neatly to design life or service life (how long the asset performs technically and structurally) and legacy life (how long the asset continues to serve, uplift, adapt, and remain meaningful).

Many assets have a long service life but a very short legacy life. They stand, but they aren't serving. They exist, but they aren't loved. They function, but they don't enrich. Others have a long legacy life but a short service life - temporary structures that become cultural anchors, community symbols, or sources of identity long after they've disappeared.

The goal of Legacy by Design is simple: to build systems with both a long service life *and* a long legacy life.

Not just structures that last - structures that last well.

A Framework for Designing With Time, Care, and Responsibility

Most engineering tools measure whether something will last. Far fewer measure whether it will remain *healthy* - useful, loved, safe, adaptable, and light on the generations who inherit it.

To bring that distinction to life, the Legacy by Design Scorecard combines two simple tools:

1. **The Long Life/Good Life Matrix** - a way to map where a project sits
2. **The Legacy Ledger** - a way to reflect on the assets and liabilities we hand forward

Together, they help teams see not just *what* they are building, but *what it will become*.

The Legacy By Design Matrix

It's worth being clear about what the Legacy by Design Matrix lens actually is. The aim isn't another checkbox or form to fill in, and it's definitely not a new compliance requirement. It's a way of thinking - a shift that sits beneath the work we're already doing.

The legacy lens helps us see what our usual processes tend to hide. Every project is full of decisions that make perfect sense in the short term but create invisible burdens for the future: brittle materials that are cheap today but expensive to maintain later, systems no one will be able to repair in 50 years, or choices that meet specifications but slowly chip away at community trust. Legacy thinking brings those long-term risks into view.

It also invites teams to consciously choose the future they are shaping. Many engineering projects never have a conversation about what they want to leave behind - not the deliverables, but the inheritance. The legacy lens opens that conversation in a natural, practical way. It moves our mindset from "delivery" toward "stewardship", reminding us that engineers finish a project, but communities live with it for generations.

It also has a wonderfully grounding effect on decision-making. Sometimes all it takes is a small shift - a better material, a more repairable system, a cooling shade tree instead of decorative concrete - and suddenly you've added decades of value. The legacy lens helps teams see those opportunities before they disappear into the design drawings.

Most importantly, it expands the conversation. We still ask the necessary questions - does it meet requirements, does it comply, does it perform? - but alongside them sits a silent companion: will people still be grateful for this in a hundred years?

The legacy lens doesn't slow a team down. It sharpens their judgement.

That's the spirit behind the Legacy by Design matrix: a simple tool that helps us pause, look ahead, and design with the people of the future in mind - even the ones we will never meet.

Service Life × Legacy Life

Imagine a simple 2×2 matrix with two sides:

Legacy Life

How long the asset continues to make a positive impact for people, culture, ecology (how long it stays loved, valuable, regenerative).

Service Life

How long the asset physically lasts (structural longevity, durability, maintainability).

Legacy by Design Matrix

The Four Quadrants

1. Short Service Life x Short Legacy Life

The Disposable Project: Fast-built public buildings that fail standards early, cheap surfacing that must be rebuilt every few years, or short-lived infrastructure that produces more waste than value.

2. Long Service Life x Short Legacy Life

The Monolith: Brutalist civic buildings that remain structurally sound but alienating; oversized transport interchanges built for yesterday's patterns; dams or bridges that no longer align with environmental realities.

3. Short Service Life x Long Legacy Life

The Temporary Icon: Doesn't last physically, but lingers in memory, culture, and meaning.

Examples: World Expos, pavilions, pop-up parks, participatory installations, protest infrastructure, or ephemeral architecture with meaning that outlives its materials.

4. Long Service Life x Long Legacy Life

The Living Legacy: The ideal quadrant. Built to endure and designed to evolve. Loved enough to be maintained. Rooted enough to matter across generations.

Examples:

- Sydney Harbour Bridge
- The Roman aqueducts
- Nightingale Housing (Melbourne)
- Hammarby Sjöstad
- Budj Bim aquaculture

The Legacy Ledger

If the Legacy by Design matrix helps a team see *where* their project sits in the long arc of time, the Legacy Ledger helps them see *how they got there* - and where they might choose to go next. It's the companion to the matrix: less about categorising, more about noticing.

The Ledger isn't a scorecard or an assessment. It's closer to a running journal - a place to map the consequences of decisions, almost like leaving breadcrumbs for your future self.

Every engineering project is full of trade-offs: a material chosen for cost, a feature removed for time, a planting decision delayed for procurement. None of these are right or wrong on their own. But over the life of a project, the pattern they form becomes the legacy.

The Legacy Ledger helps make that pattern visible.

It holds both sides of the story: **the legacy you gained** - the moments where the project grew stronger, more adaptable, kinder to people or place - and **the legacy you compromised** - the decisions that will leave a future team with maintenance, remediation, or cultural repair. Often, we never see these side by side. The ledger lets them sit in conversation.

What I love about this tool is how simple it is. You don't need a meeting, a facilitator, or a workshop. You can open it on your laptop after a design review, make a few notes, and close it again. It becomes a habit - a way to stay honest with yourself about the long-term consequences of the work.

Over time, something shifts. Teams stop thinking only in terms of milestones and start thinking in terms of inheritance. They begin to notice when a decision nudges the project toward endurance or pushes it toward future strain. They see when something they add will live well into the next century - and when something they remove might cost a future community dearly.

The ledger doesn't judge. It simply asks: what are we leaving behind? That question, asked gently and regularly, has a way of changing the work.

This is where engineering becomes leadership - not because we get everything perfect, but because we stay conscious of the ripples our decisions create long after we're gone.

A Simple Tool to Place Your Project on the Matrix

With the current idea or design you have, score the project from 1-5 on both axes:

Service Life

1 = fragile, short-term
5 = robust, durable, maintainable, repairable

Legacy Life

1 = low cultural/ecological/community value
5 = deeply rooted, regenerative, loved, adaptable

Plot the scores on the matrix.

The conversation becomes clear:

- Are we building something that only survives?
- Or something that lives well?
- Where do we want to be?
- What must we change to get there?

The Living Proof of Long Thinking

Legacy is not about building things that never change. It's about creating things that can keep living.

Every engineer I've ever known has, deep down, wanted to build something that works in the world, something that matters. I know personally in my career I wasn't trained, like most of us, to stop to ask: what actually lasts and what can we learn from them?

Sometimes, it's not the biggest projects or the most expensive. It's the ones built with care, meaning, and connection. The ones that adapt, invite affection, and belong to their place. You see it everywhere, when you look. These are some examples I return to often when I think about what true legacy means - not just technically, but humanly.

Case Study: The Pont du Gard (France) - Legacy by Design Through Precision, Restraint, and Patience

If you ever stand beneath the Pont du Gard, you feel its presence long before you understand its mechanics. An almost humbling kind of intelligence rests inside its stonework.

Built around 19 BCE by Roman engineers, this three-tiered aqueduct once carried water across valleys to the ancient city of Nîmes. Its gradient - just a 17-metre drop over 50 kilometres - was so precise that water flowed smoothly without pooling or stagnation. No mortar. No reinforcement. Only proportion, gravity, and a huge respect for the craft.[325]

The remarkable thing is not that it still stands; the remarkable thing is *how* it still stands. The Romans built it slowly, with patience and intention - choosing stone that local people understood, shaping joints repairable for future generations, and designing an asset that could be maintained without advanced tools.

Over two thousand years later, it remains not because it was indestructible, but because it was lovable. People cared for it because it was built with care.

Why the Pont du Gard Is Legacy by Design

Legacy by Design Matrix

- **Service Life:** Remarkably long - thousands of years - because it was engineered for stability and maintainability.
- **Legacy Life:** Even longer - culturally protected, restored multiple times, beloved by community and nation.

Legacy Ledger Highlights

- Legacy Gains:
  - Repairable materials
  - Harmonious relationship with landscape
  - Resilient geometry still effective under modern analysis
- Legacy Costs:
  - Vulnerable to modern tourism pressures
  - No redundancy for contemporary water supply needs

Practical Lessons for Engineers

- Precision is not the enemy of efficiency; it is the foundation of longevity.
- Repairable materials age far better than "high-performance" ones that demand specialist intervention.
- When a structure harmonises with its surroundings, people instinctively protect it.
- Care during construction becomes care during maintenance - and that becomes legacy.

Case Study: The Pantheon (Rome) - Legacy by Design Through Beauty, Structure and Emotional Resonance

The Pantheon is one of those rare buildings that dissolves the distance between past and present. Completed around 128 CE, it holds the world's largest unreinforced concrete dome, perfectly proportioned and still performing after nearly two millennia.

Roman engineers instinctively understood how to lighten the dome as it rose - mixing heavier basalt in the lower levels and lighter pumice near the top - creating a structural gradient we now teach in engineering schools.[326]

But the true endurance of the Pantheon comes from something harder to define: it moves people. When sunlight pours through the oculus, the building feels alive. It is this emotional coherence - not just its engineering - that ensured its survival. Empires fell. Languages changed. Religions shifted. Yet the Pantheon remained, because generations found it worth saving.

Why the Pantheon Is Legacy by Design

Legacy by Design Matrix

- **Service Life:** Extraordinary - the dome still performs its structural role.
- **Legacy Life**: Off the charts - preserved continuously because of cultural reverence.

Legacy Ledger Highlights

- Legacy Gains:
    - Materials optimised for long-term behaviour
    - Universally resonant beauty
    - Cultural continuity across centuries
- Legacy Costs:
    - Maintenance complexity due to unique historical materials
    - Cannot be modernised without risking heritage integrity

Practical Lessons for Engineers

- Beauty is not a luxury; it is a maintenance strategy.
- Emotional connection strengthens stewardship - assets that move people get cared for.
- Designing with gradients (materials, loads, light) creates systems that age gracefully.
- Engineering is not only about resisting forces - it's about creating meaning.

Case Study: The Great Stupa of Sanchi (India) - Legacy by Design Through Meaning, Ritual and Human Connection

High on a hill in central India sits the Great Stupa of Sanchi, first built in the 3rd century BCE and expanded over centuries. Its hemispherical form may appear simple, but the engineering logic is sophisticated: a solid core stabilised by compacted stone, surrounded by pathways that encourage pilgrims to walk clockwise in meditation.[327]

The reason it has survived invasions, monsoons, and political shifts isn't because it is the strongest structure in the region. It's because it was never abandoned. Its meaning outlived its materials. People kept returning - generation after generation - to walk, pray, repair, and pass on the story.

Why the Great Stupa Is Legacy by Design

Legacy by Design Matrix

- **Service Life:** Long-lasting stone structure with low mechanical complexity.
- **Legacy Life:** Essentially continuous; meaning sustained maintenance for millennia.

Legacy Ledger Highlights

- Legacy Gains:
    - Ritual-based engagement continuous community care
    - Simple, repairable materials

  - Spatial logic that supports wellbeing

- Legacy Costs:
  - Preservation limits adaptation for accessibility
  - Requires protected status to avoid modern degradation

Practical Lessons for Engineers

- People care for what connects them to meaning.
- Simple, robust forms often outlast complex systems.
- Ritual is a design tool: when a place invites people into a rhythm, they return - and repair.
- Legacy is strengthened not by size, but by significance.

Case Study: The Grand Canal (China) - Legacy by Design Through Adaptability and Evolving Purpose

The Grand Canal is one of the most astonishing feats of civil engineering in human history. Begun in the 5th century BCE and extended across dynasties, it stretches over 1,700 kilometres - linking the fertile south with the political capitals of the north. It moved grain, people, ideas, and governance.[328]

What makes it remarkable from a legacy perspective is not just its scale, but its adaptability. Unlike rigid modern megaprojects, the canal evolved constantly. Sections were widened, narrowed, re-routed, dredged, or reinforced depending on political need, climate shifts or technological change.

Its identity was never fixed; it remained responsive to context.

Why the Grand Canal Is Legacy by Design

Legacy by Design Matrix

- **Service Life:** Extremely long - still functional in sections today.
- **Legacy Life**: Sustained through cultural, political, and economic relevance.

Legacy Ledger Highlights

- Legacy Gains:
  - Modular, repairable, region-specific adaptations
  - Flexible purpose economic and cultural vitality
  - Integration of water, transport, and governance systems
- Legacy Costs:
  - Environmental impacts in later dynasties
  - Vulnerability to modern pollution and industrialisation

Practical Lessons for Engineers

- Adaptability is a form of durability.
- Systems designed to evolve survive far longer than systems designed to stay the same.
- Infrastructure becomes legacy when it can be reinterpreted by future generations.
- Projects that integrate multiple functions (water, trade, culture) build resilience.

Case Study: The Sydney Harbour Bridge (Australia) - Legacy by Design Through Connection, Identity and Continuous Care

We return to the Sydney Harbour Bridge again here, because its legacy also reveals something else: the economic and regenerative value of designing for more than the minimum.

For many Australians, the Sydney Harbour Bridge is more than steel and rivets. It is a symbol of identity - a piece of engineering woven into the heart of the city. When it opened in 1932, it did something both practical and emotional: it connected a growing Sydney, knitting two halves of a fractured geography into a unified whole.

But what truly makes the bridge a legacy asset is the care that followed. For more than ninety years, maintenance crews have repainted, re-riveted, inspected, and upgraded it - not because they were forced to, but because the community values it. Its endurance is an expression of affection as much as engineering.

Why the Sydney Harbour Bridge Is Legacy by Design

Legacy by Design Matrix

- **Service Life:** Very long - designed for heavy loads and continuous maintenance.
- **Legacy Life:** Enduring - culturally iconic, protected, cherished.

Legacy Ledger Highlights

- Legacy Gains:
  - Designed with inspection and maintenance in mind
  - Symbolic value consistent investment
  - Robust structural detailing for longevity
- Legacy Costs:
  - High maintenance demands
  - Increasing climate exposure risk

Practical Lessons for Engineers

- When infrastructure becomes identity, its lifespan increases dramatically.
- Maintenance must be designed into the asset - not added later.
- Beauty, symbolism and cultural connection all strengthen stewardship.
- The best legacy assets are never "finished"; they are continuously renewed.

Case Study: Arafura Building Systems - Building a Legacy in Cherbourg

When you drive into the Aboriginal community of Cherbourg in regional Queensland, the story is written into the land before you ever reach the old abattoir site where Arafura Building Systems now stands.

Cherbourg is one of the most significant communities in the history of Queensland's protection era - a place where Aboriginal people from dozens of Nations were horrifically and forcibly removed and relocated throughout the 20th century. Families were brought here from all over Queensland, separated from Country and language, and placed under strict government control.[329, 330]

The legacy of that history remains visible today in systems that still limit opportunity - including land tenure. Most residents cannot legally own the land beneath their homes. Much of the housing stock has long been low-quality, temporary, or in disrepair. Rent is often paid for decades with nothing to inherit.

So when co-founders of Arafura Building Systems and long-time advocates for Indigenous-led construction solutions, Barry Johnson and Trevor Woodhead, looked at Cherbourg, they didn't see a challenge to avoid; they saw an opportunity and a future worth building. The two had begun Arafura on the Sunshine Coast but the more they grew, the clearer it became that the company's mission - to create affordable, durable, dignified homes for regional and Indigenous communities - could only be fully realised if manufacturing happened *in the communities they were trying to serve.*

In 2025, the company moved its entire manufacturing facility to Cherbourg. The move wasn't symbolic; it was structural - jobs, skills, training, and local capability. Every panel fabricated here now represents employment for a local family, income circulating within the community, and young people seeing engineering and construction as a pathway instead of something that happens elsewhere.

In Cherbourg, the statistical picture prior to the factory move is stark. Only 30% of adults were in the labour force, nearly 20% of those were unemployed, and two thirds of residents aged 15+ were not even registered as part of the workforce.[331]

The arrival of Arafura Building Systems in this context is not a marginal development. It represents an opportunity to rewrite the local economy: manufacturing jobs, up-skilling pathways, wages circulating within the community - all aligned to place, to people, to future-generations. That shift, from being a town defined by constraints to one defined by capability, is the very essence of engineer-led legacy.

The Importance of the Design - Indigenous-Led Thinking, Fire-Safe Systems, and a New Kind of Housing Logic

What makes Arafura Building Systems remarkable isn't only the jobs or the relocation of the factory to Cherbourg - it's the design philosophy behind the homes themselves. The system was shaped in partnership with Indigenous founders and advisers, ensuring the design responded to cultural realities as much as to climate, safety standards, and construction constraints. In many remote and regional Aboriginal communities, conventional houses fail not because people don't care for them, but because the designs were never created with Country, culture, family structures, mobility, or climate in mind. ABS homes flip that script.

The panels used in the system are engineered for high fire performance, critical in regions where fire risk is increasing and where conventional lightweight housing has proven vulnerable. The homes are designed to be cyclone-rated, thermally efficient, mould-resistant, and robust enough to withstand harsh environmental conditions. What I love most is that they combine strength with dignity - the materials feel solid, safe and intentional, not temporary or "second-rate", which is sadly the norm in many government housing programs.[332]

Another defining feature is mobility without disposability. The homes can be relocated without demolition, which matters a lot in communities where people may need to move for cultural, familial, or economic reasons. A family can buy a home, pay it off in four years (for less than they were spending on rent), and - for the first time - own something that remains theirs. If circumstances change, the home comes with them. This is culturally aligned engineering: designing with mobility in mind because mobility is a living part of Indigenous life.

What sets ABS apart is that the design is not simply "affordable." It is dignified, durable, climate-smart, and culturally responsive. It meets safety standards not as a minimum benchmark but as a foundation. And because the system is Indigenous-led - from ownership to manufacturing to deployment - it avoids the classic pattern of external providers "importing" housing to communities that never asked for it.

This is engineering not just as construction, but as repair - of autonomy, of opportunity, of legacy.

A Housing Model Designed for Cherbourg, Not Just for Australia

What makes this model so powerful is that it doesn't just build houses - it builds possibility and pride in a place where the usual rules of home ownership simply don't apply. Cherbourg's land tenure system means that residents cannot buy freehold land in the way most Australians can. Even people who have lived there for generations cannot easily own the ground beneath their feet. Traditional housing programs have repeatedly tried to force a southern, suburban model into a community where the legal, cultural and historical conditions make that impossible.

Arafura approached the challenge differently. Instead of asking, "How can Cherbourg fit the housing system?" they asked, "How can the housing system fit Cherbourg?"

The homes are engineered as relocatable, structurally-sound, fire-safe dwellings - designed to withstand heat, mould, weather extremes, and the realities of remote living. But their real brilliance is in what this mobility makes possible.

A family doesn't have to own the land to own their home. They can lease a block, place the home on it, and own the asset itself.

And because the houses are affordable - often around A$100,000 - the economics begin to shift in a way not often seen in First Nations housing:

- After four to five years, the money a family would have spent on rent equals the full cost of the home.
- From that point onward, the home is entirely theirs - an asset that they can move if they choose to.
- If they need to move for work, family, health, or cultural reasons, they can simply take their home with them.

In a community where intergenerational disadvantage has been reinforced by systems that limit ownership, this model creates a different kind of pathway - one that honours mobility, autonomy, and cultural realities.

For Cherbourg residents, this isn't just housing innovation. It is dignity restored through design.

Arafura didn't "solve" home ownership by forcing residents to fit a rigid model found somewhere else; they designed a model that fits the lives, rights, and rhythms of the people it serves. That's why the system is spreading beyond Cherbourg to other towns - because it meets people where they are and gives them something the mainstream system has rarely offered: a foundation to build a future on.

Walking through the new Cherbourg factory, you see something else emerging: pride. Young workers learning fabrication and installation skills. Families building homes for other families. A sense of agency returning to a place where it was taken for generations.

This story isn't simply about construction. It's about restoring choice, dignity, and economic power - and doing it through engineering that listens, adapts, and serves.

Why Arafura Building Systems Is Legacy by Design

Legacy by Design Matrix

- **Service Life:** Long - These homes are built from durable, thermally efficient, cyclone-rated SIPs. Their modularity means components can be repaired, replaced, or moved at low cost. The factory itself is built for longevity, expanding local technical capability for decades.
- **Legacy Life:** Very Long - The model strengthens community autonomy, creates pathways to ownership, supports Indigenous employment, and roots economic opportunity in place. The social impact continues far beyond the life of a single house.

Legacy Gains - What This Work Gives the Future

1. A pathway to home ownership where none existed

Families can own their homes without needing freehold land, breaking a cycle of perpetual renting.

2. Local employment and skills development

Indigenous workers in Cherbourg gain fabrication, carpentry, logistics, and engineering-adjacent skills.

3. A housing model designed for Country, not imposed on it

Fire-safe panels, passive thermal performance, and relocatability reflect lived realities, not generic typologies.

4. Community self-determination in infrastructure

Housing is built by locals, for locals, restoring agency in a domain historically controlled by external authorities.

5. Economic resilience

Every home becomes a long-term asset held within the community rather than profits flowing out to distant contractors.

Legacy Costs - What the Future Inherits

These are not failings - just truths to remain conscious of:

1. Panel production requires continued specialist equipment

Future-proofing will depend on maintaining machinery and training.

2. Relocatable homes require thoughtful planning approvals

The model works best when councils collaborate proactively.

3. Cultural safety must remain central in company leadership

Scaling nationally must not dilute the Indigenous-led ethos rooted in Cherbourg.

Arafura's ledger is overwhelmingly positive because every choice is made with long-term effects in mind. It is legacy by design, not by accident.

Practical Lessons for Engineers

One of the most powerful lessons from Arafura's work is that engineering can influence economic dignity, not just structural performance. Every design choice has a financial ripple effect. In Cherbourg, the question wasn't only whether the houses met building standards - it was who would be able to build wealth because of them, and who wouldn't. When engineers ask, "Who is economically uplifted by this design?" the project begins to shift from infrastructure to empowerment.

The decision to manufacture the homes locally shows how place-based production can transform a community. Moving the factory to Cherbourg didn't just shorten supply chains; it strengthened capability, created new forms of employment, and turned construction into a local ecosystem rather than an external service. It's a reminder that where we choose to build is often as important as what we choose to build.

Arafura also demonstrates the importance of designing with the social system, not against it. The land tenure arrangements in Cherbourg make traditional home ownership almost impossible, but instead of treating that as a barrier, the design team reimagined the housing model entirely. They adapted the built form to the community context, rather than expecting the community to bend to an unsuitable system.

The relocatable design of the homes reveals another insight: mobility can be a form of resilience. In many engineering contexts, a building that moves is seen as a compromise. But in communities where cultural obligations, seasonal work, or family care require mobility, the ability to take your home with you is a form of stability. It reframes value through a different lens.

Arafura's approach also reminds us that the strongest legacy assets aren't always the ones that last the longest - they are the ones that people can *own*. Ownership builds agency, pride and intergenerational benefit. In Cherbourg, these homes allow families to build wealth in a system where land ownership is constrained. That is engineering serving legacy, not just function.

Finally, this Case Study highlights something engineers don't always talk about: cultural repair. Sometimes the technical elements are only the surface. Beneath them sits something far more significant - the restoration of choice, dignity, and a sense of future possibility. In communities that have carried the weight of structural injustice, the impact of regenerative design is not just physical. It is deeply human.

Case Study: Beatrix Potter & The Lake District (UK) - Legacy by Design Through Stewardship, Land Management & Ecological Care

When most people hear the name Beatrix Potter, they think of Peter Rabbit, Jemima Puddleduck, Mrs Tiggy-Winkle, and the gentle illustrations that filled childhoods across generations. But behind the stories was a remarkable woman whose real legacy wasn't just literary - it was ecological, cultural, and profoundly engineering in spirit.

Potter was not trained as an engineer, but she understood systems the way great engineers do: through observation, curiosity, and long-term thinking. In the early 1900s, she became concerned about the industrial pressures threatening England's Lake District - rapid land sales, unsustainable farming practices, and the creeping loss of traditional ways of working with the land.

So she did something extraordinary.

She began buying farms. Not as a collector - but as a custodian.

Potter acquired more than 4,000 acres of land, including 15 working farms complete with sheep herds, stone barns, ancient woodland, and historic water systems. She restored degraded soils, protected native hedgerows, reintroduced traditional Herdwick sheep (saving the breed from near extinction), and ensured that every property she managed upheld practices rooted in generations of local knowledge.

When she died, she left almost all of it - the land, the farms, the livestock, the cultural practices - to the National Trust. Today, it forms the ecological heart of the Lake District National Park.[333]

Why This Is Legacy by Design

Potter's contribution embodies legacy thinking in its purest form:

- She designed for people she would never meet. She understood that land cared for now becomes a living inheritance for later generations.
- She valued cultural continuity. Her preservation of Herdwick sheep wasn't sentimental - it sustained a whole farming tradition tied to the ecology of the fells.
- She saw land as a relationship, not a resource. Every decision reinforced the system rather than extracting from it.

Practical Lessons for Engineers

- Restoration is a form of engineering - one that requires humility and respect for place.
- A system persists when it is loved and understood, not simply controlled.
- Cultural knowledge is often the most durable design tool we have.
- Long-term stewardship sometimes requires someone brave enough to say: "This land is not for sale. It's for the future."

Beatrix Potter never built a bridge or designed a pipeline, but her work still protects water systems, soil health, biodiversity, and cultural identity. It is, unmistakably, legacy by design.

Case Study: Pilbara Seawater Desalination Plant – A New Model of Legacy in the Pilbara

If you travel across the Pilbara, you can feel the weight of competing forces: rapid industrial expansion, large-scale decarbonisation projects, and deep living histories of Country that stretch far beyond any map.

Water sits at the centre of all of it. It is the resource that underpins every industrial plan, every future energy proposal, every community, every ecosystem. Yet for decades, much of the Pilbara's water has been taken from stressed groundwater systems and aquifers that Traditional Owners have long known were at risk.[334]

The Pilbara Seawater Desalination Plant (Pilbara Desal) project represents a shift. Instead of continuing a model where water infrastructure is imposed *on* Country, this is a project developed *with* Country from the very beginning.[335]

Australian water infrastructure developer Legacie, co-led by managing director Daniel Lambert AM, has formed a partnership with the First Nations Corporations, SUEZ (one of the world's leaders in desalination), Sequana (a global water strategy and desalination advisory firm) and leading global resources companies. Together, they are co-designing a next-generation seawater desalination system at Balla Balla - a project grounded in caring for Country, environmental protection, and a long-term vision for the region.[336]

A Project That Begins With Consent, Not Extraction

Unlike many large Pilbara water schemes, this one did not start with engineering drawings or yield forecasts. It began with relationship. First Nations partners have provided support in the environmental and heritage planning, and have an option to hold equity in the desalination plant over its lifetime - a structural commitment that would ensure the project creates generational benefit rather than generational cost.

This shift from consultation to ownership is not symbolic. It is a practical rebalancing of agency in a region where Traditional Owners have often carried the burden of industrial development but received few lasting benefits. In this project, governance, decision-making, and long-term value are all available as an opportunity to grasp for the First Nations partners.

Designing Water for a Decarbonising Future

As the Pilbara rapidly transitions toward clean energy, the need for reliable, climate-independent water is accelerating. Green iron, green steel, critical minerals processing and large-scale renewable energy hubs all require stable water supply. The Pilibara Desal project aims to provide up to 150 gigalitres per year in planned stages - an extraordinary scale - powered by the region's strong wind and solar resources.

By locating the project on site with sea water access at Balla Balla, the partnership is able to reduce environmental disturbance, accelerate approvals and avoid the ecological harm associated with new groundwater extraction. Seawater desalination, when designed responsibly, becomes a protective measure - relieving pressure on sensitive aquifer systems and safeguarding cultural and ecological assets.

This is water as regeneration, not depletion.

A Partnership Model That Rewrites the Rules

What makes this project genuinely different is the partnership structure:

- **First Nations partners** actively engaged to provide cultural governance, consent, land and law – not as stakeholders but as partners.
- **Legacie**, under the leadership of Daniel Lambert, co-invests early risk capital, aligns incentives, and holds the ethical frame of "only build with consent."
- **SUEZ** provides world-leading desalination technology and operational expertise.
- **Sequana** supports the project with global advisory capability in strategy, planning, and desalination systems design.
- **Industry partners** contribute renewable energy alignment, helping the plant move toward low-carbon operation.

This isn't a procurement model. It's a values model - and a blueprint for how complex infrastructure can be developed when Traditional Owners lead.

Why The Pilibara Desal Is Legacy by Design

Legacy by Design Matrix

- **Service Life:** Very Long - The infrastructure is designed for multi-decade operation, electric-powered desalination, staged expansion, and robust environmental monitoring. It favours long-term resilience over short-term output.
- **Legacy Life:** Generational - Equity for First Nations people, cultural governance, environmental protection, and employment pathways mean the project has the potential to continue to deliver value far beyond its physical lifespan.

Legacy Gains - What This Work Gives the Future

Cultural Authority Built Into Infrastructure

Traditional Owners have the option to be equity partners, embedding cultural governance for generations.

Environmental Regeneration Through Water Security

Desalination avoids aquifer depletion and reduces pressure on sensitive ecosystems.

Economic Participation, Not Extraction

Revenue, training, contracting, and employment flow to First Nations people, not away from them.

A Foundation for the Pilbara's Net-Zero Transition

Reliable water is essential for future green industries; this project becomes part of the region's long-term climate strategy.

Approval Pathways That Respect Country

By leveraging existing footprints and avoiding new disturbance, the project sets a new standard for low-impact planning.

Legacy Costs - What the Future Must Manage

These are not drawbacks - they are simply the realities that require ongoing care.

Operational Complexity

Desalination assets need continuity of skilled workforce, energy stability, and sophisticated maintenance.

Marine Management

Brine dispersion and coastal impacts must be monitored and managed over the plant's life.

Governance Continuity

Ensuring First Nations cultural leadership remains central as generations change.

Dependence on Renewable Energy Integration

To remain low-carbon, the project must maintain strong alignment with regional development of renewable energy supply.

Overall, the ledger is overwhelmingly positive because the project makes its long-term responsibilities explicit.

Practical Lessons for Engineers

The Pilbara Desal project offers a number of powerful insights for engineers:

Start with consent, not compliance.

Infrastructure is cultural long before it is technical. When Traditional Owners are given the opportunity to lead, the engineering becomes more grounded, more resilient, and more ethical.

Partnership models matter as much as technology.

Equity, ownership, and shared risk redefine what "good engineering" can look like.

Regenerative water systems are essential to decarbonisation.

Net zero industries fail without water security. Climate-independent systems like desalination become the backbone of the transition.

Environmental protection can be engineered into the system from the beginning.

Avoiding groundwater impacts is not a trade-off - it is a design principle.

Legacy is not scale - it is stewardship.

Pilbara Desal is one of the biggest water projects in Australia, but even more significantly it may be one of the most important because of how it has been developed

This project shows what becomes possible when engineers choose to build a water future that communities want to inherit - not endure.

Case Study: Dubai - Engineering a Modern Legacy on the Edge of the Desert

Legacy by Design Through Vision, Scale & Reinvention

Few cities on Earth have transformed as dramatically as Dubai. In the space of a few decades, it evolved from a small fishing and pearl-diving village into a global metropolis of more than 3 million people.[337] Its story is often framed as ambition - skyscrapers, islands, highways - but beneath the headlines is a bigger narrative about legacy, identity, and long-term strategic engineering.

Dubai's modern foundation began in the 1960s and 70s, when oil revenue allowed investment into ports, airports, and water infrastructure. But unlike many oil-rich regions, Dubai quickly realised something critical: oil could not be its future. At its peak, oil made up over 50% of government revenue; today it is less than 1%. The city intentionally engineered a post-oil economy before the oil even ran out.[338]

That meant:

- Massive investment in logistics and trade
- World-class aviation infrastructure (Emirates, DXB, DWC)
- Large-scale desalination powered by industrial symbiosis
- Tourism systems built as economic engines

- Metro and tram systems integrated into long-term planning
- Mixed-use districts designed to encourage density in a desert environment

Dubai's approach wasn't perfect - sustainability critics point out issues of water, energy, and ecological footprint - but from a legacy perspective, something fascinating emerges:

Dubai didn't design for what it was. It designed for what it needed to become.

Why This Is Legacy by Design

Despite its complexity and contradictions, Dubai offers powerful lessons:

It thought beyond a single generation.
Leaders explicitly planned for a post-oil economy decades in advance.

It embraced adaptability.
Whole districts have been redesigned and retrofitted as needs shifted.

It invested heavily in infrastructure that multiplies value over time.
Ports, airports, logistics hubs, and free zones continue to serve millions.

It used engineering as a tool for identity.
The Burj Khalifa, Dubai Metro, and Palm Islands are all part of a long-term narrative about place-making and global positioning.

Practical Lessons for Engineers

- Designing for the future sometimes means designing beyond present constraints.
- Legacy requires anticipating the world your project will exist in - not just the world that exists today.
- Big visions need robust systems beneath them; grand architecture means little without functioning water, energy, and transport.
- Cities are stories. Infrastructure shapes those stories for centuries.

Dubai's legacy is still unfolding. Its challenges are real - climate, density, ecological cost - but its early decisions created a platform for reinvention. That is the hallmark of a legacy mindset.

Case Study: Wangari Maathai & The Green Belt Movement (Kenya)

Legacy by Design Through Community-Led Ecological Engineering

In the 1970s, Kenya was facing eroding soils, declining food security, and widespread deforestation. The impacts were everywhere: rivers drying, crops failing, women walking further each year for firewood and water. It was a systems failure - ecological, social, and economic all at once.

Then came Wangari Maathai.

A biologist by training, she did something extraordinary in its simplicity: she taught women how to grow trees. Not symbolically, not ceremonially - but strategically, as a form of ecological engineering.

The Green Belt Movement, which she founded, went on to plant over 50 million trees, restoring watersheds, stabilising soil, regenerating biodiversity, and strengthening local economies. Entire river systems came back to life. Villages once depleted by erosion began to thrive. Women gained autonomy and leadership through managing nurseries and restoration projects.[321]

This was infrastructure - just not the kind that gets built with cranes.

It was slow, relational, community-led engineering. And it worked.

Wangari Maathai later became the first African woman to win the Nobel Peace Prize, recognised not only for her environmental work but for understanding what engineers sometimes forget: ecological integrity and social stability are deeply intertwined.

Why This Is Legacy by Design

It addressed root causes, not symptoms.

Tree planting was a mechanism to rebuild entire ecological systems.

It centred community as the engineer.

Local women designed, managed and maintained the system - ensuring longevity.

It rebuilt natural capital across generations.

Watersheds restored today will nourish children not yet born.

It showed that legacy is not scale; it is direction.

Tiny seedlings became forests. Small actions became a movement.

Practical Lessons for Engineers

- Regeneration begins with listening - and with locals as partners.
- Sometimes the most powerful engineering solution doesn't look like engineering at all.
- Communities maintain what they co-create.
- Legacy often comes from patience, not efficiency.

Maathai's work demonstrates that stewardship is a form of engineering - one that shapes landscapes and futures with extraordinary power.

Case Study: Sir James Dyson - Legacy by Design Through Invention, Education, and Stewardship

When most people hear the name Dyson, they think of the bagless vacuum cleaner - the invention that disrupted a stagnant industry and made household technology feel modern again. But Dyson's real legacy is not the vacuum. It is the way he thinks about time, stewardship, and invention. It is the culture he built, the young people he invests in, and the systems he leaves behind for a future he will never get to see.

Before any of this was obvious, James Dyson spent four years developing over 5,100 prototypes in a backyard workshop. There were no guarantees, no investors cheering him on, no market begging for a cyclone-driven vacuum. There was just the conviction that if he built something with integrity - something engineered properly, something that solved a real problem - it would eventually be worth the effort.[74]

This is legacy thinking.

A Culture of Long Horizons

Once Dyson founded his company, he doubled down on this long-term approach. The business refused to outsource manufacturing to cheaper labour markets, choosing instead to keep engineering close to production so that learning could constantly feed innovation. That decision wasn't efficient in the Jack Welch sense - but it built a culture that values mastery, craft, durability, and care.

And then came the part few companies even consider: what legacy do we leave beyond products?

Dyson's answer has been extraordinary:

1. The Dyson Institute of Engineering and Technology

He founded his own engineering university - a fully accredited institution where students work inside Dyson teams while earning their degree. No debt. Paid while studying. Industry-led, academically rigorous, built entirely to lift the next generation of inventors. This is not corporate philanthropy. It's long-term stewardship of the profession.

2. The Dyson Farming Estate

Across thousands of acres in the UK, Dyson has been restoring soil health, regenerating landscapes, and experimenting with technology-led sustainable agriculture. The farm produces food, renewable energy, and even research outputs that influence national policy. It is a powerful engineering legacy.

3. The Engineering Boxes in Schools

In classrooms across the UK and Asia, Dyson sends out hands-on engineering kits that teach students about mechanisms, turbines, airflow, motors, and design thinking. Millions of children have touched real engineering hardware because of this program. This is legacy life in action: influence that will outlive the hardware itself.

Why Dyson Is Legacy by Design

Service Life:

Long and deliberately engineered. Dyson's products - from vacuums to fans to hair dryers - are built to be repaired, not replaced. The engineering school and the Dyson farming estate extend this thinking into systems with decades of lifespan built into them: soil that improves over time, graduates who carry skills across generations, and research that matures slowly but meaningfully.

Legacy Life:

Very long - and still expanding. The Dyson Institute, Dyson Farming, and Dyson's global education programs give the enterprise a cultural and societal impact far beyond its products. These are institutions designed to shape future engineers, regenerate ecosystems, and build knowledge long after James Dyson himself is gone.

Dyson's work is not only built to last; it is built to evolve. It continues to generate value - technical, cultural, ecological - generation after generation.

The Dyson Legacy Ledger

Legacy Gains - what Dyson's work gifts to the future

- **Regenerative agriculture at scale**, restoring topsoil, building biodiversity, and proving that large farms can heal landscapes, not just extract from them.
- **A debt-free engineering degree**, removing a barrier that blocks so many talented young people from the profession.
- **A new model for engineering education**, one that blends academic learning with hands-on industrial experience from day one.
- **STEM pathways for millions of school children**, thanks to Dyson's engineering boxes and global outreach.
- **A culture of long-term invention**, proving that iterating thousands of times on a single idea isn't obsession - it's discipline, patience and legacy thinking.
- **Products built for longevity**, resisting the cultural norm of planned obsolescence and disposable technology.

Legacy Costs - what the future inherits and must care for

- **Proprietary parts and systems** that may require specialist knowledge or specific supply chains to maintain.
- **Large-scale land holdings** that need consistent stewardship long after Dyson's lifetime.

- **A unique leadership culture** that must be intentionally transitioned to remain true to its values.

Dyson's ledger leans heavily toward long-term benefit - not because his choices are perfect, but because they are intentional. The compromises are known, documented, and considered. The legacy gains are designed to stretch outwards across time.

Practical Lessons for Engineers

- **Legacy is more than product innovation - it is the ecosystem that surrounds the work.**
  Dyson shows that you can build machines, and you can also build minds, landscapes, and cultures.
- **Long-term thinking doesn't happen by accident.**
  It comes from pausing, stepping back, and asking what your work will enable a century from now.
- **Education is one of the strongest forms of engineering.**
  When you teach a generation to think differently, you change the future more than any single invention ever could.
- **Regeneration belongs in every sector.**
  If a vacuum engineer can regenerate soil and inspire millions of young inventors, then every engineer can look beyond their immediate discipline.
- **Legacy grows when values are embedded in the organisation.**
  Dyson's insistence on patience, precision, and curiosity now lives in thousands of people, not just one..

Why Dyson Embodies Legacy by Design

Dyson's story is not a tale of a single breakthrough or a clever product. It is the story of someone who treated engineering as stewardship - of knowledge, of land, of people, and of possibility. His legacy sits not only in what he built, but in who he enabled, who he educated, and the landscapes he restored along the way.

This is the heart of Legacy by Design: building things that keep giving long after our part of the story is finished.

When Legacy wasn't Considered and the Impact

It's just as important to look at the projects that *didn't* think in centuries, because they show us the other side of the ledger. When legacy isn't part of the conversation, the consequences don't arrive immediately - they accumulate, year after year, until someone else has to carry them.

These examples are not about blame. They are reminders of why legacy thinking matters, and how easily it can slip through the cracks when the focus is only on delivery, budgets, or short-term performance.

When we study where things went wrong - the failures, the collapses, the clean-up costs, the communities left with the burden - we're really studying where future engineers can make different choices. These stories help us see the long shadow a project can cast when legacy isn't part of the design.

And in doing so, they reaffirm the value of stepping back, slowing down for a moment, and asking the question that changes everything: what will this decision feel like for the people who inherit it?

That is the heart of legacy thinking - learning not only from what we build well, but also from what we wish we had built differently.

Case Study: The Fundão Tailings Dam Collapse, Brazil - When Legacy Was Not Designed For

On 5 November 2015, the Fundão tailings dam near Mariana in Minas Gerais, Brazil, collapsed, releasing approximately 43 million m^3 of mining waste into the Doce River basin.[339] Nineteen people died. Entire villages were buried. Rivers were polluted for hundreds of kilometres. The surrounding ecosystems and thousands of livelihoods suffered immediate and long-term damage.[340]

Where it missed the mark on legacy

Legacy by Design Matrix

- Service Life: The dam was technically built to perform, but ignored a layered understanding of risk, long-term maintainability, and systemic consequence.

- Legacy Life: Far lower than assumed. The project did not strengthen community trust, ecological health, or future adaptability - it became a liability for generations.

Legacy Ledger

- Gains ignored: The assumption that the dam would "just hold" rather than be repaired, adapted, cared for.
- Costs incurred: Enormous social, environmental, and financial burden - compensation estimated at tens of billions of dollars.

Practical lessons

- Designing for short-term extraction, rather than long-term stewardship, invites legacy debt.
- Risk isn't only the technical failure of structure, but the human, ecological, financial, and cultural consequences of failure.
- Maintenance, remediation and adaptation must be core to design, not afterthoughts.
- If you cannot answer "Will this asset be safe, maintained, and valuable in 100 years?", then perhaps you need to redesign.

Case Study: Infrastructure Without Adaptability - When Service Life Exists, But Legacy Life Doesn't

While not as single-headline as the Brazilian disaster, many large infrastructure assets around the world have been built with minimal consideration for change-over time, circular economy, cultural relevance, or ecological health. The result: technical performance may hold, but usefulness, adaptability, and public value decline.

Where it missed the mark on legacy

Legacy by Design Matrix

- Service Life: Often long - the assets perform technically for decades.
- Legacy Life: Short - because community, ecology, culture, maintenance burden, and climate change resilience are ignored.

Legacy Ledger

- Gains ignored: Designing for repair, adaptation, multi-purpose usage, evolving social and ecological context.
- Costs incurred: Projects become rigid, expensive to maintain, obsolete in function, and disconnected from people and place.

Practical lessons

- Durability without adaptability means "lasting but not living".
- Designing for a narrow performance window instead of a changing world invites future irrelevance.
- For legacy thinking: embed flexibility, culture, ecology and repair from the start, not as extras.

Case Study: The Love Canal (USA) - When Waste Became an Inheritance

In the 1940s and 50s, a chemical company buried more than 20,000 tonnes of toxic waste beneath what would later become a suburban neighbourhood in Niagara Falls, New York.

The land was sold to the local school district for a symbolic $1. Homes were built. Playgrounds were built. Families moved in, unaware of what lay beneath.

Two decades later, chemicals began seeping into basements, gardens, and the school grounds. Children developed unexplained burns. Miscarriage rates rose. Residents fell sick. The community eventually had to be evacuated and the neighbourhood declared uninhabitable.[341, 342]

Why this is a legacy failure

- The project performed "well enough" at the time.
- But there was no Legacy Life thinking: no consideration of how this land would one day be lived in, inherited, or cared for.
- The health impacts lasted generations, and clean-up costs extended into the billions.

Legacy by Design lesson

What we bury, ignore, or hide beneath a project never stays hidden. Future communities inherit every decision we make today - good or bad.

Case Study: The Flint Water Crisis (USA) - A Cost Saving That Became a Generational Debt

In 2014, Flint, Michigan, switched its water source to cut costs. The new water was not properly treated, corroding pipes and leaching lead into drinking water. For nearly two years, families drank, bathed in, and cooked with contaminated water, while officials insisted it was "safe".[343]

Children - especially those in low-income neighbourhoods - suffered permanent neurological harm. Thousands of pipes had to be replaced. Trust in government collapsed.[344]

Why this is a legacy failure

- Service Life was technically met: water flowed.
- Legacy Life collapsed entirely: the system harmed the very people it was meant to protect.

Legacy by Design lesson

Short-term savings can create generational harm. Legacy requires thinking beyond budgets: it is about protecting the health, dignity, and trust of future citizens.

Case Study: The Samoan Coastal Roads (Pacific Islands) - Infrastructure That Didn't Listen to Country

Across parts of Samoa in the 1990s and early 2000s, coastal roads were widened and reinforced to improve access. But local engineers and village elders warned that the alignment would disturb traditional coastal buffers - mangroves, reefs, and sacred sites that had long protected villages from storms.

The advice was dismissed as "non-technical."

When Cyclone Evan hit in 2012, several sections of roadway washed away. Flooding moved inland, and villages suffered severe damage.

The very infrastructure meant to protect communities amplified their vulnerability.[345, 346]

Why this is a legacy failure

- The project met engineering requirements, but it ignored Traditional Knowledge and environmental rhythms.

Legacy by Design lesson

Infrastructure that ignores Country imperils future generations. Cultural continuity and ecological intelligence are not "optional inputs" - they are risk mitigation.

Case Study: The Three Gorges Dam (China) - World-Class Engineering, Fragile Legacy

The Three Gorges Dam is an engineering marvel. It generates enormous renewable energy and controls flooding for millions of people. But the legacy picture is mixed.[347, 348]

- Over 1.3 million people were displaced.
- Ecological changes destabilised riverbanks and contributed to landslides.
- Sediment flows altered ecosystems downstream.
- Cultural sites were submerged forever.

Why this is a legacy failure (and success)

It delivers incredible Service Life, but its Legacy Life is contested: ecological strain, cultural loss, and social fragmentation.

Legacy by Design lesson

Scale does not equal legacy. Legacy requires honouring culture, ecology, and community - not just producing megawatts.

Case Study: American Suburban Sprawl - A Design Pattern That Borrowed From the Future

After World War II, the US experienced rapid suburban expansion built around cheap land, cheap fuel, and the assumption that every family would drive. The pattern created enormous hidden legacy costs.[349, 350]

- Massive infrastructure maintenance burdens for future generations
- Car dependence and increased emissions
- Social isolation and reduced walkability
- Financial instability in local governments unable to maintain dispersed networks

This wasn't one project failing; it was a whole design paradigm.

Legacy by Design lesson

If you build something that future generations cannot afford to maintain, it isn't an asset - it's a liability with good lighting.

Case Study: The Fast-Tracked Schools of the 1970s - Built Quickly, Falling Apart Quickly

In Australia, the UK, and the US, thousands of school buildings were constructed rapidly during population booms. Many used experimental materials, minimal insulation, and low-cost construction techniques.

Within decades, they developed mould, structural issues, and health hazards. Some were demolished early; others became expensive maintenance burdens.

Legacy by Design lesson

Speed without stewardship becomes a generational retrofit program. Fast and cheap becomes slow and expensive.

Lessons Learned: What Failing Legacies Teach Us

When you look across these case studies - from Flint to Samarco, from Love Canal to whole patterns of suburban sprawl - you start to see a pattern. None of these failures were caused by a lack of technical intelligence. They were caused by a lack of *temporal* intelligence. A lack of *relational* intelligence. A narrowing of engineering vision to the next quarter, the next political cycle, or the next budget window.

They all had engineers. They all worked - at first. And then they didn't.

What failed was not mathematics or modelling. What failed was legacy.

The collapsed tailings dam in Brazil wasn't just a structural failure - it was the collapse of a mindset that believed short-term productivity was worth more than long-term safety.

The Love Canal wasn't merely a waste mismanagement issue - it was a failure to imagine that land would one day become a home.

Flint wasn't just about water chemistry - it was about valuing cost over children.

Suburban sprawl wasn't a mistake in design - it was a pattern created in a world that didn't yet understand ecological or economic limits.

When we step back, these examples reveal a few realities that engineers don't always say out loud:

1. Short-term efficiency is often long-term fragility.

Most of these systems were designed to "work" cheaply in the beginning - and they did. But the bill always arrives eventually, and it's always paid by people who didn't choose the system in the first place.

2. A project can meet every requirement and still fail its community.

Flint met standards on paper. The tailings dam met industry guidelines. The coastal roads met transport briefs. one of that protected people.

3. Cultural knowledge and environmental rhythms are not "soft factors".

They are often the difference between resilience and collapse. Where Indigenous knowledge was dismissed, vulnerability followed.

4. Legacy - good or bad - is cumulative.

It doesn't show up on opening day. It shows up 10, 20, or 50 years later. It shows up in lives changed, in ecosystems broken or restored, in trust strengthened or shattered.

5. Communities inherit what engineers create.

And they inherit it long after we've moved on to other roles and other projects.

That is why legacy is not abstract philosophy. It is a practical, ethical responsibility.

These cases show us what happens when engineers design only for Service Life - the minimum performance window - instead of Legacy Life, which asks whether the system will still be safe, loved, adaptable, and repairable long after we're gone.

They remind us that engineering is not simply about making things stand. It is about making things worth standing for.

And they point us firmly, toward the work that matters now:

- Designing systems that won't one day appear on a list like this.
- Designing infrastructure that future generations will thank us for.
- Designing in a way that remembers the people who inherit our decisions.

If the previous chapter showed how design can heal, this section shows the other side of the equation: design can harm just as easily.

But the lesson isn't guilt - it's clarity. We now know what happens when legacy is ignored. And because we know, we can choose differently.

Legacy thinking is leadership thinking. It is engineering in its most mature form. And it's how we make sure our work becomes part of someone's inheritance - not their burden.

Legacy Reflections

These projects - ancient or modern, monumental or modest - all have one thing in common: **they were designed to serve, not just to stand.** They breathe because they were built with continuity in mind - materials that could be maintained, purposes that could evolve, and meanings that people refused to let die.

When I think about the future of engineering, I think about these lessons. Because the reality is, every bridge, library, tunnel, and footpath we design today will one day belong to someone else.

The real measure of legacy isn't just whether it lasts - it's whether it's loved enough to keep living.

From Inspiration to Application

Standing back from these examples, it's easy to be in awe of them - the sheer endurance of Roman concrete, the elegance of a stone aqueduct, the living intelligence of Budj Bim.

Legacy isn't only something we inherit; it's something we build every day in the small, almost invisible decisions we make. How thick a pipe should be. What material we choose. Who we consult. Whether we ask what the next generation might need from this, long after we're gone.

When you strip it back, legacy isn't about size or budget or prestige. It's about **care** - care for people, care for Country, care for time itself.

And while those qualities might sound abstract, they can be measured. They can be designed for. They can be built into contracts, specifications, and drawings.

That's where the Legacy Scorecard comes in. It's not another compliance tool. It's a thinking tool - a way of checking if what we're creating today will still serve, still adapt, and still be loved tomorrow. A way to bring that same precision, empathy, and foresight that shaped the Pont du Gard or Budj Bim into the everyday decisions of modern engineering.

Because legacy isn't a noun. It's a verb.

It's not what we leave behind - it's what we set in motion.

The Inheritance Test

Every profession has its oath. Doctors have *First, do no harm.* Maybe ours should be:

Leave something worth inheriting.

When I think of the young engineer who once said to me, "You designed this... and now we're left to maintain it," I hear both frustration and hope. A longing for us to think further ahead. A plea to remember that the timelines of our projects don't stop when the ribbon is cut - they continue on in someone else's hands.

She was asking us to remember that our work doesn't end when the paperwork does.

Legacy is the understanding that the decisions we make today will outlive us - sometimes for decades, sometimes for centuries.

It's the recognition that someone, long after we're gone, will inherit the consequences of our choices: the materials we selected, the shortcuts we took or resisted, the intentions we infused into the design.

Legacy is designing for people who will never know our names - but who will still be walking across the bridges we built, or sitting under the trees we planted, or breathing cleaner air because of choices we made with care. It is the knowledge that future communities will live inside the systems we create, whether those systems serve them or constrain them.

And this is where a bigger truth sits:

Engineers are, whether we realise it or not, stewards of the future.

Every drawing we sign, every specification we write, every detail we approve becomes part of the world that others will inherit. Our work shapes the daily rituals of people we will never meet. How they move. How they breathe. How they feel in their bodies. How safe they are. How connected they feel to place and to one another.

We don't always talk about it this way, but engineering is a generational profession. We inherit the work of those before us, and one day our work will become the inheritance of those who come next.

To design with legacy in mind is to honour that lineage - to acknowledge that we stand in a long chain of hands shaping the world, each one leaving something for the next.

That's what generational engineering really is. Not grandeur. Not perfection. Just care, stretched across time.

And if we do this well - if we choose endurance over convenience, reciprocity over extraction, story over speed - then perhaps the future will look back and whisper something extraordinary: "Thank you. You left us something worth keeping."

That is the hope at the heart of legacy by design. That is how the future will thank us.

The Takeaway

When we begin to design for people we will never meet, legacy stops being an abstract idea and becomes a practical one. Every assumption, shortcut, and optimisation choice carries forward, shaping whether future generations inherit assets that support them or systems that drain them.

The uncomfortable reality is that legacy is not determined by technical competence alone. It is shaped by the values that guide our decisions, especially when trade-offs are required. When time, cost, risk, and performance are in tension with connection, care and legacy, what we prioritise reveals what we truly value.

Engineering has long relied on inherited values, embedded so deeply in standards, contracts, and professional norms that they feel unquestionable. But designing for long-term health and inheritance requires something different: a conscious choice about what we stand for.

The next chapter turns to values directly, exploring the difference between core values and chosen values, and why the future of engineering depends on making those choices deliberately, rather than by default.

Chapter 7: Core Values and Chosen Values

Every engineering decision is guided by values, whether we acknowledge them or not. Some are inherited through standards, systems, and professional culture. Others are chosen consciously, often in moments where trade-offs must be made and no option is perfect.

This chapter explores the difference between core values and chosen values. The core values here are, I believe, essential to the Future of Engineering and have the potential to be the defaults we carry. Chosen values are the additions we can choose to include, to deliberately step into when the defaults are no longer sufficient for the world we are designing for.

Designing for the future requires more than technical skill or good intentions. It requires clarity about what we are willing to protect, prioritise, and pass forward. This chapter examines how engineers can move from inherited values to intentional, deliberately chosen ones, and why that shift is essential if we are serious about legacy, wellbeing, caring for the planet and long-term impact.

The Hidden DNA of Values

Every profession inherits its own DNA - a set of assumptions, habits, and hierarchies so deeply embedded they become invisible. We know that engineering's DNA was not written overnight. It was shaped across centuries, through revolutions both industrial and intellectual.

In the beginning, this DNA was built for a different world. The Industrial Revolution demanded scale - bridges, railways, factories, power grids. The focus was speed and standardisation. Human systems began to mirror mechanical ones, and in that transformation, a certain way of thinking took root.

Frederick Taylor - The Scientist of Efficiency

At the dawn of the 20th century, mechanical engineer Frederick Winslow Taylor introduced what he called Scientific Management.

He spent years with a stopwatch in his hand, observing steelworkers, bricklayers, and machinists, measuring every motion to find "the one best way" to perform a task.

"In the past, the man has been first; in the future, the system must be first." - Frederick Taylor, 1911

Taylor's ideas revolutionised productivity. They made work measurable, predictable, and controllable.[1, 2] But they also subtly changed what engineers were taught to value. The worker became a variable; efficiency became a virtue.

In a sense, Taylor's stopwatch still ticks in the background of modern project management - every schedule, every milestone, every earned-value chart is an echo of his pursuit of the perfectly efficient system.

Henry Ford - The Architect of Scale

Taylor's theories came alive on Henry Ford's factory floors. In 1913, Ford introduced the moving assembly line to mass-produce the Model T. The result was astonishing: a car that once took twelve hours to build could now be made in approximately ninety minutes.

"*The man who places the part does not fasten it... That is his only work.*" - Henry Ford, *My Life and Work*[351]

Ford didn't just make cars. He made a new kind of world - one where scale equaled success and uniformity became a kind of progress. It was a triumph of systems thinking but a narrowing of imagination. Ford himself famously said,

"Any customer can have a car painted any color that he wants so long as it is black."

That statement - half joke, half manifesto - captures the paradox at the heart of industrial engineering. By mastering control, we began to lose diversity. Efficiency became a universal language, but it left little room for nuance, creativity, or care.

Jack Welch - The Era of Extraction

We have seen in previous chapters the impact Jack Welsh had on the business world. By the late 20th century, this mindset reached its corporate zenith under Jack Welch at General Electric. Welch became a symbol of industrial-era success: relentless, data-driven, uncompromising.

Under his leadership, GE's market value soared. He preached "fix, sell, or close" - if a division wasn't first or second in its market, it didn't deserve to exist.

"*Control your own destiny or someone else will.*" - Jack Welch, 1989

This was efficiency as survival - not just for machines, but for companies. Welch popularised "rank and yank" systems, cutting the bottom 10% of performers annually. For decades, his model became the template for management everywhere, even far beyond manufacturing.[231, 232]

It worked. Until it didn't.

The same principles that made companies efficient also made them brittle. The short-term rewards of optimisation came at the long-term cost of adaptability, creativity, and trust. When Welch stepped down, GE was worth hundreds of billions. Two decades later, it was a fraction of that.

The Legacy of the Efficiency Gene

These thinkers - Taylor, Ford, Welch - weren't villains. They were visionaries solving the problems of their time. Their DNA was built for expansion, industrialisation, and rapid global progress.

And it delivered: cities grew, bridges spanned rivers, homes filled with light.

But every design has a design life and a design era - the time that suited them the best to be the most efficient. The operating conditions of their world no longer match ours and they are no longer effective for our world.

What they couldn't have foreseen was a planet under strain, a society craving connection, and systems so complex they defy linear logic.

We've inherited their brilliance - but also their blind spots.

Our engineering DNA is still optimised for speed, certainty, and control. But the future requires something else: adaptability, empathy, and balance.

When I was a consultant seconded onto a multi-billion-dollar tunnel project, I could feel those conflicting values colliding in real time. Il chaired one meeting with twenty people - engineers, government officials, contractors, operators - who were all arguing for different priorities: public trust, buildability, budget, long-term maintenance.

Everyone wanted the "right" answer, but each was drawing from a different value system, all pulling in different directions.

I realised then: we weren't fighting over data; we were fighting over values.

No one was wrong. We were simply optimising different parts of the same system aligned to the DNA and the values of that organisation.

And that's why it's time to consciously evolve that system - to rewrite our professional DNA and uncover our true values for the world we're actually building now, for the time we are living in now and for future generations.

Rewriting the DNA for the Future

This isn't about rejecting our past. It's about acknowledging that the genes of efficiency, control, and precision - the ones that built the modern world - are only half of the story.

To meet the challenges ahead, we need to switch back on the genes we've silenced: care, creativity, connection, curiosity, and courage. We need to measure success not just by what we build, but by what remains after us.

That's the next evolution of engineering - and the purpose of this chapter: to map the DNA repair sequence that restores integrity, balance, and humanity to how we design.

The story of Taylor, Ford, and Welch doesn't just live in history books; it lives on silently in how we manage projects today. Their philosophies became the invisible architecture of our daily decisions.

In engineering, we don't talk about "Scientific Management" anymore, but we live it through key performance indicators, dashboards, project control group meetings, gateway reviews, the Project Management Book of Knowledge (PMBOK), and earned-value charts. Every task is tracked, every milestone measured, every deviation explained.

Precision is prized - and for good reason. Accountability matters. But when every conversation is reduced to data, the things that can't easily be measured - trust, care, creativity, connection, meaning - slowly disappear from the room.

Procurement systems, too, carry the imprint of this lineage. They were designed in the image of Ford's production line: inputs, outputs, deliverables, contracts. Each stage has been optimised for control and efficiency. We separate designers from constructors, constructors from operators, people from the places they shape.

The result is a system that functions beautifully on paper and delivers in terms of bringing projects to fruition, but often struggles in reality - because the feedback loops between intention and impact are broken.

Even our culture of risk aversion is part of this inherited DNA. We've been trained to believe that uncertainty is failure, that deviation from the plan is dangerous. Yet the world we're designing for - the climate, the economy, our communities - is becoming more unpredictable, not less.

The paradox is that our fixation on certainty now creates risk, because it prevents us from learning and adapting fast enough.

Walk into almost any project office and you'll feel this heritage at work. Mantras said in meetings and media releases like "On Time, On Budget, High Quality". Spreadsheets tracking defects per thousand units. Risk registers that grow thicker with every review. These are the tools of control - vital for safety and accountability - but they're also the legacy of a century built on efficiency above all else.

None of this is wrong. It's simply incomplete.

We are living through a time when the conditions that once defined success - control, certainty, replication - no longer match the complexity of the world we serve. The next generation of engineers will need new metrics: measures of trust, wellbeing, regeneration, adaptability, and legacy.

That's why it's time to look beneath the spreadsheets and schedules and ask the question: what values are driving the way we design?

Because when we change those values - when we rewrite that DNA - everything else follows.

The DNA Repair Sequence: Core Values

If the industrial age encoded the DNA of efficiency, control, and certainty, then the repair sequence for our profession begins with the values that restore balance: legacy, care, integrity, and systems thinking.

These are not new ideas - they are old ones we have forgotten how to listen to. They have always been part of the human story, and in fact, part of engineering's story too. We have just been so focused on delivering, which we are exceptionally good at almost to a fault, that we stopped asking what for.

Time and time again, projects are delivered but miss the point. It's like someone forgot to ask the clarifying question upfront and jumped straight into problem solving. I know I have been guilty of this myself many times. My hunger for finding the answer and figuring it out takes over, and I miss the essence of the question.

When these four values are switched on together, something powerful happens. Projects move from being technically successful to being hugely successful. They endure. They build trust. They heal as much as they build.

Let's walk through them one by one - the four core and three optional genes of a new engineering DNA.

1. Legacy over Short-Term Gain - Measure Success by What Remains After Us

Every engineer knows the pressure of deadlines and delivery cycles. Projects are designed to be delivered within election terms, budget years, or corporate reporting periods - all short windows in the life of an asset that may last a century.

The tension is clear: our systems reward short-term milestones, but society will live with our work for generations.

When I think about legacy, I often return to a question from the poet Drew Dellinger:

"What did you do once you knew?"[316]

Legacy is not about monuments or awards - it's about responsibility. It's the knowledge that the decisions we make today ripple forward into the lives of people we'll never meet.

There are Roman aqueducts across southern Europe, some of which have remained structurally intact for centuries and, in certain cases, have been adapted for continued use. They still stand, their beauty is undeniable, but their endurance came not just from material strength, but from the mindset of those who built it. They weren't designing for compliance; they were designing for continuity, for the future.

By contrast, we've all seen projects that technically succeed but fail the legacy test: infrastructure that decays too quickly, that can't be adapted, or that isolates rather than connects. The Paradise Dam in Queensland is one such reminder - an engineering feat built quickly and cheaply, now failing structurally and costing billions to repair.

Case Study:

From the late 1990s onward, engineers and designers began a long-term process of strengthening, upgrading, and carefully adapting the Sydney Opera House. They noticed that the marine life around Bennelong Point was struggling. Rather than ignoring the issue, they partnered with marine ecologists to install artificial reefs under the harbour's edge.[315] Decades after its completion, the Opera House has become a site of regeneration, not just performance. That's legacy thinking: when a structure becomes part of the living system it inhabits.

Reflection Prompt:

If your project were rediscovered 100 years from now, what would people say it stood for?

2. Care for Country - The Umbrella Value Across All Decisions

In First Nations knowledge systems, care is not a sentimental act - it's a structural one. "Country" is not land; it's a living system of relationships between people, place, and spirit. The health of one depends on the health of all.

"We don't own Country; Country owns us. Our role is to look after it, so it can look after us." - Yuin Elder, Uncle Max Dulumunmun Harrison[352]

Indigenous Australians developed sophisticated systems of land, water, and food management over tens of thousands of years - practices that many scholars now recognise as forms of advanced ecological engineering. They built fish traps, aquaculture systems, and land management practices that worked in harmony with natural cycles. It is often so hidden today because it integrates so perfectly with Country and the world we live in here in Australia.

The Budj Bim fish traps in Victoria, dating back at least 6,600 years, are one of the oldest examples of engineered landscape systems on Earth[248] - built from volcanic rock, designed to feed communities without exhausting the ecosystem. That's regenerative engineering in its purest form.

Today, Associate Professor Shannon Kilmartin-Lynch, a proud Yowong-Illam-Baluk and Natarrak-Baluk man, continues this lineage. His research explores how traditional plant resins can be incorporated into modern concrete to make it more durable and sustainable. It's a living example of "care for Country" expressed through science - proof that ancient wisdom and modern innovation are not opposites but partners.[313]

In the Blue Mountains in New South Wales, there is evidence that mixing the resin from grass trees and kangaroo poo creates a super glue that can be used to create axes. Some of these axes still exist decades after their creation.

Reflection Prompt:

If your project were part of an ecosystem - not separate from it - how would that change your decisions?

3. Integrity and Accountability - The Courage to Tell the Truth in a Safe Space

Integrity is often described as "doing the right thing when no one is watching". In engineering, it's more than that - it's having the courage to tell the truth when everyone is watching.

Every project reaches a point where something goes wrong: a safety concern, a budget blowout, a design flaw, a stakeholder dispute. In those moments, our integrity defines not just the outcome, but the culture that follows. Whether you are the leader or not, it is the response to these situations that sets the standard for the whole team, and demonstrates how accountability works in our profession.

During my career, I've seen both sides of this. On some projects, people were afraid to raise issues, worried they'd be blamed or sidelined. They would put off telling people bad news and it would often end in a damaged relationship, an argument, and a much worse situation. On others, leaders created an environment of openness and trust - where problems were shared early and solved collaboratively. The difference wasn't process. It was culture.

Engineers' projects impact people and the general public too, so sometimes these lapses in integrity are aired in a hugely public context.

Integrity is the DNA of public trust. And accountability doesn't mean punishment; it means ownership. It means seeing mistakes as data, not disgrace.

Case Study:

In the aviation industry, safety reporting systems shifted decades ago from blame to learning.[59] Pilots can report errors confidentially, and the entire system learns from them. That mindset could transform engineering, moving us from fear of failure to a culture of collective care.

Reflection Prompt:

What would it look like if your team measured success not just in terms of delivery, but in terms of honesty?

4. Connect to the System - Optimising the Whole, Not the Part

The final core value is the one that ties all the others together. Systems thinking reminds us that no part of a project exists in isolation. A bridge affects traffic, air quality, and local businesses. A water plant affects biodiversity and community health. A housing development affects social cohesion and transport networks.

Yet too often, we optimise for the part we can control - our budget, our schedule, our piece of the system - and forget to look up, sideways, or over the fence to what is next to the system.

The future of engineering depends on our ability to think like ecologists and sociologists as much as we need to think like builders. Every technical choice is also a social and environmental one.

Case Study:

In Stockholm's Hammarby Sjöstad, engineers designed an entire urban district as a circular system. Energy, water, and waste were treated as interconnected loops. Biogas from waste powers buses. Treated wastewater heats homes.[353, 354] This is what happens when systems thinking moves from theory to practice: complexity becomes coherence.

Reflection Prompt:

What would you design differently if you optimised for the whole ecosystem - not just your part of it?

Reactivating the Core DNA

Legacy. Care. Integrity. Systems thinking.

Together, these four values form the repair sequence of our profession's DNA.

They don't replace the old genes of precision, discipline, or efficiency - they rebalance them.
They remind us that we are not just builders of things, but custodians of time, place, and trust.

These values aren't asking us to completely overhaul every aspect of our work, but nudge us into new directions and ways of thinking. Even just one of these values or reflection questions could change the trajectory of a project. It's like a plane taking off for a ten hour flight. If the flight path changes just one degree at take off, then it will land in a different location. That is the power of these core values.

Once these values are consciously activated, they act like a compass. They guide decisions when data is incomplete. They reveal the social and ethical dimensions of technical work. They expand our thinking. And they help engineers see themselves not just as problem-solvers, but as stewards of the future.

Optional Values - Contextual, Collective, Chosen

If the core values are our shared DNA - the genes that make us recognisably engineers - then the optional values are where we express individuality, context, and culture.

Every project is unique: a tunnel in Sydney, a hospital in Darwin, a desalination plant in the Pilbara, a coastal protection system in Fiji. Each serves a different community, environment, and legacy. The values that guide those decisions should reflect that diversity.

These contextual values give teams the freedom to customise their ethics - to express what matters most in their specific place and moment. They act as fine-tuning for the moral compass of a project.

But there's one rule: they must sit inside the boundaries of the core values - legacy, care, integrity, and connection to systems thinking. That way, local choices add richness without losing coherence.

5. Community First - Prioritising Wellbeing and Empowerment of People

If legacy is about what we leave behind, community first is about who we build for.

Too often, we design projects for "users", "stakeholders", or "the public". These are polite technical terms, but they strip the humanity out of our purpose. Communities aren't abstract entities; they're people with histories, hopes, and everyday lives that our projects shape - sometimes profoundly.

Community-first thinking doesn't mean endless consultation or populism. It means shifting the power dynamic from telling to listening. It's about seeing engagement not as risk management, but as co-creation.

Example:
The Medellín Metrocable is one of the clearest examples of community-first engineering. Rather than bulldozing new infrastructure through informal settlements, planners and engineers built cable cars that connected hillside barrios to education and jobs. They paired transport with libraries and public spaces. The outcome wasn't just mobility - it was dignity.[83, 84, 85]

Reflection Prompt:
Who benefits most from this project, and who carries the burden? How can we rebalance that equation?

6. Curiosity & Courage (with Kindness) - Asking Better Questions, Not Just Finding Faster Answers

Curiosity is what turns an engineer into a learner again. It's what lets us see problems from angles no one expected.

But curiosity without courage achieves little, because curiosity often means questioning established norms or raising uncomfortable realities. And courage without kindness risks turning conviction into conflict.

The balance of all three - curiosity, courage, and kindness - is where innovation flourishes.

I think back to the manager who noticed that I tended to ask a lot of questions. She said, "You know, this can make some people uncomfortable. But don't lose that. Keep asking questions - it's your way of seeing differently."

That conversation changed how I saw myself. Curiosity wasn't a flaw in my engineering DNA; it was a strength.

Example:

James Dyson's relentless experimentation - 5,127 prototypes before his first bagless vacuum worked - embodies curiosity and courage. He wasn't driven by profit, but by frustration with inefficiency and the want for excellent design. His motto could have been: "ask why until something better exists".[74]

In today's complex world, curiosity and courage might look like challenging a standard that no longer fits, or proposing a solution that doesn't yet have a line item in the budget.

Reflection Prompt:

When was the last time you changed your mind because of something you learned?

7. Kindness to Planet, People, and Communities - Making Compassion Operational

Kindness may sound like an odd word to use in engineering. But when we strip it back, kindness is simply the act of recognising interdependence - that what we design affects others, and what affects others eventually comes back to us. A sentiment widely attributed to Mahatma Gandhi tells us: "a civilisation is often judged by how it treats its most vulnerable members."

Kindness in practice is empathy made structural. It's the bridge with a shaded pedestrian path for elderly walkers. The road surface that reduces tyre noise in residential areas. The design choice that prioritises biodiversity corridors rather than concrete drainage.

In our work, kindness isn't about being soft - it's about being smart. It reduces opposition, improves long-term outcomes, and builds trust.

Example:

Vienna's gender-mainstreaming approach to urban design - often referred to through initiatives like the 'Female City' projects - re-imagined urban spaces through the lens of women's lived experience. Planners mapped how women and children moved differently through the city and designed lighting, seating, and pathways accordingly. The result was a safer, more inclusive city that benefited everyone, not just its intended group.[86]

Reflection Prompt:

Where in this project could kindness reduce conflict, stress, or harm without reducing performance?

Putting Optional Values into Practice

Optional values are not decorative. They shape decisions, budgets, and behaviours just as much as technical standards do.

Now you have started to look behind the curtain of the hidden DNA and embedded values, that there is more work to do here. You may find as you unpick the values that are within you, your teams, your projects, and organisations that there are other values that need to be embedded to deliver on the desired project intent.

But unlike standards, they must be chosen consciously and collectively. A value only matters when it's shared.

That's why the next section - Chosen Values and Collective Alignment - focuses on how to identify, prioritise, and agree on these guiding principles as a team before a single design decision is made. When values are chosen together, they become more than words on a wall - they become the code that governs how a team thinks, acts and behaves, and cares.

Chosen Values and Collective Alignment

When a project begins, there's usually a rush to define scope, budgets, and timelines. The brief is written, the standards referenced, and the program locked in. But rarely - if ever - do we stop to ask: "what values will guide the way we work together?"

Yet the value conversations are the ones that define everything that follows.

In complex projects, conflict rarely comes from poor technical work. It comes from misaligned values. From differing interpretations of "good". From unspoken expectations about speed, safety, risk, and reputation.

If we don't make those assumptions visible at the beginning, they surface later - in tension, rework, poor behaviour, and mistrust. This is why alignment on values is not "soft skills" work. It's strategic project risk management at its highest level.

The Power of Alignment

When I worked as a design manager on a major tunnelling project, we had more than twenty stakeholders in one room regularly to make decisions on the project - each with different priorities. Our fortnightly technical meetings were usually a minimum of fourteen people. And I was the design manager for just one of three zones on the project.

The state government was focused on public accountability and political commitments. The contractors wanted to build efficiently and safely. The operations team was already thinking decades ahead to maintenance.

All of them were right in their own way - but their values collided. The meetings weren't just about design; it was about meaning.

It struck me then how powerful it would be if, at the start of every project, and when new team members were onboarded, if we mapped these values openly - before the first drawing, before the first milestone - and treated them like design criteria.

That's what this section is about: making values a shared framework for decision-making, not a background philosophy.

When Values Clash - The Windscreen Wiper Story

A few months ago, our car's windscreen wipers needed replacing. The closest shop to our house was Supercheap Auto - and perhaps the clue was in the name.

My husband walked in and said, "I want the best wiper blades you've got." The young assistant handed him a pair confidently.

Back in the car park, he tried to fit them. They were the wrong kind - fiddly strips that needed to be threaded through, not the simple click-in, click-out type he wanted.

So back he went. "I asked for the best," he said.

"Yeah," replied the assistant. "These are the cheapest ones we have."

In the assistant's mind, best meant lowest price. In my husband's mind, best meant ease of installation. Two completely different definitions of "best" - each valid, but based on totally different values.

By the third trip upstairs - with two restless kids in the back seat - the frustration was real. But it was also such a simple, perfect example of how values drive behaviour and misunderstanding.

This is such an incredibly simple example of how values were misaligned. Extrapolate this out to a design decision, an interpretation of a technical standard or drawing note, or communicating in a meeting of a complex project.

We actually use this example now in leadership workshops that we run. It makes people laugh, but then it clicks: this is what happens in our teams, our projects, and our relationships every day.

When people say they want the "best outcome", they might mean the fastest, the cheapest, the safest, the greenest, the kindest, or the most beautiful. If we don't clarify what we mean - and if we don't align on values - we will talk past each other forever.

This situation reminded me of something I've seen in both my work and my relationships: most conflict doesn't come from incompetence or bad intentions. It comes from different values which drive thinking, definitions of what "good" looks like, and behaviours.

Whether it's wiper blades, washing the dishes, or billion-dollar tunnels, values shape every decision - silently, powerfully, and constantly. Once you see that, you can't unsee it.

And that's why alignment on values isn't optional. Even a simple job of getting windscreen wipers had hidden values driving the decision and the outcome!

It's the foundation of everything that works in engineering - because all we do, ultimately, is create outcomes for people. And if the people creating those outcomes aren't aligned on the values guiding them, the results will never fully work.

The Process of Collective Alignment

The following approach has worked across disciplines, scales, and project types. It's simple, repeatable, and adaptable to a huge range of contexts.

1. Gather Broadly

Invite people who see the system from different heights:

- Engineers, designers, and constructors
- Policymakers, funders, and maintainers
- Community representatives and future users
- Environmental and cultural custodians
- Different disciples outside the usual ones involved in the project

The diversity of perspectives is what gives the alignment strength. This is where the three lenses - Human Wellbeing, Regenerative Environments, and Generational Legacy - become your framing tools. They keep the conversation expansive, ensuring that every voice relates back to one of the three: people, planet, or time.

2. Ask the Right Questions

Rather than starting with "what are our values?", begin with questions that reveal them indirectly. People often live their values long before they can name them.

You might ask:

- What do we want people to feel when they experience this project? (Human Wellbeing Lens)
- What do we want this place to look and feel like in 50 years? (Generational Legacy Lens)

- How can this project strengthen - not just protect - the natural systems around it? (Regenerative Environments Lens)
- When challenges arise, what will we choose to protect first: schedule, cost, or trust?

These questions pull values out of hiding.

3. Identify the Hidden DNA

Once the group begins to share, patterns emerge. You'll hear words like safety, innovation, community, transparency, care, and integrity. Write them down.

Then ask: "Which of these are non-negotiable - and which are contextual?"

This distinction matters. Non-negotiables are your core values - the DNA that must always be present. Contextual ones become your chosen values - specific to this team, this place, this moment.

This process doesn't create consensus through compromise. It creates clarity through contrast. You can't align values you haven't seen.

Choosing and Weighting Values with Stakeholders

Every project begins with a set of core values - the non-negotiables like Care**,** Integrity**, and** Legacy. But the next layer, the contextual values, depends on people and place. A city renewal might elevate Community First; a regional water scheme might emphasise Kindness or Curiosity.

Choosing these values isn't a branding exercise - it's an act of alignment, effectiveness, and strategic de-risking.

The most effective teams do this early, before drawings or procurement. They hold a short "Values Alignment Workshop" and invite engineers, designers, policy leads, operators, Traditional Custodians, and community voices. Together, they explore these questions through the three lenses.

Each participant rates which values matter most for this project and why. The group then maps the results into a Values Heatmap - a simple grid showing where alignment is strong and where differences need conversation. The visual becomes a shared compass throughout delivery and a place to re-orientate yourself back to.

The aim isn't consensus. It's clarity and direction. When trade-offs inevitably arise, teams can return to that compass and ask: "Which value are we serving right now - and which one are we willing to compromise?"

You can't manage what you haven't made visible. The heatmap makes the invisible DNA tangible - a living reference point for decisions, rather than a forgotten slide deck.

Making Values Contractual

Values hold power when they shape systems, not just slogans. Here's how they can live inside the instruments of delivery - the everyday levers that define what success really means.

Briefs: Write value outcomes alongside technical ones. Instead of "deliver a bridge", say "deliver a bridge that restores walkability, biodiversity, and cultural connection". Add a short Statement of Values and Intent signed by all partners at project inception.

Contracts: Translate values into key performance indicators. For example: community satisfaction, local employment, biodiversity restoration, or whole-of-life carbon savings. Make them auditable, not aspirational.

Design Reviews: Include a Values Reflection line in every review form, asking: how does this design perform under our three lenses? Require a short Values Compliance Statement with each major submission.

Procurement: Use multi-criteria analysis to reward innovation, social impact, and collaboration - not just lowest cost. Include criteria for community benefit or cultural stewardship.

Operations and Maintenance: Add Legacy Metrics to asset performance audits: not just how well the structure functions, but how well it continues to serve people, place, and planet. Ask: are the original values still alive five years on?

Continuous reminders: Add values to the first item agenda of a meeting. Ask people how a value was demonstrated that week and acknowledge their team mates. Create visual representations around the project work space with values outlined.

When values are written into the formal systems - briefs, contracts, budgets, and audits - they stop being invisible ideals. They become part of the engineering language, and part of accountability itself.

Caution: Avoid Value Overload

Values should guide, not clutter.

A project can't serve every value equally - nor should it try to. Start with three to five that best express the project's purpose and context. More than that, and values risk becoming wallpaper: visible but unread. Fewer, and they lose nuance.

The goal isn't to have more values; it's to have clearer ones that everyone can recognise in action. Like a good compass, they only work if people can read them quickly in a storm.

Translate Values into Behaviour

A value only becomes real when it guides behaviour. This is where most teams stop - at the words. The next step is what brings alignment to life. When we say we value care, integrity, curiosity, community, or kindness, what does that actually look like on a Tuesday afternoon under project pressure?

The following table is great for workshops and project kick-offs. It helps teams describe what each value looks like when it's alive - and what it looks like when it's missing.

Value	When it's alive	When it's missing
Care	We pause to listen before making assumptions.	Decisions are made without consulting those affected.
Integrity	We surface bad news early and learn from it.	We hide errors or blame others.
Curiosity	We explore multiple solutions and perspectives.	We rush to defend the first idea.
Community First	We design with - not for - people.	Engagement is transactional or performative.
Kindness	We consider how decisions affect stress, wellbeing, and trust.	Pressure overrides empathy.

Teams can revisit this table during key milestones - design reviews, risk workshops, or gateway approvals - to check whether their behaviour matches their intentions.

Below are a few examples of how these values have been brought to life on real projects around the world.

Care - Hanlon Park/Bur'uda Waterway, Brisbane, Australia

When Brisbane City Council decided to restore a 1930s concrete drain into a living waterway, the easy route would have been to patch and reline the channel. Instead, they paused - and listened.

The team collaborated with Traditional Custodians from the Nunukul Yuggera people, local residents, and ecologists to reimagine the site. Over 460 new trees were planted, and the creek was reconnected to its floodplain.[355]

Care, in this context, was not sentimentality - it was systemic listening. By caring for water, ecology, and community, the project reduced flood risk and created a thriving public space.

When care is alive: The design reflects relationships - not just requirements.

Integrity - The Citicorp Centre Story, New York

In the 1970s, structural engineer Bill LeMessurier discovered a flaw in the Citicorp Centre tower's design that could have caused catastrophic failure in high winds. He worked with the building owners and city officials to retrofit the structure without public panic.

LeMessurier's decision to come forward - risking reputation but saving lives - remains one of the most powerful examples of engineering integrity.[48]

Integrity means surfacing bad news early and acting with transparency. It's about doing what's right, not what's easy.

When integrity is alive: Trust becomes your strongest material.

Curiosity and Courage - Co-designing with Communities, Uganda

A group of engineers and students in rural Uganda discovered that the most innovative designs often begin with questions, not answers. Rather than imposing technical solutions, they co-designed with local communities - learning how people used water, gathered resources, and maintained systems.

The result was infrastructure that was simpler, cheaper, and far more resilient because it fit the rhythms of local life.[356]

Curiosity led to courage - courage to admit not knowing, and to let others teach.

When curiosity is alive: You stop designing for people, and start designing with them.

Community First - Frauen-Werk-Stadt, Vienna

In Vienna, planners asked a question few cities had ever asked: "What would our city look like if it were designed by and for women?"

Through listening sessions, workshops, and on-site observations, they learned how daily mobility was shaped by caregiving, prams, groceries, and safety at night.

The results - wider pavements, better lighting, local childcare, and community gardens - benefited everyone.

Putting community first didn't slow the project down. It made it smarter.

When community comes first: Inclusion becomes an innovation strategy.

Kindness - OzHarvest, Australia

When OzHarvest - Australia's leading food-rescue organisation - first began reflecting on its internal culture, they realised something subtle but important: people were "nice", but not necessarily kind. Niceness smoothed things over. Kindness made things better. Niceness avoided discomfort. Kindness leaned into truth with care.

So they made a deliberate shift.

Instead of a culture built on politeness, they cultivated one built on kindness, curiosity, and honest feedback. Staff were encouraged to check in with one another, name stresses early, offer help without ceremony, and acknowledge small acts of care as part of everyday work - not as extra, but as essential. What emerged wasn't softness, but strength.

The change didn't stay within HR policy documents. It showed up in how teams collaborated, how challenges were surfaced, and how people responded to high-pressure moments. Independent reporting showed increases in engagement, stronger performance outcomes, and a healthier, more trusting workplace.[357, 358]

The lesson translates directly to engineering. Kindness isn't a distraction from performance - it's the cultural infrastructure that makes high performance sustainable. People do their best work when they feel respected, supported, and safe enough to think clearly, speak honestly, and collaborate generously.

When kindness is alive: People don't just complete their tasks - they look after each other. They notice. They lean in. And the work gets better because the people doing it can bring their full humanity to the table.

Anchor Values in the Project DNA

The final step is to build those values into the systems of the project - so they're not dependent on goodwill or memory. Values need to live where decisions live: in briefs, contracts, design reviews, procurement, and operations.

When values live in documents, budgets, and governance - not just posters - they become part of the engineering language. Below is how this can look in practice, with examples from real projects.

1. Briefs and Contracts: Setting Intent in Writing

When you set up a project brief, that's the DNA moment - it defines the system before it forms. Embedding values here means moving beyond outputs ("a bridge") to outcomes ("a bridge that connects communities and restores ecosystems"). It moves from the how and what to the why - a common element missing in our everyday language as engineers.

Example: The UK Government's Five Case Model for business cases requires not only financial and management justification, but also social, environmental, and strategic cases - a values-based shift from "how much?" to "why and for whom?"[359]

How to embed:

- Write KPIs that measure inclusion, trust, and sustainability, not just time and cost.
- Include a Statement of Purpose and Values signed by all partners at project inception.
- Ask, "How will success be defined in 50, 100 years, or even 250 years?"

2. Design Reviews: Reflection as Risk Management

Design reviews are the heartbeat of any project. They can become value checkpoints - not just technical interrogations.

Example: The Citicorp Centre crisis again reminds us that ethics and foresight matter as much as geometry.[48] Some organisations now include a Values Review in every design stage, asking:

- Does this align with our Care and Integrity principles?

- Have we tested this design through the three lenses (Wellbeing, Regeneration, Legacy)?
- Who isn't in the room that should be?

How to embed:

- Add a values reflection column in design review templates.
- Document lessons and share "ethical wins", not just errors.

3. Procurement Frameworks: Rewarding What We Say We Value

Procurement has long been the place where values go to die. The lowest-cost mindset often overrides everything else, but a few organisations are rewriting this rulebook.

Example: The City of Melbourne's Social Procurement Framework weights bids based on community benefit, sustainability, and social enterprise engagement.[360] Suppliers who demonstrate how their approach creates shared value score higher, even if their bid isn't the cheapest.

How to embed:

- Include evaluation criteria for collaboration, innovation, and social impact.
- Use "multi-criteria analysis" to reflect the true cost of poor design and the benefit of trust.
- Reward partners who contribute to legacy, not just delivery.

4. Operations and Legacy: Living the Values Beyond Handover

Operations is where values either endure or erode. A brilliant design loses its integrity if maintenance neglects the people or place it serves.

Example: Working on an existing tunnel that was nearing 20 years into its design life, I was responsible for leading the consulting team to create the Enhanced Asset Management Plan. When we started, the process for documenting maintenance was pen and paper. We created a digital maintenance and monitoring system to overhaul asset and legacy management - documenting learnings, identifying improvements, and embedding transparency across decades. We even took a digital photo of the entire tunnel to create a memory of where it was at that time for future operators to reference.

Example: At the Sydney Opera House, maintenance and renewal have been guided by a "living heritage" philosophy[361] - preserving not just the structure, but the culture and creativity it symbolises.

How to embed:

- Require legacy statements at project close-out, capturing lessons and future stewardship goals.
- Include "values in operations" audits - are care, community, and curiosity still alive five years later?
- Measure whole-of-life performance, not just delivery success.

Anchoring values in the project DNA is not about adding bureaucracy - it's about making meaning operational. When done well, this not only creates better outcomes for people and the plant. It also reduces risks and long term operational costs.

When care, integrity, curiosity, community, and kindness live in every review, clause, and handover, they shift from ideals to infrastructure. Ultimately, values are the specifications that shape everything we build.

From Alignment to Action

Teams that start with values alignment spend less time in conflict and more time in creativity. They build trust faster, communicate better, and make decisions more confidently.

Values alignment isn't about everyone agreeing - it's about everyone understanding what they're agreeing from, so they can cohesively work towards a common goal.

When we align our values before we align our plans, we create the conditions for the kind of engineering that lasts - not just in concrete and steel, but in trust, care, and memory.

That's how we begin to rewrite the DNA of our profession: together, transparently, and with purpose.

When we begin to see values not as words on a wall but as living systems, something profound shifts. A team that once debated cost versus time starts to ask the questions: "How does this decision care for people?" "What will this mean in 50 years?" "Are we proud of how this will be remembered?" Those questions mark the moment when engineering stops being just a profession and becomes a practice of stewardship.

The Takeaway

This chapter has shown that values are the invisible infrastructure behind every bridge, policy, and design review. They shape what gets measured, who gets heard, and how success is defined. When values remain implicit, they reinforce the status quo. When we make them explicit, name them, test them with stakeholders, and embed them into the documents that guide decisions, we begin to rewrite the DNA of engineering itself.

But values do not live in statements or frameworks alone. They only come to life through behaviour. They are expressed in the daily choices teams make: how setbacks are handled, how dissent is received, how urgency is balanced with care, and how trade-offs are navigated when there is no perfect answer.

That is where the next chapter begins. If this chapter was about alignment, Chapter 8 is about activation. Seeing our values clearly is only the first step. Practising them consistently is the real work. The next chapter translates these shared values into twelve working principles, practical habits that bring care, curiosity, integrity, and community into the meetings, drawings, and decisions where culture is built and legacy is shaped.

Chapter 8: The Twelve Working Principles - From Alignment to Action

Clarity about values is only the beginning. What ultimately shapes engineering outcomes is not what we say we believe, but what we consistently do, especially in the ordinary moments where decisions are made and culture is formed.

This chapter is about action. It bridges the gap between alignment and practice, translating shared values into habits that guide everyday engineering work. The Twelve Working Principles offer a practical way to bring care, curiosity, and integrity into the places where they matter most: meetings, designs, site decisions, procurement, and operations.

These principles are not an alternative to technical rigour. They build on it, expanding what good engineering looks like in a world facing planetary, social, and generational challenges. By operationalising values, they help ensure that every project, no matter its scale, becomes an expression of responsibility to people, place, and future generations.

From Values to Behaviour

Values sit in the foundations of every decision, often unnoticed, until they translate into action and behaviour. Principles are how that happens. They are the daily rituals, meeting habits, and design instincts that make values visible in practice.

This chapter is about activation - the translation of care, curiosity, and integrity into the ordinary moments where culture is formed: in the design meeting, the options report, the site inspection, or the procurement clause.

For decades, the engineering profession has had principles - safety, cost-efficiency, compliance, quality. They've served us well.

But the challenges we face now - planetary instability, loss of trust, human disconnection - require principles that extend beyond the purely technical. The world no longer needs engineers who can only make things work; it needs engineers who can make things work well for life and many generations into the future.

That's what these twelve working principles are for: to operationalise values so that every project, no matter its size, becomes a living expression of care for people, planet, and generations to come.

Each principle has three qualities:

- It's **practical** - something you can actually do or check for in your work.
- It's **systemic** - it applies across the project lifecycle, not just one stage.
- It's **relational** - it brings connection back into the core of engineering practice.

And because ideas only matter when they're embedded, each principle comes with a simple "where it lives" guide - showing how it can be embedded into projects, systems, and problem solving, and therefore carried through to operation. Of course there are endless ways to embed these principles and these are suggestions. Take what fits for you in your project, system, or context.

These are not commandments. They're not compliance tools. They're a set of working habits for a profession in transition - habits that build trust, reduce risk, and leave legacies we can be proud of.

The Twelve Working Principles

1. Co-create with people and communities to strengthen trust, reduce risk, and design with shared intention.

Engagement isn't a checkbox - it's design intelligence. When people and communities shape the process, the questions, and brief, the risks you usually fight downstream dissolve upstream. Co-creation builds legitimacy, strengthens trust, and reveals insights few models or manuals can show you. Consider using the word "communities" rather than "stakeholders" to create connection, as the latter can create an "us" and "them" separation.

2. **Learn from nature's systems to regenerate, restore, and redesign with life.**

Nature has already solved most of the problems we face - elegantly, efficiently, and without waste. Its systems show us how to design for interdependence, renewal, and resilience. When we learn from Country and natural cycles, we design with life, not against it.

3. **Use human experience and planetary health as core metrics alongside cost and time.**

Success is more than delivering on budget. It's improving people's lived experience and strengthening ecological health. Build metrics that capture safety, belonging, wellbeing, biodiversity, and future resilience - and let them guide decisions from concept to completion.

4. **Practice legacy thinking: design for future generations, not just today's demands.**

Every design is a time capsule. Your decisions will outlive you - sometimes by centuries. Build as though future generations are in the room, because they're the ones who will carry what we create. Completion is short-term. Contribution is legacy. Build not for completion, but for contribution.

5. **Build teams with diverse archetypes and perspectives - challenge, creativity, and follow-through.**

Diversity isn't just demographic - it's cognitive and cultural. Great teams need challengers, visionaries, finishers, and stabilisers. When these archetypes work together, we uncover blind spots, unlock creativity, and deliver solutions that stand up in the real world.

6. **Remember beauty is not optional - design must soothe, inspire, and foster belonging.**

Beauty is functionality for the human nervous system. It fosters pride, connection, and care. When spaces feel welcoming and calming, people treat them with respect and feel a sense of belonging. Beauty isn't a luxury - it's a design requirement for human wellbeing.

7. **Design to support nervous systems, not overstimulate them.**

Light, sound, flow, colour, texture, rhythm - every design decision affects how people feel in a space. Good engineering calms. Overstimulation and stress is a design flaw. Protect attention, safety, and comfort by designing for the full spectrum of human experience.

8. **Balance global collaboration with deep listening to local ecosystems.**

Global knowledge is powerful, but every place has its own story, climate, culture, and custodians. Listen to Country, First Nations people, and community. Let local wisdom shape the solution before you apply global expertise. Listen to place before applying a template.

9. **Be a guardian of public trust - legitimacy is earned through transparency and care.**

Engineering exists because the public believes we act in their best interests. That trust is fragile - and once broken, difficult to rebuild. Honour it through transparency, humility, and decisions guided by care, not convenience.

10. **Redefine engineering: not just making things work, but regenerating for future generations.**

Repairing damage is no longer enough. Our designs must restore ecosystems, strengthen social fabric, and create possibility for future generations. Regeneration is the new standard - for systems, communities, and hope.

11. **Count the real cost of bad design - what do we harm, forego or leave behind?**

Efficiency means nothing if it erodes wellbeing, destroys biodiversity, or fractures community and connection. Every omission has a cost. Expand your cost lens to include the human, social, and ecological impacts of design decisions - seen and unseen.

12. **Connect then lead with values and intention - purpose guides every design decision.**

Clarity of purpose transforms uncertainty into courage and enables people to push through even the biggest challenge. When you're anchored in your values, you design with intention, you question outdated assumptions, and you create outcomes that reflect who you are and what you stand for. Knowing why you build shapes how you build.

Applying the Principles on Projects: A Living Framework

These twelve principles form the bridge between values and practice. They are the actions that rewrite our inherited engineering DNA from the inside out - making care intentional, curiosity courageous, designs intentional for people and planet, and legacy something we design for, not stumble into.

Before we go any further, it's worth naming something openly: we all know we don't live in a perfect world.

Engineering is full of constraints, competing priorities, and projects that already feel stretched to capacity. The aim here isn't to overwhelm anyone with an impossible ideal or suggest that every project can - or should - transform overnight.

Instead, think of what follows as a living framework: guidance to explore, adapt, borrow from, experiment with, and grow into. Not a checklist. Not a mandate. Just a clearer picture of what becomes possible when we allow our work to evolve.

Engineering has never been static. It has always been a living thing - constantly evolving, adapting, and responding to new knowledge and needs. These principles are simply the next step in that evolution.

In the following sections, we'll explore where these principles could live across the lifecycle of a project, beginning long before the brief is written and continuing long after the ribbon is cut.

Remember, principles don't live in policy; they live in people. And every project, system redesign, or problem to solve is a chance to practice them. Even small shifts can change the trajectory of what we build.

We also know that a project is never truly finished. Once something is built, it begins its next life-operating, evolving, aging, adapting to the people and place it serves.

The same principles that shape good beginnings are the ones that sustain good guardianship.

Later in this chapter, we'll explore how these ideas extend beyond design and construction into the ongoing care, renewal, and regeneration of the systems we create - because engineering doesn't end at completion; it endures through stewardship.

Stage 0 - Project, System, or Problem Conceptualisation

Principles

- Connect then lead with values and intention, so that purpose guides every design decision
- Practice legacy thinking and plan for generations ahead
- Use human experience and planetary healing as metrics alongside cost and time
- Be a guardian of public trust in every project

Before a brief is written, a drawing is sketched, or a model is run, something triggers the brief to be created. We decide what we are really trying to change, do, and make happen. What is the point of what we are doing here?

Stage 0 is all about asking the right questions to powerfully set the intention. This is the phase that can often be overlooked because we are focused on receiving the brief, but it is important to acknowledge and sometimes question the conversations, meetings, and strategy documents that came before where we choose the problem, frame the opportunity, and set the intention. This is where we decide *why* the project exists, whose needs it serves, and what "better" actually means.

In traditional engineering culture, this stage can feel informal or "someone else's job" - left to planners, politicians, executives, or business cases. Engineers are often invited in once the decision is made: "Here's the project, now make it work." But as we've seen throughout this book, the biggest failures rarely come from calculations; they come from solving the wrong problem brilliantly.

This is why we also need engineers in the board room, strategic sessions, and leading businesses - so that they can have input here, as well as when the brief is received.

In every project, we want to make sure that we have a Stage 0. This is the moment to:

- **Set the intention** - What future are we trying to enable, not just what asset are we trying to deliver?

- **Understand the why** - What is the real need behind this request, and whose stories and data define it?
- **Create the values** - Which values will guide trade-offs when things get hard: speed, profit, safety, connection, Country, carbon, care?
- **Apply the principles** - Out of the twelve, which three to five will we actively design to in this project, and how will we know we're honouring them?

You don't need a perfect world or a blank slate to do this. Stage 0 isn't about rewriting a whole program; it's about being as rigorous with your intention as you are with your design. A few early conversations, a short values workshop, or a single page of project intentions and values can shift the trajectory of everything that follows.

When you receive a brief, it will be clear and obvious if this stage happened or not.

When it worked

Here is a real-world example of intention at Stage 0 reshaping an entire city.

As we have seen in previous chapters in the early 2000s, Medellín was considered one of the most dangerous cities in the world, with homicide rates reaching 381 deaths per 100,000 people in 1991. Vast parts of the city - particularly the hillside comunas - were disconnected physically, socially, and economically.

City leaders realised that infrastructure alone wasn't the problem. The failure was a lack of access, dignity, and connection.[83, 84, 85]

At Stage 0, the intention was defined clearly and powerfully: "Engineering must connect the people who have been historically disconnected." This wasn't a slogan - it became a design requirement.

Values Chosen at Stage 0

- Equity first: serve the most marginalised communities before the affluent ones.
- Access as dignity: transport should deliver opportunity, not just mobility.
- Integration over isolation: projects must link to education, health, and public space.

- Design for human experience: create calm, safety, and belonging in areas long associated with fear.

What they built from that intention

- Metrocable gondola system connecting remote, steep hillside communities to the city's metro.
- Outdoor escalators replacing 28-storey climbs, reducing travel time from 35 minutes to 6 minutes.
- Library parks combining public space, education, and greenery.
- Urban acupuncture projects - small but high-impact interventions targeting community needs.

Outcomes

- Travel times dropped by up to 80% for hillside residents.
- Employment accessibility increased by 30-40% for cable-served communities.
- Tourism grew by more than 400% over a decade.
- Homicide rates dropped by over 80% city-wide.
- Public trust increased dramatically - Medellín was named "Most Innovative City in the World" in 2013.

Most importantly, the projects worked because they stayed anchored to the Stage 0 intention: infrastructure as social integration and human dignity.

This wasn't perfect-world engineering. Budgets were tight, constraints were enormous, and risks were real. But the intention was clear enough to guide every subsequent decision.

When it didn't work - and what it taught

Brisbane's Clem Jones Tunnel (Clem7) stands as one of Australia's most significant examples of a project that failed not because of engineering capability, but more likely because of misaligned intention, incorrect assumptions, and cultural blind spots at Stage 0.

When conceptualised in the mid-2000s, the tunnel was positioned as a transformational north-south link intended to reduce congestion, shorten travel times, and stimulate Brisbane's growing metropolitan mobility. The business case projected 100,000+ vehicles per day soon after opening. This was a number borrowed heavily from modelling templates and datasets used in Sydney and Melbourne - cities where toll-road culture was mature and tunnel usage relatively normalised.[362]

The tunnel was needed for a growing city and a great solution to set up Brisbane for the future, but the assumptions and therefore the business model behind it failed.

I remember my friend from engineering and her husband, also an engineer, had invested in the RiverCity Motorway along with their parents. All lost their investment. These were smart people who understood engineering and Brisbane locals, but they were let down by incorrect assumptions and an organisation that didn't understand the local context.

The Potential Stage 0 misstep

For this project, it could have been that the "why" was built on a set of assumptions that didn't fit Brisbane's culture, behaviour, or transport patterns.

Instead of understanding the cultural reluctance of Brisbane drivers to use tolls, or the local history of preferring surface roads even when congested, demand forecasts were imported, not interrogated. Did the project team deeply explore what Brisbane drivers valued, how they made decisions, or how pricing elasticity would influence behaviour?

When Clem7 opened in 2010:

- Daily traffic volumes were less than one-third of those forecast in the first year.
- Instead of the expected 100,000+ vehicles per day, usage hovered around 30,000-36,000 vehicles per day.
- Revenue was insufficient to cover debt repayments.
- RiverCity Motorway, the operator, entered voluntary administration within 12 months.
- Investors, including many retail shareholders, lost approximately $1.1 billion collectively.
- The tunnel was later sold for about $618 million, a fraction of its construction cost.

All of this stemmed from an intention that was never clearly or locally defined. The problem Clem7 aimed to solve - "reduce congestion through a high-use toll tunnel" - was never validated against the cultural and behavioural reality of the city.

What it taught us

Clem7 is a textbook Stage 0 failure for engineers:

- Wrong assumptions wrong problem wrong business model.
- Imported models replaced lived knowledge.
- Behavioural insights were ignored.
- Values (cost sensitivity, distrust of tolls, preference for surface routes) were not explored.
- Success metrics (high initial demand) were never realistic for Brisbane's transport culture.

The engineering was world-class. The tunnel performs exactly as designed. The failure wasn't technical - it was related to project intention.

The Stage 0 "why" was misaligned, and the project never recovered.

What makes this stage unique

Stage 0 is the least visible phase but the most leveraged. It's where you make decisions that won't show up on a drawing but will echo through the calculations, procurement choice, and maintenance budget.

What makes this stage unique is that:

- You're not yet constrained by a specific solution. You still have permission to ask, "Is this the right problem?"
- You can choose the metrics of success before they choose you - adding human experience, Country, carbon, and regeneration alongside cost and time.
- You can define the values and principles that will guide trade-offs later, when time is short and pressure is high.

This is also the moment to acknowledge constraints honestly and have the courage to be curious and challenge the brief. Stage 0 isn't about imagining a frictionless world; it's about making the invisible

visible - the politics, legacy assets, community history, climate risks, cultural significance - so you can design with them, not around them.

If later stages are where you pour concrete and work out the "what" and the "how", Stage 0 is where you pour the "why" into the work.

Where to focus in Stage 0

You don't need a three-day offsite to do this. Even in a busy project environment, you can focus on a few high-leverage moves:

- Name the intention clearly
    - Write a one sentence intention statement that answers: What are we really here to change? For whom? Over what timeframe? Make it easy to remember so people can recall it easily throughout the project
    - Check it against your organisation's strategy, community commitments, and climate goals. If it clashes, pause.
- Interrogate the "why" before locking in the "what".
    - Ask: Why this project, in this form, now? What problem data and whose stories are we relying on?
    - Map who benefits, who bears the impact, and who isn't in the room yet.
- Co-create values and principles for this specific context
    - Choose 3-5 of the twelve principles that are most relevant to this project. Translate each into one practical behaviour or decision rule. (For example: "Use human-experience as a metric" becomes "We will run at least one design review focused solely on how this feels to users, not just how it performs.")
    - Agree how you'll measure them - even if the measures are qualitative at first.
- Define "non-negotiables" early
    - Identify the commitments you will not trade away later: e.g. a maximum heat-island threshold, guaranteed accessibility, or a defined community engagement minimum.
    - Write them into early scoping and governance documents, not just hopeful slide decks.

- Decide how you'll revisit intention
    - Set a cadence (e.g. at each major gateway) to ask: Are we still solving the problem we set out to solve? Has our context or our understanding of the "why" changed?

These are not extra tasks on top of engineering; they are good engineering applied upstream.

Reflection prompt

How could we be as rigorous with our intention and values as we are with our design and cost estimates?

Stage 1 - Brief & Concept Definition

Principles

- Co-create with communities to reduce risk and increase trust
- Connect with values and purpose
- Be a guardian of public trust

Every project starts with an act of translation - turning the vision and concept into a brief. It's usually the first moment where the DNA of the project is written.

In engineering, we've often treated the brief as a technical artefact: scope, budget, deadlines, risk. But in reality, it's also a moral document that encodes what we believe success looks like. The tension isn't between good and bad practice; it's between complete and incomplete definitions of value.

When a brief considers only cost, scope, and time, it produces clarity but can often miss connection and meaning. When it includes connection, care, and legacy, it creates alignment that lasts long after handover.

When it worked

The Sumida Hokusai Museum in Tokyo offers a powerful example of a brief defined not only by function, but by human experience. Before a single line was drawn, the project team - including architects, engineers, urban designers, and community groups - studied the site as a living system within one of Tokyo's densest wards. They mapped pedestrian flows, sun paths, visual corridors, and microclimates to understand how the building could become a connector rather than a blockage.[363, 364]

The resulting brief contained equal weight for technical requirements and emotional purpose. It called for:

- Seamless pedestrian permeability across the site
- Glare-free façades to avoid blinding surrounding residents
- Universal access and intuitive wayfinding
- A building that "reflects the neighbourhood back to itself" through mirrored geometry
- Disaster resilience and low-energy performance

Engineers used these intentions to guide the structural and environmental design. The museum's faceted aluminium panels reduce heat load and glare. The building's geometry creates natural wind corridors. The structure exceeds Japan's strict seismic standards.

As a result:

- The building achieved significant reductions in operational energy through passive façade performance.
- It met universal design benchmarks, improving accessibility for elderly and mobility-impaired residents.
- Post-occupancy surveys showed high levels of perceived safety, comfort, and ease of navigation.
- The museum rapidly became a civic landmark despite its modest footprint, demonstrating how a values-led brief shapes community identity.

The project succeeded not just because the plans were good, but because the brief captured purpose as precisely as it captured performance.

When it didn't work - and what it taught us

In 2014, Scotland opened the Edinburgh Trams network - five years late, massively over budget, and with a route far shorter than originally promised. Technically, the project succeeded: it delivered a functioning tram line. But taking a step back, the whole story reveals what happens when a brief is written with technical clarity but without community alignment, shared purpose, or a trusted narrative.

The original brief prioritised cost control, schedule, and political optics over community connection and system-wide value. Stakeholder engagement was minimal in the early stages and the brief masked complex risks - particularly the relocation of underground utilities, which was far more extensive than assumed. As these issues surfaced, trust eroded. Residents lost patience, businesses suffered, and public confidence collapsed.[365, 366]

What the inquiry found:

- The brief was incomplete, unrealistic, and inadequately informed by evidence or stakeholder insight.
- Risk assessments failed to consider ground conditions and utility complexity - the eventual relocation effort involved over 1,200 separate utility diversions, far beyond early estimates.
- Business impacts were underestimated. A local study recorded revenue losses of 20-50% for shops along the corridor during prolonged street closures.
- Public trust plummeted, requiring expensive re-engagement efforts later in the project.

In other words: the brief answered what and how, but didn't explore deeply the why or for whom. The misalignment echoed through the entire project. Teams battled over priorities. Decisions were reactive rather than intentional. The public narrative spiralled into frustration and, at times, anger. All because the project's "moral document" - the brief - encoded the wrong definition of success.

This story repeats around the world. Perhaps you've even seen it firsthand. What it teaches us is simple but powerful: clarity of intent isn't a one-time event. It's a continuous practice. A brief that is technically complete but emotionally and socially incomplete cannot hold a project steady through complexity. But a brief grounded in values can.

Success for many failed projects becomes measured in kilometres delivered, not trust maintained - and the gap between the two becomes the real cost to society, now and into the future.

What makes this stage unique

This is the phase where the foundations are laid - not just for the design, but for the foundation of the values that will shape every decision that follows. Arguably, it's the most important stage of the entire lifecycle, because this is where the project's DNA is written.

Here, the goal isn't simply to "engage more", but to embed trust, purpose, and care into the brief itself. It is the perfect time to practice embedding the values in all the documentation, conversations, and meetings and continually align to the project values and continually reflect on these.

This is where we define not only what we intend to build, but why it matters and how people should feel when they encounter it: calm, safe, connected, and considered. When the values are clear at this point, alignment becomes easier, risk reduces naturally, and the project gains a steady compass that guides it all the way through to delivery and stewardship.

Where to focus in Stage 1

Here are a few high-leverage actions that bring this stage to life:

1. Build the brief with communities, not for them

Hold early listening sessions before the brief is finalised:

- Ask residents, users, operators, and Traditional Owners what success looks like.
- Capture lived experience data (heat, safety, access, noise, flooding, delight).
- Translate these into requirements, not "nice-to-haves."

2. Write values into the brief explicitly

For each chosen principle, articulate:

- How it informs design decisions
- What behaviours it commits the team to
- How it will be measured

Example: "We design for human experience" "The brief requires at least two design reviews centred exclusively on user experience, accessibility, and emotional response."

3. Define success in human terms

Alongside cost, schedule, scope, include:

- Community trust indicators
- Access, safety, and wellbeing metrics
- Ecological or regenerative targets
- Cultural or local identity outcomes

4. Make the brief a living document

Revisit the brief at milestones:

- Are we still aligned with the intention?
- Has the community's context or needs changed?
- Are we drifting toward technical optimisation at the expense of purpose?

When the brief is treated as a moral as well as technical document, the entire project strengthens.

Reflection prompt

How does our brief describe success and fulfill on the intention - in human terms, not just technical ones?

Stage 2 - Options Analysis & Design Development

Principles

- Learn from nature's systems to repair, rebuild, regenerate
- Remember beauty is not optional
- Design to support nervous systems, not overstimulate them

By the time we reach design development, the conversation changes. The "why" becomes "how" and the potential for creativity is at its peak. This is where engineering becomes artistry - where we balance function, form, and feeling in the reality of a living system.

The old DNA of engineering taught us to solve for efficiency: "How can we make it work and optimise this?" The new DNA adds a deeper question: "How can it calm us, be adaptive, and be generous to the future?"

When it worked

The Øresund Bridge between Denmark and Sweden remains a landmark example of design development grounded in ecological intelligence. Spanning 16 km across a sensitive marine corridor, its designers faced two complex challenges:

1. Building a motorway and rail link across the Øresund without disrupting bird migration routes, and
2. Avoiding ecological harm to the shallow seabed and fish nurseries below.

Instead of default mitigation, the engineering teams adopted a regenerative approach. They shaped bridge piers and submerged foundations to function as artificial reefs, providing textured surfaces and cavities that support mussels, algae, and small fish. The submerged tunnel island, Peberholm, was intentionally left undeveloped so natural succession could occur unhindered.

The results:

- Biodiversity on Peberholm has increased dramatically, with more than 700 plant species now recorded - none artificially planted.
- Marine surveys show the bridge's foundations support growing populations of crustaceans, molluscs, and juvenile fish - turning structural elements into habitat rather than obstruction.
- Bird migration pathways remained intact, confirmed through multi-year monitoring programs.
- The project met stringent EU environmental conditions while delivering a resilient transport corridor used by tens of thousands daily.[367, 368]

This is engineering that heals while it connects, proving that design development can serve both nature and people through the same intervention.

When it didn't work - and what it taught

In too many projects, beauty and wellbeing are treated as optional - the first things trimmed when budgets tighten. But when we ignore human experience, the costs return elsewhere: in maintenance, in stress, in disconnection or infrastructure that isn't used.

We've all seen designs that functioned perfectly yet left people confused, fatigued, or alienated. These moments aren't failures of intelligence; they're symptoms of an incomplete definition of performance - one that misses the nervous system or a focus on people, not just the structure.

A clear example is Melbourne's Docklands precinct. During rapid development in the early 2000s, building orientation and massing were determined primarily by commercial yield, views, and floor space efficiency. Wind tunnel modelling for downwash and corner effects was limited in rigor, and the harbour-edge microclimate was not fully considered.

The result was a now well-documented wind tunnel effect, with gusts exceeding comfortable thresholds for pedestrians and far surpassing acceptable speeds for outdoor seating or cycling.

Measured consequences:

- A 2019 urban climate study recorded wind speeds in Docklands up to 4-6 times higher than comparable Melbourne streets.
- Certain laneways frequently exceed the Lawson comfort criteria, making them effectively unusable on windy days.
- Businesses reported reduced foot traffic and limited ability to use outdoor dining areas due to unpredictable gusts.
- Cyclists regularly avoid affected corridors for safety reasons.

The district "worked" as a property development, but failed as a liveable, comfortable neighbourhood. Ignoring natural forces - especially wind, shade, heat, and human sensory experience - created environments that were technically compliant but physiologically hostile.[369]

The lesson is simple: if we design in opposition to nature and human nervous systems, nature and nervous systems always win.

What makes this stage unique

This is where values meet physics. The principles here aren't abstract; they live in choices about materials, form, flow, and sensory impact.

Design reviews can ask:

- Does this system mimic, complement, or fight its environment?
- Does it create connection and comfort before it tries to create control?
- Does it invite calm, even in moments of movement or complexity?

In Singapore's Punggol Digital District, engineers and designers are applying this logic to an entire precinct. Cooling is achieved through a shared district energy system integrated with greenery and shade - not just for sustainability, but for human comfort and resilience in a warming climate.

Where to focus in Stage 2

Here are practical ways to embed the principles into design development:

1. Map natural systems before drawing solutions

- Study sun, wind, water, soil, shade, noise, and biodiversity patterns.
- Use these as design partners, not constraints.
- Incorporate regenerative gestures: habitat-supporting structures, water-sensitive design, microclimate moderation.

2. Treat beauty as a technical requirement

- Include sensory impact (light quality, material texture, shade, sound, colour) in design reviews.
- Allocate budget early to elements that create calm and belonging - not as add-ons, but as core performance factors.
- Evaluate public experience using walkthrough simulations or prototypes.

3. **Model human comfort with the same seriousness as structural loads**

- Wind, shade, heat, acoustics, and wayfinding all affect safety and usability.
- Use human comfort standards as design constraints.
- Test design iterations for nervous system impacts, especially in transport hubs, public spaces, and dense precincts.

4. **Design with "future repair" in mind**

Ask early:

- How will this design age?
- How will it feel after 10, 20, or 50 years?
- Does it enable renewal without complete replacement?

The most elegant designs are those that remain generous to future users and future environments.

Reflection prompt

Are we designing for control and compliance - or designing for calm and connection?

Have we compromised on values for short-term optimisation and efficiency?

Stage 3 - Procurement & Partnership Formation

Principles

- Balance global collaboration with local ecosystems
- Redefine engineering: not just making things work or better, but regenerating for future generations
- Count the real cost of bad design now and into the operational future - what do we harm or forego

Procurement is a place where values can quietly disappear. Perhaps this is because this is where commercials are particularly triggered. The systems designed to ensure fairness and control sometimes erase the very things they're meant to protect: creativity, trust, community wellbeing, and long-term stewardship.

I've been to many value engineering workshops which are not about embedding values like we have been discussing in the book, but rather optimising projects to create "value"'. Often in these meetings, it is more about cutting costs, saving money, and squeezing out efficiencies from the old DNA and extraction or productivity mindsets. There usually isn't a strategic document in sight here and it is assumed that everyone knows what the design intent is: we are building a road or upgrading an intersection to make it safer.

The challenge here isn't corruption or neglect; it's that our traditional systems and procurement frameworks, which are very clever and have helped us deliver extraordinary engineering feats, were created for an era of predictability and linear growth. Today's complexity - climate risk, social fragmentation, cultural nuance, generational accountability - demands flexibility, curiosity, and deeper listening.

It is easy to forget intention and a project's values here, as a new team is often brought in to deliver or they aren't carried through in the project documentation. Procurement, at its best, is not about buying outputs, but enabling outcomes that endure from the initial project stages right through to construction.

When it worked

Let's look again at the Øresund Bridge and the project's procurement strategy.[367] This bridge remains one of Europe's most successful cross-border collaborations - not just because the engineering was world-class, but because the procurement itself was engineered for long-term stewardship.

Instead of separating design, construction, and maintenance into fragmented contracts, both governments created a single, binational joint venture - Øresundsbro Konsortiet - responsible for the full lifecycle: design, construction, financing, and ongoing operation.

This meant:

- One team held fully integrated accountability for safety, performance, and whole-of-life cost.
- Designers worked directly with constructors and maintainers from the beginning.

- Decisions were judged not by lowest cost, but by 100-year durability, ecological performance, and maintenance efficiency.
- Procurement rewards were tied to long-term resilience, not short-term milestones.

The results:

- The bridge consistently outperforms traffic forecasts, carrying more than 20 million passengers annually.
- Structural health monitoring shows deterioration rates far below original projections, extending predicted service life.
- The governance model accelerated construction: completed ahead of schedule and within the joint venture's financial framework.
- The socio-economic benefits have exceeded early projections - including labour mobility and cross-border economic integration.

The Øresund procurement model is still studied today because it aligned contracting with intention, to build something that would serve generations, not just meet a milestone.

When it didn't work

The Sydney Light Rail project is an example of procurement misalignment creating long-term cost, conflict, and community frustration.

The contract used an availability Public Private Partnership (PPP) model that heavily transferred risk - especially utilities risk - to the private consortium. But the risk allocation didn't reflect reality: underground utilities were far more complex, older, and poorly mapped than the contract assumed.

Instead of fostering collaboration, the procurement framework created an adversarial environment where each party tried to protect themselves rather than solve problems together.

The result was cascading disputes, cost blowouts, and delays. A project originally costed at $1.6 billion ballooned beyond $3 billion. Contractors sued the state, local businesses faced years of disruption, and the community lost trust.[370, 371]

Measured impacts

- Construction delays of 12-18 months across major sections.
- Over 2,000 compensation claims from local businesses due to prolonged disruption.
- Legal dispute between Acciona and NSW Government: claim of $1.2 billion, settled only after extensive negotiation.

The problem was not engineering capability. It was a contract that:

- Assumed an unrealistic understanding of ground truth
- Financially punished collaboration
- Rewarded parties for avoiding risk instead of addressing it
- Embedded values of control rather than partnership

Technically, the procurement met the rules. But it failed the project, the people, and the city because it prioritised cost control over system understanding and partnership.

This project taught that when procurement is misaligned with the complexity of place, the real cost shows up later - in litigation, community harm, ecological impacts, and eroded trust. A contract that doesn't reward collaboration, curiosity, and shared problem-solving will punish them.

What makes this stage unique

Procurement isn't just paperwork - it's delivering on values in reality for people. Every clause, KPI, and risk allocation is a design choice that can shape behaviour. This is the moment where long-term values can be embedded or erased with decisions and actions on delivery.

Embedding the project's purpose here means asking:

- How will this contract reward collaboration - not competition?
- Does it measure long-term performance or short-term delivery?
- Are we buying innovation - or eliminating it?
- Does this take away from the original intention?
- Are we honouring the people who we are building this for?

- Are we honouring local ecosystems, community needs, and future generations - or optimising only for price?

A powerful example of procurement aligned with values is Scotland's Borders Railway, completed in 2015 - the longest new domestic railway built in the UK in over a century, reconnecting Edinburgh with communities that had lost rail access since the 1960s. What makes this project exceptional isn't only the engineering, but how procurement was used deliberately to deliver social, cultural, and economic regeneration alongside transport outcomes.[372]

When Transport Scotland released the tender, they embedded social value as a core evaluation criterion - not as an optional add-on, but weighted meaningfully within the commercial assessment. Bidders were required to demonstrate how they would:

- Create employment pathways for local residents, including long-term unemployed workers.
- Support apprenticeships and skills development, especially in regional communities with declining industry.
- Engage local businesses and suppliers rather than importing labour and capability.
- Minimise disruption to villages along the alignment and maintain early, transparent communication.

This shifted procurement from buying construction services to buying regional renewal.

The winning consortium (BAM Nuttall) delivered on these commitments:

- More than 900 people were employed over the course of the project, with a significant portion drawn from the Borders region.
- Over 100 apprentices, trainees, and graduate engineers received placements that became pathways into long-term industry careers.
- The project invested £89 million directly into local supply chains, strengthening small and regional businesses.
- Community engagement began early, with drop-in events, school outreach, and clear communication about the program and its impacts - fostering trust rather than resistance.

Despite its complexity, the line was completed ahead of schedule and opened in September 2015. Passenger numbers exceeded forecasts immediately - more than one million journeys in the first year alone, nearly double the original projection.[373]

But perhaps the most telling indicator was that public confidence in rail grew to such an extent that discussions began almost immediately about extending the line further south toward Carlisle - a direct outcome of a project that not only delivered infrastructure, but delivered trust.

The Borders Railway shows what becomes possible when procurement is used not to extract value, but to create it - socially, economically, and generationally.

Where to focus in Stage 3

Here are practical, high-leverage actions for embedding values into procurement:

1. **Write ecological, cultural, and social principles into procurement documents**

- Require bidders to demonstrate how they will partner with local ecosystems, communities, and Traditional Owners.
- Include regenerative performance criteria - not just "do no harm".

2. **Align risk to reality**

- Map uncertainty early: utilities, ecological systems, ground conditions, cultural heritage
- Allocate risk to the party best equipped to manage it - not the one easiest to offload onto.
- Incentivise joint problem-solving instead of claims culture.

3. **Reward collaboration, not low price**

- Include metrics for team behaviours, culture, openness, and shared learning.
- Use procurement scoring that values partnership capability as much as technical capability.
- Apply alliance or relational contracting models where appropriate.

4. **Make whole-of-life value visible**

Ask:

- How will this asset perform in 30, 50, or 80 years?
- Does this procurement decision increase long-term stewardship or undermine it?
- What maintenance burden are we creating or preventing?

5. **Keep the project intention alive**

- Present the intention statement and values at bidder briefings.
- Require bidders to respond directly: "How will your approach honour these values?"
- Embed intention-based KPIs into the contract.

Reflection prompt

What does our procurement framework teach people to optimise and value - trust, collaboration, or cost?

What behaviour does this contract make inevitable?

Stage 4 - Delivery & Construction

Principles

- Build teams from diverse archetypes
- Practice legacy thinking and plan for generations ahead
- Be a guardian of public trust in every project

A huge amount happens in delivery. It is where the project biomes delivery and the moment where engineering meets human nature. Even the best designs depend on how people collaborate under pressure - how they decide, communicate, and course-correct when things go wrong.

You know what it's like when unexpected things pop up on site - when a new risk appears, the quality of materials is lower than expected, the government changes, or a key stakeholder suddenly changes their mind. In many ways, this is where values are stress-tested.

A common theme I see with a lot of early career leaders in construction is how much time they spend putting out fires and reacting rather than proactive tasks that are important but not urgent. Over time, we see them shift from a reactive to a proactive state. This mindset is crucial if you want to be able to actually embed these principles in delivery and construction.

Traditionally, delivery culture has rewarded control, precision, and speed. Those instincts are critical - yet they can overshadow empathy, inclusion, and reflection. A "firefighting" mindset solves today's problem, but often ignores tomorrow's consequences.

Legacy thinking asks: what will this decision look like in 10, 20, or 50 years, when the public inherits it? It reminds us that delivery doesn't just build structures; it builds reputation, legitimacy, and trust. This can be exceptionally difficult as often projects don't reward this behaviour. Instead, they reward the old DNA of cost cutting and efficiency - not the new DNA we need.

When it worked

Crossrail - one of Europe's most ambitious transport projects - involved 42 km of new tunnels travelling beneath some of London's most sensitive heritage buildings, live rail corridors, and geotechnically fragile zones. The technical complexity was staggering: 10 new stations, enormous underground caverns, +£20 billion in investment, and a construction environment involving more than 10,000 workers and hundreds of subcontractors at peak.[374]

What distinguished Crossrail was not only its engineering sophistication but its delivery culture. The project explicitly required every contractor to participate in a shared learning ecosystem:

- A centralised lessons-learned platform captured innovations, near misses, engineering insights, and constructability improvements in real time.
- Weekly multidisciplinary learning forums brought geotechnical experts, tunnellers, safety specialists, and project managers together to reinterpret emerging risks.
- Cross-contractor collaboration was mandatory, not optional. If one team learned how to reduce settlement risk near historic foundations, every other team received the insight within days.

- More than 500 formal lessons and thousands of field-level innovations were shared across the program, accelerating safe tunnelling and reducing duplication of errors.

This wasn't compliance - it was institutionalised curiosity. The outcome?

- Crossrail achieved an exceptional safety performance, with a lost-time injury rate far below industry benchmarks.
- Complex station caverns such as Liverpool Street and Whitechapel were delivered with remarkable precision, despite the constraints of Victorian-era tunnels and live tracks above.
- The project created one of the world's most detailed digital engineering knowledge bases - now used by megaproject teams globally.

Crossrail's legacy is more than a rail line. It is a model showing that when delivery teams are diverse, communicative, reflective, and connected by purpose, complexity becomes manageable - and sometimes even elegant.[375]

When it didn't work - Boeing 737 MAX - when delivery culture overshadowed diverse voices

The Boeing 737 MAX tragedy remains one of the clearest modern examples of what happens when delivery culture privileges speed, schedule, and internal pressure over diverse perspectives, psychological safety, and legacy thinking.

As Boeing raced to deliver the MAX on an accelerated timeline - competing with Airbus and facing commercial pressure - multiple engineers raised concerns about the MCAS flight-control system, single-sensor dependency, pilot training assumptions, and insufficient redundancy. Yet internal communication structures and cultural norms meant these concerns were muted or dismissed.[376]

Investigations found:

- Engineers felt unable to escalate safety concerns without reprisal.
- Internal emails revealed pressure to "fit the schedule" above all else.
- Safety-critical design changes were not transparently shared with regulators.

- The organisation lacked empowered "challenger archetypes" - those who ask difficult questions, anticipate failure modes, and widen perspective.[377]

The result was catastrophic:

- Two crashes in Indonesia and Ethiopia
- 346 lives lost
- A 20-month global grounding of the aircraft
- Billions in financial losses
- A fracture in public trust that will take generations to repair[378]

The lesson is stark. When dissenting voices are sidelined, when curiosity is punished, when delivery becomes a race rather than a responsibility, engineering loses its moral centre. And when commercial pressure outruns engineering integrity, the cost is counted not just in money - but in lives, legitimacy, and trust.

What makes this stage unique

Construction is the moment of truth where leadership becomes visible and values become behaviour. It's also where public trust can be won or lost in real time.

Embedding the principles here means:

- Encouraging teams to name tensions early rather than conceal them
- Celebrating integrity, curiosity, and adaptability as performance metrics
- Recording why decisions were made, not just what was done
- Ensuring diverse archetypes are empowered, not sidelined
- Choosing actions today that a future engineer could proudly explain

The Sydney Opera House Renewal Program embodies this mindset. After decades of maintenance challenges, engineers now work alongside historians, Indigenous advisors, acousticians, and conservation specialists - ensuring upgrades honour the original intent while adapting to modern performance and future needs.

Delivery is the phase where intentions either hold - or silently fall away. It's where pressure is highest, complexity is most visible, and the human element can make the difference between a project that simply gets built and one that is remembered with pride. Even small habits here can reinforce the principles and anchor the project to its purpose.

Here's where to focus

1. Strengthen team composition - and protect the challenger archetype

Make space for diverse thinkers at the table: the detail-minded, the risk-hunters, the vision-holders, the connectors, the cultural navigators. Delivery culture often unintentionally silences the very people who see emerging risks early.

A simple practice is to name who is playing each archetype on the team, and invite each person explicitly in meetings. This normalises diverse input and prevents narrow decision-making under pressure.

2. Build a "name it early" culture

Create psychological safety by encouraging people to raise concerns *before* they crystallise into issues. Two high-leverage rituals:

- Daily 60-second tension check: "What's one thing making you uneasy today?"
- Weekly assumption review: "Which assumptions no longer feel true?"

When teams feel safe to name concerns, risks surface early, solutions come faster, and trust increases.

3. Make legacy part of every decision, not just documentation

Delivery often narrows focus to deadlines, defects, and daily coordination. To keep legacy thinking alive, ask at every major decision point: "What will this look like in 10, 50, or 100 years - and will someone thank us for this choice?" This can shift decisions on materials, sequencing, temporary works, and finishes without adding cost - simply by reframing what "good" looks like.

4. Keep a clear, simple decision rationale log

Not a compliance document - a cultural tool. Capture:

- What decision was made
- Why it was made
- Which values or principles guided it
- What trade-offs were accepted

This helps future engineers, operators, and maintainers understand the story of the asset. It also protects public trust by showing that decisions Ωwere thoughtful, accountable, and aligned to intention.

5. Centre community and worker wellbeing - even when it's inconvenient

Delivery environments can be overwhelming: lights, noise, shifting schedules, public disruption. Two simple practices are to:

- Hold a "community lens" review weekly: "If I lived here, how would this feel today?"
- Protect sensory wellbeing onsite: quiet areas, light management, and predictable routines for workers and clear communication for residents.

Calm construction creates calm communities - and earns trust without needing a single extra dollar in budget.

6. Reward behaviours that strengthen trust, not just those that hit KPIs

Often the unsung heroes of delivery are the people who:

- Walk across site to resolve issues face-to-face
- Call a resident to explain a noisy night shift
- Document a lesson that prevents a future mistake
- Pause work to protect a culturally significant feature

These actions are rarely captured in dashboards, yet they shape reputation and legacy. Build them into performance reviews, toolbox talks, and team celebrations to make them visible and valued.

7. Protect time for reflection, learning, and knowledge transfer

Even in the busiest stages of construction, five minutes of reflection per week changes outcomes. Ask:

- What surprised us this week?
- What did we learn that another team could use?
- What principle guided a decision - or should have?

This maintains alignment with intention and turns delivery from firefighting into adaptive, thoughtful engineering.

8. Keep the purpose visible - literally

Re-display the Stage 0 intention and Stage 1 values in site offices, briefing packs, and meeting rooms. When physical reminders of purpose disappear from the environment, they quickly disappear from decision-making too.

This stage is where care becomes visible - in how teams communicate, how issues surface, how decisions are justified, and how people are treated. You don't need to overhaul your whole delivery system to embed the principles. A few deliberate habits, repeated consistently, can shift a project from "delivered" to "stewarded".

Reflection prompt

What part of today's decision will tomorrow's engineer - or tomorrow's community - have to explain?

Stage 5 - Operation, Handover & Legacy

Principles

- Practice legacy thinking and plan for generations ahead
- Use human experience and planetary healing as metrics alongside cost and time
- Connect with values and purpose

Operation is often treated as the final phase - but in reality, it's the beginning of a project's real life. This is where engineering leaves the hands of makers and enters the daily rhythm of cities and communities.

Systems breathe, age, and adapt. Legacy is not about what survives; it's about what continues to serve - in human, ecological, and cultural terms.

When it worked

The High Line in New York is a remarkable case of reinvention. Once an abandoned rail line, it was reimagined through collaboration between engineers, landscape architects, ecologists, and local residents. The result wasn't merely a park - it became an ecosystem of social connection, biodiversity, and economic renewal. Engineers measured success not only in structural stability but in belonging, safety, shade, and wellbeing.[208, 209]

Another example is the Roman aqueducts, still standing after millennia. Their endurance isn't just material; it's philosophical. The Romans designed for inspection, repair, modularity, and reuse - principles that remain as relevant today as they were 2,000 years ago.[217]

When it didn't work

Grenfell Tower, which we have already discussed in this book, is a clear example of how operation and legacy fail when long-term thinking, community voice, and transparent maintenance are sacrificed.[31, 32]

The tower's original structure (1974) was robust. The failure was not in the initial engineering but in decades of governance breakdown, poorly designed operational manuals, and a retrofitted solution gone wrong. What happened:

- Resident safety concerns repeatedly ignored
- Cost-driven refurbishment (2012-2016) introduced highly flammable cladding
- Fire compartmentation issues left unresolved
- Insufficient holistic safety review, despite known vulnerabilities

The result was catastrophic: 72 people died, a national reckoning unfolded, and trust in housing authorities and regulatory systems was profoundly damaged.

Grenfell teaches us that:

- Compliance is not care.
- Operation is never passive.
- A building can be functional and still fundamentally unsafe.
- Legacy is determined by the accumulation of stewardship decisions - or the absence of them.

Another example are the many deindustrialised power plants left without closure planning. Across the world, hundreds of coal and gas power stations have been decommissioned without regenerative closure. Research from the IEA and global environmental regulators shows recurring issues:

- Unlined ash ponds leaking heavy metals into waterways
- Asbestos and hazardous materials left sealed but not remediated
- Communities excluded from the planning of future land uses
- Sites left economically stranded instead of being transformed into renewable hubs, public spaces, or innovation precincts

These failures impose enormous financial liabilities. The US EPA estimates more than $100 billion in unfunded remediation obligations across aging thermal plants.[379]

When regeneration is not planned early, the operational legacy becomes an environmental and cultural debt passed to future generations. It begs the question: do these designs even stack up in the first place, even on a basic cost benefit ratio when we calculate the cost to people and planet after their life has come to an end?

What makes this stage unique

This is where stewardship replaces ownership. When legacy principles are embedded early, operations teams inherit not just an asset, but a philosophy - a coherent story of why it was built, who it serves, and what future it's meant to enable.

That can look like:

- A legacy statement handed over with the as-built documentation
- A digital twin that tracks human experience and ecological performance over time
- A governance model that keeps communities involved after the ribbon-cutting
- Clear visibility of the building's "nervous system" - how it responds to light, heat, flow, and human use

Global parallel

In Amsterdam's De Ceuvel, a former shipyard polluted with hydrocarbons, heavy metals, and industrial waste has been reborn as a circular innovation precinct. Every building is designed for eventual disassembly and every material chosen for reuse. Even the landscape itself performs phytoremediation, slowly cleansing the soil through carefully selected plants.[380] Operations here are an act of regeneration - not just upkeep. The site demonstrates what it looks like when legacy isn't an afterthought, but a design constraint from day one.

Where to focus in Stage 5

Operation is where decisions echo the loudest across time. You don't need to overhaul entire systems to embed the principles - just a few intentional practices that honour stewardship, people, and place.

1. **Begin with a "legacy handover", not just a technical handover. Include:**

- The project intention from Stage 0
- Design principles and values
- Decisions that shaped the asset
- Known vulnerabilities and future considerations

It ensures the operations team inherits not just drawings, but wisdom.

2. Measure what people feel - not just what the asset does

Track human-experience indicators:

- Shade and comfort
- Perceived safety
- Accessibility
- Noise, heat, flow
- Emotional resonance

If an asset feels good to use, it will be used - and protected.

3. Treat operation as continual renewal, not slow decline

Schedule proactive upgrades that extend life, regenerate ecology, and enhance wellbeing. Ask quarterly: how can this asset give more than it takes this year?

4. Keep community dialogue alive

Hold an annual community review or "state of the asset" session. Invite residents, users, Traditional Owners, and local businesses. Ask:

- How is this serving you?
- What's not working?
- What possibilities do you see?

Legacy grows when people feel part of the story.

5. Use maintenance as an opportunity for ecological repair

When resurfacing, replanting, upgrading lighting, or renewing stormwater systems, look for regenerative options:

- Permeable pavements
- Native plantings
- Shade trees
- Biodiversity corridors
- Energy-efficient retrofits

Small choices, repeated over years, shift entire ecosystems.

6. Document the lived behaviour of the system

Capture not just faults or failures but patterns:

- How the site is really used. Does this differ from what was originally intended?
- Where people naturally gather
- Where they avoid
- How weather shapes experience

This is a great learning opportunity and becomes invaluable intelligence for future upgrades and the future purpose of the asset.

7. Honour the story of the place

Keep interpretation, cultural narratives, and historical memory visible. Places with meaning are cared for more deeply and more consistently.

8. Protect the future by making today's decisions reversible

Where possible, choose materials, assemblies, and systems that can be adapted, repaired, or replaced without major disruption. Reversibility is a gift to tomorrow's engineers.

Reflection prompt

If this project could speak in 50 years, what would it thank us for - and what would it question?

Practical Pathways: What Renewal Looks Like in Practice for Maintaining Assets Within Design Life

A few years ago, my husband and I had the opportunity to visit the Sandhurst Leadership Training Academy in the UK - a place where the British Army has trained leaders for more than 250 years. The director of the academy shared a story that stayed with me.

He said: "People often think we train soldiers just for battle - specifically the last hundred metres of battle, when they need to use everything they have got in them. We do train them for this. But actually, we train them for peacekeeping - for the years between conflict. Most of a soldier's life is not spent in combat but in community - building, maintaining, and protecting peace."

I've often thought about that story in the context of engineering.

We focus so much of our time, attention, and resourcing on designing and building the project - our last hundred metres - that we forget the preparation going into the race, the intention values and purpose, the system we are in, and everything we have discussed in this book. And we often forget the marathon that comes after: maintaining the asset, caring for it, learning from it, and ensuring it continues to serve.

Like the army, engineering isn't just about readiness for the moment of action; it's about stewardship in the long quiet that follows. Every bridge, tunnel, or power station spends most of its life not being built - but being used, cared for, and adapted. And yet, our systems and metrics rarely reflect this.

If design and construction is the battle, maintenance is peacekeeping. It's where we prove the integrity of our design - not just its strength but its spirit. The true measure of good engineering is not how well it begins, but how gracefully it endures.

Beyond Projects: Engineering as Stewardship

Most of the time when we talk about engineering, we talk about projects. New things being built. Bridges rising and tunnels dug. Buildings topped out and systems commissioned.

But when I think about the engineers I've known - the ones who really made a difference - their legacy wasn't just in what they made. It was in what they maintained.

Engineering isn't just about making; it's about maintaining the conditions for life. Projects are only one moment in a system's story. Once something exists, engineers become its stewards - maintaining, adapting, and sometimes letting go. We are custodians of the built world, just as First Nations peoples have always been custodians of Country.

In Aboriginal and Torres Strait Islander cultures, care for Country isn't about ownership or extraction - it's about ongoing relationship, reciprocity, and renewal. It's knowing that what you inherit, you must also care for. That idea - of custodianship - is something we can all learn from and carry into how we engineer.

A few years ago, I worked on a project that changed how I saw this. It was one of the tunnel projects I've worked on in an Australian capital city - a huge, complex piece of infrastructure built decades ago. By the time I was brought in, the tunnel was over twenty years into its design life.

On paper, it was performing exactly as intended: structurally sound, reliable, efficient. But what fascinated me wasn't the physical system - it was the human system around it. Maintenance logs were still being kept on bits of paper, problems were tracked in people's heads, and knowledge was being passed along through informal conversations rather than systems. It was a reminder that legacy isn't just about concrete and steel - it's about information, memory, and culture.

Our work on that project was to digitise those processes, to build a record that could be handed on to future engineers who would never meet the original designers. It was, in a small way, a bridge between generations, filling the knowledge gap.
That experience made me realise something: every project eventually becomes someone else's inheritance. And yet, so much of our professional language still ends at handover.

We talk about design life as though it's an expiry date. But the reality is, most assets live far beyond their intended lifespan. Bridges, tunnels, water networks, and energy systems - they're all part of an evolving ecosystem.

The question isn't just how long will it last, but how will it live?

The same twelve principles that shape new projects also guide how we care for existing ones - but they take on new expressions as time passes. For assets still within their design life, they remind us to sustain performance while deepening value: to operate with care, integrity, and curiosity. For assets nearing or beyond their design life, they guide us to adapt, renew, or retire with dignity - to think of regeneration not just in environmental terms, but in social and cultural ones too.

In this way, engineering becomes an act of long-term stewardship. Our job isn't to control the built environment - it's to help it evolve gracefully. Just as nature renews itself through cycles of growth, decay, and rebirth, our systems too must be designed and maintained to breathe, heal, and continue serving life.

So in this next part of the book, we'll look at how these working principles apply across the whole engineering life cycle - from conception and creation, through to operation and maintenance, and adaptation and legacy.

Because the real test of great engineering isn't what we build. It's what endures - and how we care for it when it does.

Applying the Principles: Maintaining Assets Within Design Life

A lot of engineering work today isn't about building new things. It's about caring for what already exists. It's not usually the most glamourous project, or the one that goes viral on LinkedIn and gets a media release or the headlines.

But these assets matter greatly to local communities. So many of these are hidden in our daily life that we don't notice them until they stop working - when a storm turns off power, heavy rainfall caves in a road, a fallen tree stops access, or the swings are broken at the playground and it takes weeks to fix them.

Every asset - whether it's a bridge, water network, hospital, or transport corridor - moves through long periods of maintenance where small decisions shape its long-term legacy.

When systems are new, it's easy to celebrate their innovation. But it's in maintenance - the quiet, unglamorous years that follow - where our values are truly tested. Do we listen to the operators who know the system best? Do we upgrade ethically, or only when failure forces our hand? Do we learn from the small fractures before they become structural?

Maintaining assets within their design life is an act of respect - for the people who built them, for the public who use them, and for the generations who will inherit them next.

Here's what each of the Twelve Working Principles looks like when applied in this stewardship phase.

1. Co-create with communities to strengthen trust, reduce risk, and design with shared intention

When assets are operational, co-creation shifts from public consultation to continuous dialogue. Communities are not just users; they are living sensors of performance.

When it worked: Copenhagen's cycling network is guided by one of the world's most consistent feedback systems: the Copenhagen Bicycle Account, a biennial survey that gathers detailed input from residents and cyclists.

The city asks riders about lighting, surface quality, comfort, safety, and maintenance needs, using the insights to prioritise upgrades and redesigns across the network.

This ongoing feedback loop has helped Copenhagen respond quickly to issues and direct investment where it matters most, contributing to measurable rises in cyclists' perceived safety and satisfaction over time.[381]

This consistent, transparent process has strengthened trust: people see that their experiences directly shape the network they rely on.

In the literature on infrastructure and community conflict, one recurring pattern is clear: when communities are excluded from decisions - or when communication is opaque - even technically sound projects can be derailed by distrust.

For example, a 2022 case study review of community opposition to construction and demolition waste facilities in Australia found that "lack of communication or exclusion of the public in the decision-making process" was frequently the key factor driving opposition - even when environmental and technical risk assessments were favourable.[382]

The report shows that opposition is rarely rooted only in objective risk; instead, it often stems from how decisions are made and communicated. When people don't feel included - when information is withheld or the process feels remote - the result tends to be outrage, distrust, and refusal to accept even safe, rational solutions.

In other words: the problem isn't always the engineering. It's the failure to treat people as partners. Without transparent dialogue and meaningful participation, engineering becomes a force of exclusion rather than inclusion.

This is why the principle you champion - co-create with communities - matters so much. It isn't just a nice add-on. It's the difference between building trust and building resistance; between long-term success and avoidable failure.

2. Learn from nature's systems to regenerate, restore, and redesign with life.

Nature doesn't maintain by replacement - it maintains by renewal. Regenerative maintenance applies the same mindset: use each intervention to deepen resilience rather than just "reset" the asset.

When it worked: In Singapore's Bishan-Ang Mo Kio Park, a straight concrete drainage canal carrying the Kallang River was transformed into a meandering, naturalised waterway as part of the ABC Waters Program. Engineers and landscape architects replaced hard canal walls with bio-engineered riverbanks, wetlands, and floodplains that still manage stormwater and prevent flooding, while dramatically increasing biodiversity and public amenity.[281, 282, 383] Here, maintenance of flood capacity became an opportunity to restore habitat, cool the city, and reconnect people with water.

When it didn't: For decades, the Los Angeles River has been treated primarily as a concrete flood control channel, with vegetation routinely cleared to maintain capacity and non-native plants mechanically removed to keep flows efficient.[384] While this approach reduces short-term hydraulic risk, it has come at the cost of habitat loss, urban heat, and a river most residents can't safely or pleasantly engage with. This has prompted a new generation of plans and pilot projects now trying to "re-green" what earlier design and maintenance practices stripped away.

Nature doesn't separate form from function. Neither should we.

3. Use human experience and planetary health as core metrics alongside cost and time

Maintenance affects daily life - noise, dust, access, safety, sleep. If we don't measure those impacts, we end up optimising for the easiest spreadsheet, not the real experience of people and planet.

When it worked: On Tokyo's metro, upgrades like the long-running renewals of the Ginza Line are deliberately scheduled overnight. Track relaying and resignalling are carried out in a narrow window after the last train and before the first, so daytime services remain reliable.[385] This is part of a culture of frequent maintenance and continuous improvement that underpins Japan's reputation for extremely punctual, safe and clean rail systems.[386] Reliability, safety, and rider experience are treated as core performance metrics, not "nice-to-haves".

When it didn't: Across Australian cities, residents regularly complain about months of night-time roadworks that bring jackhammers, floodlights, and detours to their front doors. Councils like the City of Sydney have had to issue detailed construction noise guidelines because unmanaged night works have left communities exhausted, stressed, and angry - even when projects meet their technical KPIs.[387]

The lesson is simple: if a project succeeds on paper but frays people's nerves and wellbeing, the metrics were wrong. If maintenance causes harm, it's not maintenance; it's erosion.

4. Practice legacy thinking and plan for generations ahead

Legacy during design is visionary. Legacy during maintenance is discipline - the unglamorous, ongoing work that extends the life of what we've built.

When it worked: The Sydney Harbour Bridge, opened in 1932, is kept alive by almost continuous preventive maintenance: inspection, repainting, steel repair, and documentation. Recent plans include an eight-year repainting campaign and rust treatment as part of preparations for the bridge's 100th anniversary, backed by an annual maintenance budget of around AUD $40 million and a dedicated team of specialists.[388] Together, they've allowed the bridge to safely carry hundreds of thousands of vehicles and trains every day, decade after decade.

When it didn't: Paradise Dam in Queensland was completed in the mid-2000s using roller-compacted concrete. Within years, investigations found long-term concrete strength issues and safety concerns serious enough that the dam wall had to be lowered and storage reduced. The initial build cost was about $200 million, but the rebuild is now expected to cost at least $4.4 billion as the wall must be effectively replaced to restore safety and water security.[389] Cutting corners upfront simply shifted a much bigger bill - financial, social, and political - onto future generations.

5. Build teams from diverse archetypes

Maintenance often rewards the Operator archetype - steady, dependable, risk-aware. But long-term care also needs Creators, Challengers and Connectors who question routines, bring new ideas, and broker perspectives across disciplines and communities.

When it worked: The Netherlands' Room for the River program is a powerful example of multidisciplinary engineering in practice. Instead of relying solely on higher dikes, Dutch teams brought together hydraulic engineers, landscape architects, ecologists, planners, and communities to widen floodplains, relocate levees, and create multi-functional river landscapes that improve safety and amenity.[390, 391] That blend of technical rigour, ecological insight, and social perspective has become an international reference point for integrated flood-risk management.

When it didn't: In many infrastructure organisations, reviews of asset-renewal programs show the cost of sticking too long with "we've always done it this way". In London, for example, Transport for London's own planning documents note that underinvestment in critical renewals leads to an accumulating backlog and future reliability problems, even when operations teams are doing their best within existing routines.[392] When only one mindset dominates - optimise today, defer tomorrow - maintenance becomes preservation of the past, not preparation for the future.

I've anecdotally heard this to be true for the Melbourne City Loop. This critical piece of infrastructure is at capacity, and rail engineers I have spoken with say the issues relate to the upgrading of signals. However, it is very challenging to get public approval for this and my engineering friends claim this is why the upgrades haven't happened.

6. Remember beauty is not optional

Even during maintenance, beauty matters. It shapes whether people feel pride in a place or turn away from it.

When it worked: New York's High Line transformed a disused elevated freight line into a linear park now home to more than 500 plant species, many of them native. The City owns the structure, but Friends of the High Line - a non-profit partner - is responsible for daily operations, horticulture, and maintenance, employing specialist gardeners and rangers to care for the planting in all seasons.[393] Maintenance here isn't just about safety; it's about keeping an evolving, wild-yet-crafted landscape that seven million visitors a year feel proud to inhabit.

When it didn't: We've all seen the alternative: pedestrian bridges and underpasses left to peel and rust, covered in tagging and grime, with broken lights and no sense of care. Technically, they may still "work" as crossings. But they communicate neglect - people feel less safe, less welcome, and less inclined to look after them. When beauty is treated as optional, the social fabric around an asset unravels.

7. Design to support nervous systems, not overstimulate them

Light, sound, flow, texture, rhythm - every design and maintenance decision affects how our bodies feel in a space. We have discussed in previous chapters good engineering calms while overstimulation is a design flaw.

When it worked: Maggie's Centres - a network of cancer support centres in the UK and beyond - are designed around a simple idea: architecture and landscape can reduce anxiety and restore a sense of dignity. Their buildings use natural materials, soft lighting, domestic-scale spaces, and therapeutic gardens. Research on Maggie's has found that people experience the centres as non-clinical, welcoming, and relaxing environments that help them cope with the emotional load of cancer.[96, 394] These are nervous system-aware spaces, designed as much for emotional regulation as for function when it is needed at a critical time in people's lives.

When it didn't: On many 24-hour worksites, from construction zones to industrial plants, it's still common to see harsh floodlighting, constant noise, no quiet breakout spaces, and little thought for sensory overload. Work gets done - but at the cost of fatigue, stress, and error rates. Contrast that with airports now introducing sensory rooms with controlled light, acoustics, and tactile features to support neurodivergent travellers, and you can see the gap between "compliant" and truly caring design.[395]

Wellbeing isn't a bonus - it's a performance requirement for infrastructure and us as a society.

8. Balance global collaboration with deep listening to local ecosystems.

Global knowledge is powerful, but every place has its own story, climate, culture, and custodians. Templates travel poorly when they ignore Country.

When it worked: Along the Birrarung/Yarra in Melbourne, the Burndap Birrarung burndap umarkoo (Yarra Strategic Plan) recognises the river as a single living entity and embeds Wurundjeri Woi-wurrung Traditional Owners in governance and decision-making.[396] River health, cultural values, ecology, and urban planning are treated as intertwined. Technical interventions - from bank stabilisation to access paths - are guided by listening to Country first, then applying engineering expertise.

When it didn't: In ports and coastal infrastructure around the world, climate-risk assessments show what happens when global design rules are copied into tropical or monsoon climates without adaptation. Corrosive marine conditions, rising sea temperatures, and changing storm patterns accelerate damage to quay walls, piles and equipment, driving up maintenance costs and shortening asset life.[397] The engineering was "by the book" - just not by this place's book.

Always listen to place before applying a template.

9. **Be a guardian of public trust - legitimacy is earned through transparency and care.**

Engineering exists because the public believes we act in their best interests. That trust is fragile - and once broken, hard to rebuild.

When it worked: In the United States, the Federal Highway Administration's National Bridge Inventory makes condition data on more than 600,000 public road bridges available to governments and the public, including ratings of key structural elements and summaries by state.[398] States such as Minnesota and New York build on this with public dashboards and datasets that show inspection ratings, explain what "poor" condition means, and track progress over time. Transparency doesn't eliminate risk - but it shows people that inspections are systematic, data is shared, and action is traceable.

When it didn't: The 2018 collapse of the Morandi Bridge in Genoa, Italy, which killed 43 people, became a symbol of what happens when maintenance and oversight fail. Investigations and court proceedings have highlighted long-standing concerns about corrosion in the bridge's load-bearing cables, allegations of neglected or falsified maintenance to preserve dividends, and serious design and monitoring shortcomings.[399] The physical structure failed - and so did public trust in the institutions meant to keep it safe.

10. **Redefine engineering: not just making things work, but regenerating for future generations.**

Repairing damage is no longer enough. Every intervention is a chance to give back more than we take.

When it worked: At Sydney Opera House, a modular artificial reef has been installed along the seawall at Bennelong Point as part of a research project with universities to restore marine habitat and boost biodiversity in the harbour.[400] Hexagonal reef pods beneath the waterline provide structure for marine life, improving ecological conditions without compromising the asset's primary function.

A routine seawall renewal became a chance to regenerate the underwater neighbourhood.

When it didn't: All over the world, "like-for-like" replacement is still the norm - asphalt for asphalt, pipe for pipe, concrete for concrete - even where nature-based or low-carbon alternatives exist. It keeps projects simple in the short term but foregoes opportunities to cool cities, restore habitat, cut emissions, or improve health. Regeneration starts the moment we ask, "How could this repair leave people and Country better off than before?"

11. Count the real cost of bad design - what do we harm, forego, or leave behind?

Efficiency means nothing if it erodes wellbeing, destroys biodiversity, or fractures community. The cheapest line item on a project sheet can become the most expensive decision in human and ecological terms.

When it worked: London Underground's renewal plans frame steady, ongoing reinvestment as the baseline required to keep assets in a "state of good repair" - explicitly linking track and systems renewals to safety, reliability, and customers' experience of a "good public transport" network.[401] Their own analysis shows that deferring critical renewals doesn't save money; it simply creates backlogs that drive future operating costs and reliability problems. Designing for cost *and* performance over the whole life of the asset is treated as a strategic necessity.

When it didn't: The Flint water crisis in Michigan began when officials switched the city's water source in 2014 from treated Lake Huron water to the more corrosive Flint River to save money, without adding required corrosion control treatment.[402] Lead leached from aging pipes into homes, exposing tens of thousands of residents - many of them children - to toxic contamination and triggering a public health emergency, lawsuits, criminal investigations, and years of pipe-replacement works. A decision framed as a short-term cost saving became a multi-billion-dollar financial, social, and moral catastrophe.

Every omission has a cost - especially the ones we don't count.

12. Connect then lead with values and intention, so that purpose guides every design decision.

Long-term care work is often invisible. That's what makes it sacred. Our values show up not in the ribbon-cutting, but in the records we keep, the repairs we choose, and the stories we tell about what we're here to protect.

When it worked: The Japanese art of kintsugi repairs broken pottery with lacquer and gold, deliberately highlighting the cracks rather than hiding them - a philosophy that treats repair as honouring history, not erasing it.[401] Maggie's Centres carry a similar spirit into architecture. Founded on the belief that people facing cancer deserve spaces that feel like home, not institutions, they've built a network of centres where natural light, gardens, warm materials, and kitchen tables are as essential as any clinical treatment next door.[404] Here, values - dignity, care, beauty, agency - are not slogans on a wall, but design constraints that shape every beam, window, and pathway.

When it didn't: In systems where maintenance is treated purely as a contract to be outsourced at the lowest price, the work is often stripped of purpose. Local knowledge disappears when contracts change hands; documentation is seen as overhead, not stewardship; and decisions get made far from the communities who live with the consequences. The physical assets may stand for a while, but the deeper infrastructure - trust, memory, care - corrodes.[58, 405]

Clarity of purpose turns maintenance from a cost centre into a moral practice. When we remember why we build and why we keep caring, we design and maintain in a way that future generations would thank us, not forgive us for.

The Quiet Years Matter Most

Maintenance is not the middle act between innovation and obsolescence. It is the heartbeat of engineering - the steady rhythm that keeps everything else alive. It is where our design, ideas, and outcomes are lived in real life with real people and communities every day. When we treat maintenance as mere compliance, we flatten its potential and the opportunity of what we can provide as engineers. When we treat it as care, we extend both function and meaning.

The best engineers I've ever met weren't just people with great ideas or builders; they were stewards. Because legacy isn't what we make - it's what we maintain on an ongoing basis.

Even small shifts in how we care for assets can create impact for people and places. You don't need to apply every principle at once or overhaul your entire approach. You just need to use your engineering judgment to decide how you can embed these principles in practical, achievable ways that work best for the context you are in.

Embedding the Principles in Everyday Maintenance Practice

So what does this look like, practically? How do these principles show up not as abstract ideals, but as daily habits in maintenance teams, operators, and engineers?

Look at these principles as lenses - simple, practical ways of seeing and deciding that can be woven gently into what maintenance teams already do every day. You don't need a restructure or to do tons of additional work. You just need small, intentional habits that align daily choices with long-term legacy. The following are some simple ways to embed the principles into maintenance practice - quietly, consistently, and meaningfully.

If you're reading this as someone working on an asset already deep into its maintenance phase, this next shift is for you. Your job now isn't just to preserve design life or stretch service life. It's to begin transitioning the asset into legacy life - the phase where care, culture, values, and community shape what the asset ultimately becomes for the next generation.

No doubt that your project already has a rhythm of systems, processes, and meetings that help it run. The below ideas are suggested as tweaks to these systems and procedures to be embedded in rather than creating something new.

It might be worthwhile to gather a team and workshop what ideas you want to implement and embed into existing systems, or a community of practice that will implement this and champion the core values and principles in the project or asset.

1. Bring a Principle Into Operational Meetings

Choose one principle each month - just one - and ask two simple questions:

- Where did we live this principle recently?
- Where did we drift from it?

It could be "Co-create with communities" one month and "Count the real cost of bad design" the next. Or you could keep coming back to one principle every month for several months with a discussion of how you might embed a principle.

The goal isn't critique - it's awareness. A ten minute discussion could spark a simple idea or a small change. This rhythm builds reflective muscle without adding more meetings.

2. Annual Strategy and Budget Cycles

Before renewing budgets or maintenance programs, review them through the three lenses of human wellbeing, regenerative environment, and generational legacy. Ask:

- Does this budget support the nervous systems of workers, residents, and ecosystems? Does this reduce stress, disruption, or risk for people?
- Are we regenerating more than we consume? Does this repair or regenerate?
- What will this look like in 50 years?

Even one question can shift a decision from reactive to regenerative.

3. Contractor and Supplier Briefings

When onboarding maintenance contractors, embed values and principles alongside technical specs. For example:

- Instead of "Minimise disruption", say "Maintain community wellbeing during works."
- Instead of "Deliver on time", say "Deliver with trust and transparency."

Reframing metrics changes behaviour.

4. The Principles in Practice Log

Create a shared log or dashboard - digital or physical - where maintenance teams capture examples of decisions that reflect the principles. It could be a lineworker who adjusted schedule timing to reduce school drop-off disruption, or a supervisor who reused removed material for habitat restoration. Both become living proof of principles in motion.

This can demonstrate to the team:

- The principles are valued and important as they are noticed
- You can see how people are implementing ideas
- Autonomy and reward for implementing the principles

- How to think about embedding the principles

Perhaps you might even create a team award for the month where people nominate team mates demonstrating the values, so there is a fun, collaborative team building element too.

5. Lessons Learned Reviews

You don't need to wait until the end of a shut down period to do a lessons learned review, although that is a good idea. You can include a values reflection in post-maintenance reviews, in annual reviews, and as part of strategic planning:

- Which of our core values were visible here?
- Where did we trade one off against another, and why?
- How could we embed them earlier next time?
- What principle can we adopt and take forward in our next maintenance cycle that would help decrease risk or improve outcomes?

6. Annual State of the System Report

For major assets, publish a brief, plain-language report for the public summarising:

- Technical health of the asset
- Environmental impact
- Community sentiment or feedback themes
- Legacy trajectory (how it's tracking against long-term intent)

This builds transparency, reinforces trust, and invites civic pride - transforming maintenance into a shared social contract.

7. Cultural Rituals

Replace or supplement traditional "safety moments" with "care moments" - 60-second reflections on kindness, community, or legacy. It's a simple act, but rituals shape culture. When care becomes part of the rhythm of maintenance, it redefines what "good engineering" feels like day to day.

8. Long-Term Knowledge Stewardship

Record lessons, stories, and insights from operators and long-term staff. Capture community conversations and seasonal patterns they've observed. This turns maintenance from mere compliance into cultural continuity. Legacy depends not only on what is maintained, but on what is remembered. Embed this into inductions for new staff on the project or team.

9. Linking to O&M Systems

Where possible, integrate these principles into asset management software or dashboards. Prompt users to record not only technical metrics (downtime, cost) but qualitative metrics (community feedback, ecological improvement). Data alone doesn't tell the story; reflection does. Have the prompts link back into the core values or principles for the project.

10. Leadership Accountability

Finally, make these principles part of leadership KPIs. Superintendents, asset managers, and team leads should be asked in reviews: "How have you demonstrated care, curiosity, or community first in your decisions this year?" Make it matter more by measuring and embedding into existing systems.

A Quiet Revolution

Embedding values in maintenance doesn't require a new framework - just a new mindset. It's about remembering that engineering isn't only about building the future, but about caring for the present. When teams start to discuss care, curiosity, and community in the same breath as cost and compliance, something subtle but powerful shifts.

This is where the profession evolves - not through revolution, but through repetition. The steady rhythm of engineers who care enough to keep things working well for life - just as we always do, but now with deeper connection.

Applying the Principles: Assets At or Beyond Their Design Life - Regeneration, Renewal, and the Legacy of What We Leave Behind

Early in my career, I worked for a mining company on a project that left a lasting impression on me.

A pit at the mine in regional Queensland had reached the end of its operational life. The coal was gone, the pit was massive, and the water that had filled it wasn't safe for drinking. My task was to lead a team of professionals to run a desktop study to determine what could be done next to make it safe and rehabilitate the pit.

It is an eerie sight: a gaping, water-filled scar in the landscape - too deep to fill, too polluted to repurpose, too remote to justify major redevelopment. For years, it had powered cities and livelihoods. But now, the trucks stopped running and the site fell quiet.

When it was first excavated and extracted, it was clear that no one had really thought about what would happen after. Now it was after and the time to think about this.

That project showed me how frequently we start thinking about regeneration only when extraction ends. Only when the hole is in the ground do we ask what comes next.

Today, we're much better at considering this earlier - building rehabilitation plans into mining operations from the start, designing landscapes to be restored, and thinking about closure as part of the project lifecycle. But there are still a lot of projects that were started decades ago and didn't consider it.

There are two paths for assets at the end of their design life: we either rehabilitate them to give back more than we took, or renew them because they're too important to lose. You can see both of these paths clearly in the way we treat old rail corridors.

When it worked: Across Australia, disused railway lines have been transformed into rail trails - long-distance cycling and walking paths that stitch regional towns together again. There are now more than 100 rail trails in Australia, from short local links to routes over 1,000 kilometres long.[406] One of the most loved is the Murray to Mountains Rail Trail in north-east Victoria: a 116-kilometre sealed path running from Wangaratta to Bright with a branch to Beechworth. It links wineries, farm gates, bakeries, mountain views and small town main streets, and now attracts more than 315,000 users a year, including over 250,000 visitors from outside the region.[407]

Economic studies of the Murray to Mountains and other Victorian rail trails show just how powerful this kind of regeneration can be. On average, every visitor day on these trails injects around $51 into the local economy, and a later study of the Murray to Mountains alone found overnight visitors spending on average $483 per person per day on accommodation, food, wine, and experiences - with almost half of users visiting for the first time and many staying two to three days.[408] Cycling is often the main reason for the trip, but the value flows to the towns: cafés, farm stays, pubs, bike hire, shuttles, cellar doors.

In Queensland, the Brisbane Valley Rail Trail does the same kind of work. It follows the old Brisbane Valley railway line for 161 kilometres from Wulkuraka near Ipswich to Yarraman, connecting a string of small towns through farmland, forest, and river country.[409] Visitor research shows that more than 80% of surveyed users come from outside the local council areas, turning a disused freight corridor into a regional tourism spine that supports accommodation, hospitality and tour operators.[409] The Queensland Government has recognised this potential, investing in rail trails as a way to power regional economies, preserve heritage, and create safe, car-free active transport routes on old state-owned rail corridors.[410]

My husband and I have jumped on the bikes from Ipswich out along the Brisbane Valley Rail Trail in Queensland too - even putting our son Winston on the back with us when he was 18 months old! We were hugely ambitious but proud to say we did 191kms with him. We turned it into a great holiday and loved it. We met locals, stopped at tiny corner shops for a coffee, and stayed in old school pubs, visiting towns we might never have visited otherwise. You can feel, in a very tangible way, how an old piece of hard infrastructure becomes a living social and economic artery again when it's given a new purpose and creates connection.

When it didn't: On Queensland's Sunshine Coast, a very different story played out. For more than a century, the Moreton Central Sugar Mill at Nambour was the heart of the local sugar industry. To feed it, an extensive narrow-gauge cane tramway network was built from the 1890s onwards. By 1905, there were already about 26 miles (42 kilometres) of permanent tramline in use. Over time, the network extended east to the Maroochy River and Coolum Beach, carrying cane and - for a period - passengers, helping open up the coast to early tourism.

When the mill closed in 2003, its last crush brought an entire era to an end. The closure had real upsides for residents. The black soot that once fell on nearby homes stopped, and the mill site was eventually redeveloped as a shopping centre and town centre, bringing new services into Nambour. But there was also a huge missed opportunity. As part of the closure, the majority of the tramlines were removed as leases with landholders were wound up, leaving only a short section of track through the CBD on the heritage register.

That means that what was once a continuous network of cane train corridors snaking across the Sunshine Coast - linking farms, river crossings, towns and beaches - can no longer easily become a modern light rail route or a continuous cycling and walking network in the way disused railways have elsewhere. There are now small heritage tram proposals for a short section in Nambour, which have been expensive to fund and still haven't really got off the ground, but the broader corridor that could have been a climate-friendly transport spine or a regional active transport trail has largely been lost.[411]

Both stories are about assets at the end of their life. In one case, old rail lines become rail trails that connect, regenerate regional economies, support small businesses, and give people experiences of place. In the other, an equally rich network of tracks has been mostly pulled up and forgotten, extracted from the community. The contrast is not about technical feasibility; it's about imagination, planning, and values. At the end of an asset's life, we can either design for renewal and reconnection - or let the opportunity disappear.

This is what legacy really looks like in engineering. It's not just about preserving things for nostalgia's sake - it's about evolving them for the next generation and building on already great infrastructure. It's recognising that every structure, every system, every piece of infrastructure has a lifecycle - a moment of creation, a long period of care, and eventually, a point of renewal.

When forests regenerate, old trees fall. But those fallen trees become the nourishment for new growth - part of an ongoing system that sustains itself. Our infrastructure should follow the same pattern. Wouldn't that be extraordinary if what came before could feed what comes next?

That's the spirit of regenerative engineering - designing and maintaining in ways that allow assets to continually give back to people, to the planet, and to time itself.

So when we reach the end of an asset's design life, it's not the end of the story. It's the beginning of the next chapter - one that asks us to look at what remains and ask: "How can this place, this structure, this system now give life again?"

Aligning to the principles, here are some ideas and suggestions for assets at the end of life to consider what is next. The aim of these is to spark ideas and inspiration rather than a checklist of items to implement. Sometimes just one thing can make a huge impact, and that one thing might actually be doable rather than trying to do everything.

1. Co-create with communities to reduce risk and increase trust

When an asset nears its end of life, community becomes even more vital. The conversation shifts from "how to fix" to "what does this place need next?" and "is this still relevant today compared with when it was designed and built?"

When it worked: In Seoul's Cheonggyecheon Restoration Project, an aging elevated freeway was demolished - but only after years of dialogue with locals, shop owners, and historians. The result was a revitalised stream and public space that healed a city's heart.[412]

When it didn't: In contrast, when Boston's West End neighbourhood was "renewed" in the 1950s, residents were displaced with little say. The new infrastructure stood - but the community's spirit was lost.[413] End-of-life decisions can either rebuild trust or erase it.

2. Learn from nature's systems to repair, rebuild, regenerate, re-experience

In nature, nothing truly ends - it transforms. When assets reach the end of their technical life, what if we followed that same principle?

When it worked: The Brisbane Powerhouse is a perfect example of regenerative thinking applied to an industrial relic. Once a defunct, graffiti-covered power station on the banks of the Brisbane River, it sat unused for decades - a concrete skeleton of the city's energy past. When the time came to decide its future, the city didn't wipe it clean or demolish it. Instead, it embraced the principle nature models so well: transformation.

Architects and engineers preserved the original industrial structure, brickwork, and even the graffiti - treating it not as vandalism, but as part of the building's living history. Initially they were going to remove the graffiti, but when it came to working out what graffiti to keep and what do we take away, they couldn't choose. What had a right to be there more than another? Rather than sterilising the surfaces, they let the layers of paint, tags and textures remain as a record of the building's in-between life.[414]

The renewal brought new purpose (a vibrant arts and cultural centre) while retaining the emotional and material memory embedded in the place. Today, the Powerhouse is a great example of adaptive reuse: bold, raw, beautiful, and alive - proof that regeneration can be cultural as well as ecological.

When it didn't: A powerful counterexample is the Packard Automotive Plant in Detroit, once one of the world's most advanced car factories and a symbol of American industrial innovation. After manufacturing ceased in the 1950s, the 3.5-million-square-foot complex sat largely abandoned for decades. Rather than being stewarded, stabilised, or creatively repurposed, it became a slowly collapsing ruin - a site of fires, vandalism, dangerous structural failures, and environmental contamination.[415]

Across multiple decades, redevelopment attempts came and went: art precinct proposals, housing visions, manufacturing revivals. But without consistent stewardship, investment, or a regenerative mindset, the site deteriorated beyond safe restoration.

By 2022, the City of Detroit ordered demolition of the complex, declaring sections structurally unsalvageable. What could have been one of the world's great adaptive-reuse projects - an industrial cathedral turned cultural landmark - was instead reduced to rubble because the window for transformation was allowed to close.[415]

It is a reminder that when we fail to imagine new life for aging assets, we don't just lose buildings - we lose history, identity, community memory, and the ecological potential that regenerative reuse could have created.

In nature, neglect leads to renewal. In our systems, neglect often leads to demolition.

3. Use human experience and planetary healing as metrics

At the end of an asset's life, the question is not just "Is it safe?" but "Is it still serving?" and whether it could improve the lives of people and the community. The end of life for a to include the principles and as an opportunity for healing.

An asset can be structurally sound and still fail its current and future community. Likewise, an aging structure can become a vessel for cultural, social, and ecological renewal when we choose metrics that include human experience and planetary wellbeing alongside cost, risk, and durability for its future. Could this mean that assets that once would have been demolished or removed could be an opportunity for something else?

When it worked: Cockatoo Island in Sydney is a great example of how an asset at - or well beyond - its design life can continue to serve by shifting the metric from technical function to human experience.

The island has lived many lives: a convict prison in the mid-1800s, a reform school for girls, a major shipbuilding yard, a submarine base during both World Wars, and later a vast industrial shell after the dockyards closed in the early 1990s. Each era left its own physical scars - tunnels, cranes, slipways, sandstone cuts which made some of Sydney's most beautiful buildings - creating a layered, complex landscape of memory.

Instead of erasing that history, selling off the land, or defaulting to demolition, many uses have been developed for the island, including the Biennale of Sydney embracing Cockatoo Island[416] as a site for artistic transformation. They used the existing structures as canvases, allowing warehouses, shipyards, tunnels, and decaying industrial spaces to become immersive environments for contemporary art. These interventions are deliberately temporary and light-touch: installations come and go, seasons shift, and the island evolves with each festival cycle. Little is rebuilt; much is re-experienced. The island's relevance comes not from new construction but from the human experiences it hosts - awe, curiosity, reflection, and connection.[417]

Cockatoo Island proves that assets do not need permanent redesign to serve communities. They just need to be opened, interpreted, and cared for in ways that honour their layered past and invite people into new forms of meaning. It is a powerful reminder that cultural, emotional, and experiential value are legitimate engineering outcomes - and that sometimes a regenerative path is simply giving an old place new life without erasing the old one.

London's Tate Modern is another powerful example of how an asset at the end of its design life can be reimagined to serve people in ways its original engineers never predicted and make people feel good through art.

Once the decommissioned Bankside Power Station, it sat vacant on the Thames until architects and engineers transformed it into one of the world's most visited art museums. Rather than razing the building, they preserved its monumental brick façade and cathedral-like turbine hall - honouring the industrial heritage while creating vast public spaces that evoke awe, belonging, and communal experience.[418]

The museum's opening in 2000 didn't just repurpose a structure; it helped regenerate Southwark, sparking new walking paths, businesses, public life, and ecological improvements along the riverfront. The building's rebirth shows what becomes possible when we measure success not by square metres of retail or technical life, but by emotional resonance, accessibility, cultural vibrancy, and the renewal of place.

When it didn't: Contrast this with the fate of countless suburban shopping centres across the world - many built in the 1960s-1990s and now known in research literature as "dead malls". Structurally, many of these assets remain intact. Technically, they could keep operating for decades, but their social contract expired long before their concrete did.

As online retail grew and communities shifted, these centres failed to evolve, becoming sterile, dimly lit corridors of vacancy and decay. Research shows that when owners prioritised retail metrics alone - foot traffic, lease turnover, car parking - and ignored social relevance, community use, cultural identity, or ecological potential, the centres lost their purpose.[419] Some now sit abandoned; others limp along as underused spaces offering neither joy nor service. The engineering succeeded, but the human experience failed.

4. Practice legacy thinking and plan for generations ahead

When an asset nears the end of its design life, this is where legacy becomes visible. The question is no longer "how do we keep this operating?" but "what future are we creating through what we choose to repair, retire, restore, or reimagine?" End-of-life decisions are some of the most consequential in the entire engineering lifecycle, as they shape what future generations inherit as heritage, waste, or foundation.

When it worked: The Roman aqueducts are extraordinary examples of legacy thinking in action. Many remained functional for centuries because they were maintained, adapted, and valued as civic gifts rather than disposable utilities. Later societies repurposed their arches, channels, and stonework for new uses - from transport routes to urban landmarks. Their endurance was not an accident but an expression of culture: maintenance as stewardship, not compliance. Even beyond their technical life, they continued serving because they were built - and cared for - with the long view.[325, 420, 421]

When it didn't: London's Millennium Dome was built with grandeur but without a clear plan for life beyond the millennial celebrations. Once the year 2000 festivities ended, the structure sat largely empty for years - a costly symbol of ambition without foresight. Only after extensive reinvention did it finally become the O2 Arena.[422] The failure wasn't the engineering; it was the absence of a generational horizon. Legacy thinking is not about size or spectacle - it is about imagining the value of a structure long after the headlines fade.

5. Build teams from diverse archetypes and perspectives

At this phase, diversity becomes non-negotiable. Sunset decisions require more than technical judgement - they demand cultural intelligence, ecological insight, economic understanding, creativity, and storytelling. Assets that have lived long lives must be reimagined with the same richness and multiplicity that shaped their pasts.

When it worked: Barangaroo Reserve in Sydney transformed a disused concrete container port into a layered cultural landscape by assembling one of the most diverse teams ever engaged in an Australian precinct renewal. Aboriginal landscape architects led the cultural narrative; ecologists restored habitat; engineers reconstructed an 1830s headland using 10,000 sandstone blocks; artists embedded meaning into public spaces.[423] The result is a storied, regenerative place grounded in Country. It demonstrates what becomes possible when many ways of knowing come together at the end of an asset's industrial life.

When it didn't: Many post-industrial redevelopments led exclusively by commercial or technical expertise result in sterile waterfronts and placeless promenades that age poorly. When heritage, ecology, and cultural meaning are absent from the conversation, the outcome may be functional but emotionally empty. Renewal needs many hands - and many minds - or it risks flattening the richness of what came before.

6. Remember beauty is not optional

As assets age, beauty could become a form of dignity if we enable it. People fight for places they love and forget the ones they don't. At the end of life, beauty can determine whether a structure invites stewardship or slips silently toward demolition.

When it worked: The Venetian lagoon shows how beauty and maintenance can co-exist. Buildings weather, tide marks rise, colours soften - yet the city adapts, repairs, and stabilises its structures without erasing their patina. Age is not hidden but honoured. This graceful approach to decline creates emotional resonance that compels ongoing care. Venice is one of the world's most fragile environments, but also one of the most fiercely protected. Beauty, even in decay, sustains loyalty and preservation.[424]

When it didn't: By contrast, brutalist car parks and aging megamalls across the world often face demolition not because they are unsafe, but because they were never designed to inspire connection. Harsh edges, blank façades, and lifeless interiors create no emotional anchor. When beauty is absent, so is belonging - and people rarely fight to save what they cannot feel.

7. Design to support nervous systems, not overstimulate them

End-of-life work - demolition, decommissioning, renewal - is often noisy, dusty, unpredictable, and emotionally charged. Communities are already grieving the loss of familiar places. Workers carry cognitive and emotional load. Designing for calm is not a luxury; it is a safety practice, including in this phase when assets are at the end of their life.

When it worked: After the Christchurch earthquakes, engineers and urban recovery teams recognised that rebuilding was as emotional as it was technical. Mental health specialists were embedded in engineering teams, community communication was compassionate and constant, and recovery hubs supported residents navigating disruption. This approach reduced burnout, increased trust, and acknowledged that rebuilding the city required rebuilding nervous systems too. The rebuilding of the city needed to heal nervous systems, and this became an opportunity to do this.[425]

When it didn't: In Detroit, large-scale industrial demolitions often proceeded without adequate dust suppression, noise management, or communication with nearby residents. Schools and homes were coated in demolition debris, and communities already experiencing economic hardship were subjected to additional environmental and psychological stress. The technical work was completed - but at a human cost.[426]

8. Balance global collaboration with deep listening to local ecosystems

At this phase, renewal becomes an invitation to design with the place rather than impose upon it. Global expertise gives us possibilities, but local ecosystems, cultures, and stories tell us what is appropriate.

When it worked: In Fremantle, Western Australia, the old port sheds and maritime warehouses were restored and repurposed into community markets, maker spaces, and creative venues. Designers used local timber, local skills, and local craft traditions to preserve the character of the working harbour. The result is regeneration that sounds, smells, and feels like Fremantle: salty, artistic, maritime, and unmistakably local.[427]

When the Myer department store in Fremantle closed in 2013 after about 40 years in operation, its five-storey building in Kings Square was left vacant - a void in the heart of the city.

Rather than letting it sit empty while long-term redevelopment plans lingered, the building was handed a second life as a temporary pop-up creative, retail, and community hub known as MANY 6160. Just six months after Myer closed, MANY 6160 opened - activating the empty space with around 20,000 m^2 given over to independent designers, retailers, artists, creative makers, cafés, and communal gathering spaces.

Inside the former department store, visitors found a mix of retail, handmade goods, vintage clothing, artisanal furniture, art galleries, a café, even creative studios - a vibrant, diverse, low-barrier ecosystem. They didn't do extensive remodelling of the building. In fact, they left things like the flooring in, which actually gave it a warmly familiar department store feel. Local creatives paid heavily-subsidised rents and used the space to build their ventures.

Rather than seeing the building only as a future development site, this interim use turned it into a living, breathing part of the community again - drawing people into the city centre, generating foot traffic, supporting small businesses, and giving new life to architecture that would otherwise have remained vacant. The activation helped preserve the structure's value, reduced the chance of decay or vandalism, and kept Fremantle's historic heart alive while the long-term vision unfolded.[428]

In other words: the building's "in-between" phase was treated not as waiting time or fallback, but as opportunity. Its in-between life became part of its legacy.

When it didn't: Across the world, redevelopment projects that import global templates without adapting to local climate, culture, or ecology often produce spaces that feel generic and disconnected. Tropical cities receive glass-heavy buildings suited for cold climates, historic waterfronts are replaced with global mall architecture, and coastal defences copied from Europe fail under local storm patterns. When place is ignored, relevance dissolves quickly.

9. Be a guardian of public trust - transparency is care

End-of-life decisions - closures, demolitions, hazardous material removals, relocations, or temporary disruptions - are moments of heightened public sensitivity. These are the moments when communities watch closely. They want to know what is happening, why it is happening, and who will be affected. Transparency isn't merely information sharing; it is an expression of care. It signals that people matter as much as the project.

When it worked: An Australian example comes from Brisbane during construction of the Legacy Way Tunnel. As part of the early works, the project team proposed using Anzac Park in Toowong as a temporary construction compound and overflow parking area for workers. On paper, this seemed efficient and technically sound. The park was close to the tunnel portals, the terrain was suitable, and the plan included a commitment to restore the park to a better standard afterwards.

But Anzac Park holds emotional and historical significance for the local community. Residents voiced strong opposition, worried about tree removal, noise, access, and the symbolic impact of turning a beloved memorial park into a worksite. And critically, the project team listened.

Instead of pushing forward with what was easiest for construction, they shifted the compound to an alternative site. It was less convenient, more complex, and required redesign. But it honoured community sentiment, preserved trust, and demonstrated that engineering can bend toward people, not just logistics.[429] The relationship between residents and project teams strengthened because the community saw that their voice genuinely shaped the outcome. The engineering still progressed - but not at the cost of care.

A similar commitment to democratic transparency shaped the transformation of Berlin's Tempelhof Airport. When the airport closed in 2008, the city didn't predetermine its future behind closed doors.

Instead, officials opened the vast airfield to the public and held a referendum on its long-term use. Citizens overwhelmingly supported keeping the space open, wild, and free - a park at the scale of a small city.[430] Tempelhofer Feld is now one of Europe's largest and most beloved public parks - a symbol of civic trust and an example of how transparency turns infrastructure decisions into shared ownership.

Both examples show the same principle in action: public trust is built not through perfect engineering, but through respectful process. When people see themselves reflected in the decisions that shape their neighbourhoods, legitimacy grows. Transparency becomes a form of stewardship - a way of saying, "We are building this with you, not just around you."

When it didn't: The opposite of care is not conflict - it is silence. And nowhere is this more evident than in cases where infrastructure is decommissioned or altered without transparency or meaningful community involvement.

A striking example comes from the closure and partial demolition of the Hazelwood Power Station in Victoria, once Australia's most emissions-intensive coal plant. When Hazelwood shut down in 2017, locals were given just five months' notice, despite the profound economic, environmental, and social implications for the Latrobe Valley.

The community, unions, small businesses, and local councils repeatedly stated they felt blindsided, excluded from planning, and left without clear information about the long-term rehabilitation strategy, job transition plans, or environmental risks associated with the enormous ash ponds onsite.[431] Instead of feeling part of the decision-making process, the community experienced uncertainty, frustration, and a sense of abandonment at precisely the moment when clarity mattered most.

Subsequent reports, including by the Victorian Auditor-General and independent transition bodies, noted that a lack of early and transparent engagement eroded community trust and amplified fear.[432] Residents were unsure whether the site would be rehabilitated safely, how long it would take, and what would happen to the thousands of hectares of disturbed land. Confusion also surrounded groundwater contamination, dust management, and the future stability of the mine void.

The technical work of decommissioning eventually progressed, but the trust deficit remained, influencing subsequent projects and shaping public perception of government and industry. Hazelwood is now widely referenced in research on just transition planning as a case where opaque decision-making worsened social, economic, and psychological impacts, and where more transparent, participatory processes could have created far better outcomes for the region.[433]

End-of-life decisions don't just shape land - they shape relationships. Hazelwood shows us what happens when communities are treated as recipients rather than partners: trust erodes, fear fills the gaps, and even responsible engineering actions can be interpreted through a lens of doubt. That's why transparency should be seen not as an afterthought, but as an integral part of engineering.

10. Redefine engineering: not just making things work, but regenerating for future generations

End-of-life decisions reveal what we truly believe engineering is for. Do we leave land wounded or renewed? Do we close chapters or open new ones?

When it worked: Hammarby Sjöstad in Stockholm is one of the world's most studied examples of regenerative redevelopment. What began as a contaminated industrial harbour and polluted lakefront became a living model for circular urban design. The site, once home to factories and scrap yards, required extensive soil remediation and environmental cleanup. Rather than simply capping the contamination and building over it - the traditional approach - Stockholm's planners committed to restoring ecological function first.[353]

The redevelopment introduced an integrated closed-loop energy, water, and waste system now known as the Hammarby Model. Waste heat warms homes. Biogas from sewage powers buses. Stormwater is treated through green corridors and wetlands before returning to the lake, improving biodiversity.

The precinct's emissions are estimated to be half those of comparable Swedish developments, and its water use efficiency is internationally recognised.[354] Academic reviews describe Hammarby as a "global benchmark for eco-districts" and a case that shows how former industrial sites can become healthy, restorative neighbourhoods rather than environmental liabilities.

This is regeneration in its fullest expression - not merely cleaning up a problem, but transforming damaged land into a thriving community that gives more back to the environment than it takes.

When it didn't: Around the world, many decommissioned coal and gas power stations offer a stark contrast. Instead of regenerative closure, they have been left as lingering hazards - physical, environmental, and communal. Research from the International Energy Agency and environmental regulators highlights a recurring pattern: when power plants shut down without strong remediation frameworks, the burden is passed to future generations.[434]

Contaminated ash ponds, often containing heavy metals such as arsenic, mercury, and lead, are frequently left minimally lined or unlined. When operations cease and water management declines, these ponds can leach pollutants into groundwater and nearby rivers - a documented issue at multiple sites across the United States, India, and Eastern Europe.[435]

Asbestos insulation, PCB-containing equipment, unstable boiler houses, and corroding smokestacks can remain in situ for years because demolition is expensive and remediation standards vary.

A well-known example is the Widows Creek Fossil Plant in Alabama (USA), where a massive coal ash spill in 2009 exposed underlying storage weaknesses. Years after closure, the site still requires containment and remediation.[436]

In Australia, state auditors have repeatedly raised concerns about the long-term risk profiles of decommissioned or mothballed coal plants, noting that unclear ownership, insufficient funding for rehabilitation, and inadequate regulatory oversight turn these sites into ecological debts - burdens that communities and governments must eventually absorb.[437]

The pattern is sobering: when closure is treated as the end of responsibility rather than the beginning of healing, damaged landscapes persist, ecosystems degrade further, and public trust erodes.

Ending operations is not enough. Regeneration must begin the moment extraction ends - or we simply push the true cost of our designs onto those who come after us.

11. Count the real cost of bad design - what do we harm, forego, or leave behind?

Short-term decisions at the start often become the most expensive - financially, socially, ecologically - at the end. End-of-life is where shortcuts reveal their true cost.

When it worked: The Eiffel Tower, originally designed as a temporary structure for the 1889 World's Fair, survived because its beauty and unexpected usefulness as a radio tower made it worth preserving. Its flexible structural design allowed adaptation, maintenance, and renewal across generations.[438] Good design creates options. It gives the future room to decide.

When it didn't: The Grenfell Tower tragedy in London exposed the devastating consequences of cost-cutting, poor material choices, and ignored resident concerns. Flammable cladding turned a residential tower into a deadly trap.[32] The true cost of bad design is never confined to the balance sheet; it is borne in human lives, community trauma, and national reckoning. Legacy is an accounting of consequences.

12. Connect with values and purpose - purpose guides every design decision

The end of an asset's life is a moment of reflection. What story does this place tell about who we are? What do we choose to honour, restore, or release?

When it worked: The Brewarrina Aboriginal Fish Traps (Baiame's Ngunnhu) in New South Wales are among the oldest engineered structures on Earth. They endure because they were built inside a living system of culture, governance, ecology, and community. Care was inseparable from function.[439] Knowledge was passed down through story and practice. Their longevity is not an engineering anomaly, but rather an example of engineering rooted in values.

When it didn't: Modern infrastructure that is stripped of cultural meaning often loses relevance the moment its function ends. When purpose is reduced to throughput or efficiency, there is nothing to anchor a place in memory or identity. Assets become disposable and communities feel no connection to what is left behind. To care for Country - or for any built system - is to care for people and time.

Marginal Gains and the Power of 1%

When I look back on that mining project from the early days of my career - the one where we were left with a vast, empty pit and a poor demonstration of "remediation" - I often think about how easily that could have been different. Not through massive reinvention, but through a handful of small, early questions in every stage of the project:

- What happens next?
- How could this give back when it's done?
- How could we make this useful again for people or for nature?

It is more about what we didn't do than what we did, and it's the same story with so many of the assets, systems, and infrastructure we have in our world today.

Those questions don't take huge budgets or decades to answer. They do take time and they do take being intentional and having the willingness to make small, deliberate shifts. While there might be some challenging conversations involved, they really can de-risk a project and save time and money in the later phases of the asset's life and beyond it.

We really must shift our mindsets to effectively implement these changes.

The kind of change needed reminds me of the story of British Cycling. For almost a century, they were one of the most ordinary teams in the sport - no Olympic golds, no Tour de France wins, not even a strong reputation. Then, in 2003, they hired a new coach, Sir Dave Brailsford, who believed in what he called the aggregation of marginal gains.

He focused on tiny improvements - the 1% changes - across hundreds of areas: slightly better seats for comfort, better sleep for athletes, cleaner tyres, improved nutrition, even the paint inside the team truck to better spot dust that could affect bike performance. Individually, those changes seemed trivial. But together, they transformed the entire system.

Within five years, British cyclists won 60% of the gold medals at the Beijing Olympics. Within a decade, they won the Tour de France multiple times.

That's the power of marginal gains - and I think engineering works the same way. I also think this is the approach we need to take here with the principles, ideas, and concepts in this book. Just imagine if every engineer, designer, or policymaker took just one of the principles in this chapter - just one - and found a 1% way to improve it on the next project, or in the next maintenance cycle, or as an asset nears the end of its design life.

Maybe it's one more conversation with the community.

Maybe it's choosing a slightly longer-lasting material.

Maybe it's restoring one extra patch of native landscape.

The point isn't perfection - it's progression. If everyone in the profession made a 1% shift toward more care, more curiosity, and more legacy, the cumulative change would be extraordinary. Those small actions, added up over time and across teams, would transform the culture of engineering from one that fixes to one that heals.

Legacy isn't built through a single grand project. It's built through the steady rhythm of small, deliberate acts of care that compound over time in new builds and maintaining the assets we have - just like those marginal gains. The reality is, the future won't be engineered in leaps. It will be engineered in increments - one thoughtful decision, one design, one project, one percent at a time.

Bridge Forward: From Principles to the Values Beneath them

If the Twelve Principles show us *how* engineering can be practiced differently, the next step is to ask the question: why do some teams instinctively make regenerative, human-centred choices while others - working with the same data, tools, and constraints - do not? Why do some projects feel alive with care, connection, and intention while others, equally competent on paper, fall short in purpose or legitimacy?

The difference is rarely technical. It lives in something hidden, deeper, and far more powerful: the values we've inherited and the ones we choose.

Beneath every design standard, procurement model, risk workshop, and drawing package lies an invisible layer of assumptions - about what matters, what success looks like, and what can be safely ignored. These assumptions shape what gets measured, rewarded, funded, and protected. They determine whether we design for efficiency or for empathy, for extraction or regeneration, for compliance or connection.

Just like the hidden DNA we explored earlier, these values can either expand what is possible or silently constrain it.

That is where we turn next.

It's time that we reveal the cultural patterns, inherited beliefs, and invisible norms that have guided engineering for decades. Some of these values once served us. Some still do. Others now limit our imagination and narrow our capacity to design for a thriving future. This next part is about making those values visible again.

It is about understanding the legacy of the Industrial Age - the pursuit of efficiency, control, certainty, and productivity - and how those values still shape the profession today, often without being named. It is also about rediscovering the traits that have always lived in engineering's heart: creativity, care, courage, curiosity, stewardship, and service to society. These values have always been present - they just haven't always been legitimised.

Before we can redesign our systems, we must first understand the beliefs we've inherited. Before we can build differently, we must learn to *see* differently. Before we can lead the next era of infrastructure, we need to ask a more foundational question: what do we want engineering to stand for now - and for the next 100 years?

The Takeaway

Principles only matter when they are tested. Not when conditions are calm or choices are easy, but when decisions are constrained, trade-offs are real, and the pressure to fall back on familiar habits is strongest.

The Twelve Working Principles are not about adding complexity to engineering. They are about making better decisions under complexity.

When values are embedded into how choices are framed, how options are weighed, and how risk is understood, uncertainty becomes more navigable rather than more dangerous.

With this foundation in place, the book now shifts from alignment into action.

The next part focuses on how these principles show up when it matters most, in decision-making, risk, and moments where engineering judgement carries real consequence. It explores how engineering can be de-risked by design, not through more control or compliance, but through clearer intent, better questions, and choices that align technical rigour with human, environmental, and generational responsibility.

Part III: The Hidden Heart of Engineering

Seeing the problem is not the same as changing it. Awareness matters, but on its own it does not shift outcomes. The real work begins when insight turns into practice, and intention shows up in decisions.

Part III is about activation. It takes the values, lenses, and questions explored earlier in the book and translates them into action at the scale where engineering actually happens: in meetings, design reviews, risk assessments, procurement choices, and everyday trade-offs. This is where culture is built, trust is earned, and legacy is shaped, often quietly, through repeated small decisions.

The chapters in this part offer practical ways to embed care, integrity, and long-term thinking into engineering work without losing rigour or credibility. They are not about perfection or idealism. They are about doing the work differently, consistently, and deliberately, so that the future of engineering is shaped not by chance, but by choice.

Chapter 9: De-Risking by Design and Making Decisions

Every engineering project involves risk. But not all risk is treated equally. Some risks are modelled, mitigated, and reported with precision. Others remain unnamed, unmanaged, and absorbed by communities, ecosystems, or future generations.

This chapter reframes risk as a design choice rather than an unavoidable by-product of complexity. It explores how decisions are actually made under pressure, and how narrow definitions of risk can produce technically compliant outcomes that still fail people, place, or purpose.

De-risking by design is not about eliminating uncertainty or adding layers of control. It is about expanding what we pay attention to, and when. By integrating values earlier into decision-making, engineers can reduce downstream failure, rebuild trust, and make choices that hold technical rigour alongside human, environmental, and generational responsibility.

The Myth of Perfect Information

If there's one myth that often shapes engineering thinking, it's this: that the best decisions come from perfect data.

We chase the illusion of certainty, believing that if we just had one more report, one more model, one more simulation, then we could finally decide with confidence. I know because it's what I have done as an engineer - been on a quest for perfection while withering away with analysis paralysis.

Life rarely gives us that luxury of perfection, as any parent will know. And in my experience, both personal and professional, it's the waiting - the hoping for more information - that can be the riskiest move of all.

When I was pregnant with my first child, I learned more about decision-making than any professional training could have ever taught me. At my 20-week scan, something unexpected appeared on the screen. The sonographer went quiet - that silence that makes your stomach drop - and said she'd just need to fetch the doctor. The doctor came in, studied the image, and said they'd found an anomaly. They couldn't yet tell me for sure if it was accurate.

It was a Friday afternoon, just after two o'clock. My obstetrician's office had closed at midday. There was no one to call.

I walked out of the hospital with a thousand questions that no one could answer. The doctor suggested doing another scan in four weeks' time to check again and make an assessment at that point.

That weekend felt like a lifetime. My brain, trained as an engineer, went straight to analysis mode. I wanted to map the risk, define the variables, calculate the outcomes. But nothing was certain.

At the 24-week scan, the same anomaly appeared again. The doctors gave predictions about what this could mean. Sure, they had data and percentages behind this but again, it wasn't certain and they didn't know the full impact. They could tell me what they'd seen - the data - and could make estimates and educated guesses, but couldn't tell me what it meant for certain.

I read every research paper I could find, scanned medical journals, asked friends who were in the health profession. I tried to fill the gaps with logic and cried a lot. But at some point, the data ran out - and all that was left was trust that it would work out.

That's when I realised something: I was facing this decision like an engineer. I was asking the same kinds of questions I'd learned to ask on projects when the stakes were high and the information incomplete. These were great for engineering tasks, but not great for my life.

What do I know for sure? What don't I know - and can I live with that uncertainty? What are the possible failure modes? And what happens if I do nothing?

That framework - the same one I'd used years earlier managing a $45 million portfolio of road projects at 23 - became the structure I leaned on. Back then, I didn't have all the answers either. I was young, inexperienced, and surrounded by people with far more technical knowledge. But I learned quickly that leadership wasn't about knowing everything. It was about asking better questions, synthesising information from others, and deciding with humility and courage.

And so, during that pregnancy, I found myself doing the same thing. Asking questions. Listening to the experts. Running mental "failure scenarios". Testing options. And then - when I had done enough thinking, when I had gathered enough to feel grounded, not perfect - I made decisions about my baby and pregnancy.

It required enormous trust. Not just in the medical team, but in myself. Trust that even with incomplete information, I was capable of making the right call for that moment.

This was one of the most formative experiences of my life, where I had to have complete trust in my instinct and intuition, and in making the right choice with the information I had.

When he was born, it turned out that what they picked up on the scan at was accurate. There were also other things that weren't picked up. This started a lifetime journey navigating medical choices, making decisions on behalf of my child, and trusting my gut like never before, in the most high-stakes environment I have ever been in.

My son is incredible - resilient, curious, adventurous, full of life. Looking at him now, you would have no idea we faced these challenges. I often think about those days and how much they taught me about leadership, engineering, and making decisions.

In the end, making decisions under uncertainty - whether it's about a baby or a bridge - isn't about being right. It's about being responsible. It's about balancing the technical with the human, the analytical with the intuitive, and having the courage to move forward when there is no perfect data.

When Enough Is Enough

I was reminded of the challenge of getting the perfect amount of data again years later, when I began my PhD. My research method relied on interviewing people to uncover patterns - and in my engineer's mind, that meant scale. I thought I'd need to interview 40 or 50 people before I could draw meaningful conclusions.

When I explained this to my supervisor, he smiled and said, "You'll probably start seeing patterns after about 20."

I was shocked. Only 20 interviews? For a doctoral study? It seemed impossibly small. Surely we'd need more data to make a solid, evidence-based conclusion.

I learnt many valuable lessons from this. That insights don't come from quantity, but from depth, from the consistency of human experience. That data doesn't always mean numbers. That sometimes, the quality of the information matters more than the quantity. And often, the signal of truth emerges much earlier than we expect - if only we're willing to trust what we're seeing.

These experiences reshaped how I approached projects and making decisions in engineering, business, and life. The fact is, we rarely have perfect information. But when you know what's enough - when you can sense that you've reached the point where patterns and possibilities start to converge - that's when the work of judgment begins.

Uncertainty Isn't the Enemy - It's the Environment.

Engineering is, at its heart, the management of uncertainty. Every model, every plan, every risk register is a best estimate - a momentary snapshot of a world that is always changing.

Yet, we've inherited a culture that often sees uncertainty as a threat. We disguise it with spreadsheets and confidence intervals, as though wrapping it in numbers could make it go away.

But pretending perfect information exists doesn't reduce risk - it increases it. Typically, the longer we wait for certainty, the fewer options remain.

NASA learned this during the Apollo missions. The engineers had to make life-or-death adjustments with incomplete telemetry and delayed communications. They didn't wait for perfect data; they trusted disciplined judgment and teamwork.[440]

Closer to home, Infrastructure Victoria has adopted adaptive planning frameworks for this exact reason, because climate, technology, and social needs evolve.[441] Rather than freezing decisions in time, they design for flexibility, so infrastructure can adapt as the world changes.

What Happens When Data Runs Out?

One of the most empowering habits we can teach engineers is the "60% rule". Decide when you have enough information to be confident, not certain. This small shift in language can be incredibly powerful.

Waiting for 100% accuracy is a luxury the real world doesn't offer - and by the time it arrives, the window for action has usually closed.

At roughly 60%, great leaders shift from analysis to synthesis. They start using values and judgment as their compass. They talk to people, test options, and ask, "What's the cost of waiting? What's the risk of acting now?" The goal isn't perfection - it's progress with integrity.

The aviation industry applies this logic all the time. Flight operations are designed to handle uncertainty in real time - weather, human factors, mechanical issues - using training and values to guide quick, ethical decision-making.

Boeing learned early that waiting for perfect information slowed progress. Teams that made timely decisions with incomplete data, and built feedback loops through testing and iteration, were able to innovate faster and surface errors earlier than those who delayed for certainty.[442]

It's not about being reckless. It's about understanding that waiting has a cost too.

Risk as Curiosity

When most people hear the word "risk", they picture red spreadsheets and compliance forms - something to be managed or mitigated. But risk, when done well, is a creative process. It's one of the most imaginative spaces in engineering.

I will be the first one to proudly tell you that I absolutely love risk management workshops. I know, you could say I'm a total nerd, but I do. It was one of my favourite things to do as a consulting engineer. Here's why.

These workshops were often a space for creativity and curiosity - a place where you could solve multiple problems at once and really get embedded into projects. When done well and combined with strategy and budget, it can be a game changer for projects in so many ways.

When I run risk workshops, I see this transformation happen in real time. At first, people think it's about ticking boxes. But when we start talking about what could go wrong - and what could go right if we approached things differently - something shifts. Suddenly, it's not about control. Creativity and curiosity are awakened in people, and they start to see it can be about discovering possibilities you can't see when everything feels safe.

Good risk workshops aren't about compliance; they're about connection, care, and legacy. They connect commercial, design, environmental, and human factors. They bring diverse voices into the room. They ask, "Who might be affected by this?" and "What don't we yet understand?" They keep people safe and make sure they feel supported.

When you treat risk as curiosity - not control - you don't just de-risk a project; you make it more human, more resilient, and more capable of creating long-term value.

De-Risking by Design

Engineers are taught to control risk - to eliminate it, to tame it, to box it neatly into registers and probability matrices. But real-world risk has never lived inside a spreadsheet. It lives in complexity - in people, in changing weather, in unexpected human behaviour, in the unknowns that reveal themselves only when we start building.

The reality is that we can't eliminate uncertainty - but we can design with it.

That's what "de-risking by design" really means. Not pretending the unknown doesn't exist, but learning to make it part of the creative and strategic process, rather than a shadow lurking at the edges.

Here are the common ways we see risk and how we might shift this to a new way of thinking.

From Control to Curiosity

In many organisations, risk management is treated like a compliance ritual - something to be completed, filed, and rarely looked at again. Rows and columns. Heat maps and colour codes. It's neat. It's organised. It's comforting. It is usually delivered in boring meetings.

Comfort isn't the goal. Clarity is. And clarity comes from curiosity, and knowing you have asked the important questions.

Early in my career, I worked on a major infrastructure program with a risk matrix that could have covered an entire wall. Every month, we'd update it - red to orange, orange to yellow - glance at the top five risks, and copy them into a database, which would automatically go into a report for the monthly project control group meeting.

The spreadsheet looked precise, almost scientific. But behind those colours, there were assumptions no one had challenged in months. The real risks - the human and behavioural ones - never made it onto the spreadsheet at all.

So when I began running my own risk workshops, I did something different. I aimed to create a safe space, get people talking and engaged in conversations.

I started asking:

- What don't we yet understand?
- Who else should be in this room?
- What assumptions are we silently relying on?

And suddenly, risk became alive and so did the people engaging in the meeting. The spreadsheets filled with insight, not just data and then linked back into the systems of the project. Designers and commercial leads started cross-pollinating ideas. Curiosity had replaced control - and in the process, the team uncovered opportunities we'd never seen before. Curiosity, not control, made that possible.

Here are some real case studies showing how this has worked in other projects:

Cheonggyecheon Corridor, Seoul

By the early 2000s, the Cheonggyecheon corridor in central Seoul had become a tangle of elevated roads, buried waterways, and urban heat - the legacy of 1970s engineering that prioritised traffic flow over ecology and human experience. The four-lane overpass above the buried stream carried 100,000 vehicles a day, yet the surrounding businesses were declining and air quality was among the worst in the city.

When Seoul committed to removing the freeway, the project team realised this wasn't simply an engineering challenge; it was a cultural, ecological, economic, and psychological one too.

Instead of convening only traffic modellers and structural engineers, the city brought in freshwater ecologists, hydrologists, sociologists, historians, heritage groups, community organisations, and urban designers. Crucially, they were engaged not as advisers but as co-authors of the brief.[443]

Where a traditional approach might have framed ecology and social impact as "risks to mitigate," this team treated them as design parameters:

- How should the daylighting of the stream support native species?
- How will people feel moving through this space?
- What microclimates are we creating, and how can we cool the city?
- What cultural stories need to be restored, not erased?
- What economic life could re-emerge if the space invited people in instead of pushing them out?

This curiosity-driven process de-risked the project by surfacing issues upstream - seasonal water availability, historic artefacts, business access - and designing for them early. It reduced downstream surprises and reframed complexity as an asset, not a threat.

The outcomes are now globally documented. Within three years:

- Fish species increased from 4 to 25
- Insect species rose from 15 to 192
- Average corridor temperatures dropped by 2-7°C
- Particulate pollution (PM10) dropped by ~35%
- Daily public use exceeded 60,000 people
- Property values in adjacent districts increased by over 300%

Cheonggyecheon became far more than restored infrastructure - it became a restored relationship between a city, its river, and its people. It is one of the clearest examples of how curiosity derisks complexity far better than control ever could.

Room for the River, Netherlands

After catastrophic floods in 1993 and 1995, the Netherlands faced a national dilemma. Climate change was intensifying river flows, and the country's centuries-old strategy of raising levees was reaching its limits. Traditional engineering would have framed this as a water control problem. But the Dutch took a radically different approach.

They began by bringing ecologists, hydrologists, landscape architects, spatial planners, farmers, social scientists, and community residents into the earliest design sessions. Rather than asking, "How do we hold back the water?" they asked, "How do we give the river more room to behave as a river?"[390]

This shift - from controlling nature to being curious about how to live with it - unlocked an entirely different design logic:

- Move levees back instead of raising them
- Create secondary channels and floodplains
- Integrate habitat corridors into hydraulic capacity
- Support farmers and communities to transition land uses
- Turn river management into public space and ecological regeneration

The multidisciplinary team identified risks before they became emergencies: sediment movement, ecological migration paths, heritage assets, and the social impacts of land acquisition.[391] Because these insights emerged early, solutions could be tailored, collaborative, and transparent.

The outcomes have been profound and independently studied:

- Lower peak water levels during major floods
- Reduced flood risk for millions of residents
- New wetland and riparian habitats supporting biodiversity
- Improved recreational access and public amenity
- High long-term public trust and acceptance due to early and ongoing engagement
- Economic benefits when avoided flood damage, biodiversity, and recreation are included

Room for the River is now considered one of the world's strongest precedents for designing with nature, not against it. It proves that curiosity, when combined with humility, creates more adaptive, resilient, and socially supported engineering outcomes than control ever could.

Elliott Bay Seawall Replacement, Seattle

By the early 2000s, the century-old Elliott Bay seawall in Seattle was failing. The timber foundations, damaged by marine borers, posed serious risk to the seawall, the roadway above, and the waterfront economy. Traditionally, seawall replacement would focus on structural integrity and flood protection alone.

Instead, the City of Seattle initiated a design process that brought marine biologists, tribal cultural advisors, ecologists, lighting specialists, coastal engineers, urban designers, and community groups into the project before any design options were drawn.

The team asked new questions:

- How do juvenile salmon navigate shoreline shadows?
- What underwater geometry best supports kelp, algae, and small fish?
- How can a seawall act more like a natural shoreline?
- What cultural and ecological stories must be honoured?
- How can a working waterfront also be a habitat corridor?

These weren't afterthoughts; they shaped the engineering brief.

Because risks were identified early (loss of salmon habitat, poor light penetration, migration disruption, accelerated predation), the team could design for them instead of react to them:

- Light-permeable sidewalks allowed daylight to reach the water column
- Habitat benches at varying depths mimicked natural intertidal zones
- Textured seawall panels were used to encourage marine colonisation
- Navigation and cultural pathways for local tribes were formally integrated[444]

The outcomes, tracked through ecological monitoring, include:

- Increased juvenile salmon survival rates along restored sections
- 25-50% increases in marine species richness

- Improved kelp and algae growth due to daylight penetration
- Strengthened shoreline stability and lower long-term maintenance cost
- Higher public support for waterfront renewal because values were visibly embedded in the design[445]

Seattle's seawall shows what happens when technical excellence is expanded - not diluted - by ecological and cultural insight. Curiosity didn't make the project softer; it made it safer, smarter, and more enduring.

From Avoidance to Anticipation

Traditional risk frameworks teach us to avoid failure - to build buffers, contingencies, redundancies. But true resilience doesn't come from avoidance; it comes from anticipation. It's the difference between bracing against a storm and learning to sail with the wind.

One of the best examples of this is in Japan's earthquake engineering. Their infrastructure is not built to avoid earthquakes - that would be impossible - but to anticipate and respond to them.

Buildings sway. Bridges flex. The engineering isn't rigid; it's dynamic, adaptive, living.

That same mindset applies to project decision-making. Instead of trying to prevent every possible disruption, anticipate change. Ask: "What signals would tell us this assumption no longer holds true?" and "How could we adapt quickly when it happens?"

Avoidance feels safe, but it can trap us. Anticipation, on the other hand, invites movement.

Singapore offers one of the clearest real-world demonstrations of moving from risk avoidance to risk anticipation. For decades, the national water agency, PUB, managed flooding by widening and deepening drains - a classic "avoidance" strategy. But in a dense urban environment with limited land and increasingly intense tropical storms, they eventually reached a point where they could not simply build bigger channels forever.

So in the mid-2000s, Singapore shifted its philosophy. Instead of treating water as a threat to be controlled, they asked a new question: "What if our city could anticipate water, adapt with it, and even benefit from it?"

This led to the ABC Waters Program, a transformation of stormwater management that reimagined parks, rivers, and public spaces as flexible, multi-use systems. Flood resilience became something that lived within the landscape, not just in pipes.

One of the most powerful examples is Bishan-Ang Mo Kio Park, where engineers and ecologists worked together to replace a straight concrete drainage canal with a meandering naturalised river. The surrounding parkland was sculpted into floodplains designed to fill during heavy rain, slowing and storing stormwater safely. On dry days, the same spaces function as playgrounds, cycling paths, open fields, and habitats.[383]

This wasn't avoidance - it was anticipation. A deliberate choice to create a system that would flex, not fail.

The outcomes are remarkable:

- Flood risk reduced without enlarging a single drain
- Biodiversity increased dramatically - dragonflies, fish, and bird species all returned
- Park visitation soared
- Water quality improved
- Residents reported stronger connection to nature and a sense of calm
- Maintenance costs stabilised over time due to lower hard-infrastructure burden

In other words, Singapore didn't just derisk a flood challenge - they created a city that lives with water, rather than bracing against it.

This is the essence of anticipation:

- Design for change, not against it.
- Let the system move, breathe, and adapt.
- Turn a hazard into a teacher - sometimes even into a gift.

It's beautiful, functional, and resilient - risk turned into design.

From Blame to Learning

Perhaps the hardest cultural shift in engineering is this one. Things go wrong on projects. I can't think of one that ever went perfectly. Our instinct is to find fault. Who signed off that drawing? Who missed that line in the specification?

But risk isn't reduced by blame. It's reduced by learning. Blame freezes information; learning releases it.

When I run post-project reviews or lessons learned sessions, I always start with one ground rule and a question: "This is not about fault. This is about insight. How can we make this an opportunity?"

Because when we focus on learning, something remarkable happens. Teams become open, real and creative. People share mistakes early instead of hiding them. And we get better, faster.

Of course, the key here is to also have the foundations of a psychologically safe team. A key indicator of this is that teams don't hide problems and mistakes; they embrace them as a team.

One of the clearest examples of shifting from blame to learning comes from an industry where the cost of failure is immediate and unforgiving: aviation. In the 1970s and 1980s, aircraft accidents often revealed a similar pattern. Warning signs were missed, junior crew members stayed silent, and mistakes went unreported because people feared disciplinary action or loss of licence. The system punished human error, so people hid it.

A turning point came after the Tenerife disaster in 1977, which is still the deadliest aviation accident in history. Investigators identified not only technical factors, but deep cultural ones: communication hierarchies, unreported near misses, and a cockpit culture where speaking up was risky.[446] Aviation regulators realised a critical reality: you cannot learn from what people are afraid to say.

This insight drove the creation of what we now call "Just Culture".

Just Culture is a safety and leadership philosophy built on a simple insight: human error is inevitable, but system failure is optional.

Instead of treating mistakes as personal failings, Just Culture treats them as signals - clues about where systems, processes, or assumptions need strengthening. It creates an environment where people can report errors, near misses, or concerns without fear of punishment, as long as their actions were not reckless or intentionally harmful.[447]

In a Just Culture, the central question shifts from "Who is to blame?" to "What can we learn?"

It distinguishes between three types of behaviour:

1. Human error - unintentional slips, lapses, or mistakes. These require comfort and system redesign, not punishment.

2. At-risk behaviour - taking shortcuts or assumptions that seem reasonable in the moment. These require coaching, clarity, and safer systems.

3. Reckless behaviour - conscious disregard for significant risk. This still has consequences.

This clarity removes fear while preserving accountability.

Just Culture doesn't excuse harmful behaviour - it contextualises it, so organisations can fix root causes instead of simply reacting to symptoms. When people don't have to hide mistakes, leaders see risks earlier, learning becomes continuous, and safety improves dramatically.

It is a cultural promise and a structural framework - one that turns honesty into a strength, transparency into prevention, and everyday decisions into opportunities to learn rather than judge.

Under this model, pilots, engineers, and air traffic controllers can report errors and near misses confidentially or anonymously. Instead of asking, "Who is at fault?", investigators ask:

- What conditions made this mistake possible?
- Where did the system set someone up to fail?
- What can we learn before this becomes an accident?

Crucially, Just Culture distinguishes between honest human error, at-risk behaviour, and reckless behaviour. This clarity removed fear while maintaining accountability.[447]

The results have been extraordinary:

- Global fatal accident rates have fallen by over 90% since the 1970s, despite massive growth in air travel.[448]

- NASA's Aviation Safety Reporting System (ASRS) now receives tens of thousands of voluntary reports each year, creating one of the world's richest datasets on near misses and systemic risk.[449]
- Airlines that adopted Just Culture earlier - such as KLM, Qantas, and Delta - show consistently higher safety performance and lower incident severity.
- European regulators formally embedded Just Culture into EU Regulation 376/2014, recognising that learning cultures prevent accidents far more effectively than punitive ones.[450]

Aviation got safer not because humans stopped making mistakes, but because the system made it safe to surface those mistakes.

Imagine if our engineering projects did the same. If risk registers were living documents shaped by open reporting instead of workarounds. If lessons learned weren't buried in post-project folders, but treated as assets that grow in value and tools for collective growth. If raising a problem early was seen as a mark of professionalism, not a threat to reputation.

A learning culture doesn't eliminate error. It illuminates it - and in doing so, transforms the entire system.

What we can learn

When we shift from control to curiosity, from avoidance to anticipation, and from blame to learning, risk management stops being a constraint. It becomes a creative discipline - a way to design with humility and imagination.

The goal of risk isn't to stop things from going wrong; it's to make sure that when they do, we know how to recover - ethically, intelligently, and together.

How can we put this into practice? This is where the 60% Rule and the Decision Matrix comes in.

The 60% Rule

When we ran our very first Power of Engineering event, I thought everything had to be perfect before we went public. It was a free one-day event for Year 9 and 10 girls in Brisbane, and our goal was to shift perceptions of engineering and show students that it could be creative, human, and meaningful.

We dreamed up the event in November. It was scheduled for 8 March the following year - International Women's Day. That might sound like a long lead time, but schools in Brisbane have nearly two months of school holidays over December and January, and the funding was from the Queensland Government Office for Women, which meant significant reporting for the project. The window for notifying schools was also incredibly tight.

My instinct as an engineer kicked in. I wanted every detail confirmed before we announced anything. The speakers, the workshops, the site tours - every box ticked, every risk mitigated.

Then someone asked me a simple question that completely changed how I approached decision-making: "Do you really need all of that before you tell schools? What if you just shared the purpose - to inspire and shift perceptions - and let them know the details will follow?"

That was my 60% moment.

I realised I didn't need perfect information to begin; I just needed enough. So we sent out a simple flyer as soon as schools went back late January with the name, the date, and the purpose and trusted the rest would fall into place.

It did. The event filled up within three days and, by the end of the week, we had 200 students registered - double the capacity of the event. This resulted in 131 students there on the day and Queensland University of Technology, our partner and host, offering to put on a second event for the students who missed out. A local television program for kids, *Totally Wild*, which I watched growing up, even came and filmed the day.

Schools registered based on the potential outcome, not detail - and we refined everything else as we went.

This experience taught me something I've carried ever since: perfectionism delays impact, and sometimes means the outcome is missed or not achieved at all. Sometimes the best thing we can do as engineers, as leaders, and as humans is to act on purpose before everything is certain.

That's the 60% Rule at its heart - enough clarity to start, enough curiosity to adjust, enough courage to trust the process.

Acting with Confidence in the Grey Zone

Every engineer has lived this moment. You're in a design review or project meeting, and someone says, "Let's just wait until we have a bit more information."

It sounds reasonable. Responsible, even. It is often hard to argue with this, especially in a room full of other engineers.

But what happens next? Weeks pass. The team hesitates.The problem grows more complicated, or the window for influence closes.

By the time the data finally arrives, the decision has already been made - by default, through delay.

The reality is: we almost never have 100% of the information. And waiting for it doesn't remove risk - it often amplifies it.

The Principle: Decide When You Have 60 Percent of the Information

The 60% Rule is simple: make the decision when you have roughly 60% of the information you wish you had.

That's the moment when you know enough to act, but still have the flexibility to adjust. If you wait for 80 or 90%, the opportunity may be gone. If you act at 20%, you're guessing. At 60%, you're learning.

It's the sweet spot between paralysis and recklessness - the grey zone where good judgment lives. To be clear - I am not saying throw out the design standards here and only design 60% of the design. This is different.

In my own life, it has taken time and unlearning but I've come to trust this instinct deeply. Whether I was navigating difficult medical decisions during my pregnancy or making the call as a leader of a team, there was never a point when I had all the facts. But I had enough to see patterns emerging, enough to understand the trade-offs, enough to know that action mattered more than perfection.

That's the art of leadership in engineering: acting early enough to influence the outcome, and humbly enough to adjust when new information arrives.

The Cost of Waiting versus the Cost of Acting

In every decision, there's a hidden balance between two costs:

1. The cost of waiting - lost time, lost trust, lost opportunity.
2. The cost of acting - the potential need to adjust, redesign, or correct later.

Great decision-makers weigh these two costs constantly. If the cost of waiting exceeds the cost of adjusting later, the decision should be made now.

Think of it as a simple equation:

Decision Confidence = (Information Gained - Cost of Delay) ÷ Flexibility Remaining

This way of thinking turns decision-making into a dynamic process - one that adapts as projects evolve, rather than freezing in fear of being wrong.

NASA's engineers during the Apollo 13 crisis understood this perfectly. They didn't have all the data. Telemetry was incomplete.[440] But they made real-time decisions based on 60% of what they knew - because the cost of waiting was death. They trusted disciplined intuition, systems thinking, and communication - and brought the astronauts home safely.

Reversible versus Irreversible Decisions

Not all decisions carry the same weight. Some are reversible - you can test, learn, and pivot. Others are irreversible - once made, they lock in cost, time, and consequence.

Good leaders act quickly on reversible decisions to keep learning flowing, and delay irreversible ones until they have enough confidence.

As Jeff Bezos said in his 2016 annual letter to shareholders of Amazon: "Most decisions should probably be made with somewhere around 70% of the information you wish you had. If you wait for 90%, in most cases you're probably being slow."

For "two-door" decisions - those that are reversible - this pace enables learning, adaptation, and agility. In engineering design and early project phases, adopting that mindset can keep options open, encourage prototyping, and reduce sunk-cost pressure.

For the irreversible decisions - final route alignment, major structural geometry, material choice - treat them with the gravity they deserve, and step forward only when you have robust evidence and confidence.

When Data Runs Out, Values Take Over

Every engineer has a moment where the spreadsheet stops helping. Where the graph runs out, the simulation ends, and you're left staring at a blank space that says: you must decide.

That's the moment where your values kick in. Not as a philosophical ideal, but as a lived, everyday practice - in the meeting, on the site, in the brief thought before signing off on a drawing.

Engineering is not just about what we can build; it's about what we should. And when data can't give the answer, values must.

Values as the Dimension of Engineering Judgment

Data tells us what's possible. Analysis tells us what's efficient. But values tell us what's right. But whose values? The community consultation can help us determine project values.

In many ways, engineering has always been a values-based profession. We hold safety paramount. We sign drawings with our names. We know what it means to carry responsibility. But the challenges of our time - climate change, resource depletion, inequality - require ethics that stretch beyond compliance.

It's not enough to avoid harm; we have to design for good. That means asking not just "Is it safe?" but "Is it fair?" Not just "Does it work?" but "Who does it serve?"

Our values are foundational in guiding our engineering judgement. We have default values running like our background programming all the time. It is important we see and understand their influence on our judgement as engineers and on our projects.

Like our hidden, default DNA, we have hidden values too. We must be proactive in seeing these and aligning ourselves and our work to values for our future.

Case Study: The Challenger Disaster - The Silence That Spoke Too Late

Many of us know the famous and devastating story of the Challenger Disaster. On the night before the launch of the Challenger Space Shuttle in 1986, engineers from Morton Thiokol warned NASA officials that the O-rings in the rocket boosters might fail in the cold. The data was inconclusive, but the risk was real.

The engineers hesitated. Management hesitated. There were pressures - political, public, emotional. The decision to launch went ahead.

Seventy-three seconds after liftoff, the shuttle exploded, killing all seven astronauts.

In the aftermath, physicist Richard Feynman famously demonstrated how the O-rings became brittle in freezing water. But the cause wasn't material - it was cultural.[451]

As Feynman said, "For a successful technology, reality must take precedence over public relations, for nature cannot be fooled."[451]

The Challenger story isn't about technical failure. It's about what happens when a project has misaligned values and courage breaks down - when fear of consequence outweighs care for people.

Case Study: Flint Water Crisis - When Cost Replaced Care

In 2014, officials in Flint, Michigan, switched the city's water source to save money. Almost immediately, residents complained about taste, smell, and colour. Children began showing signs of lead poisoning.

But because the data wasn't yet "conclusive", the decision-makers delayed action. Months turned into years before the reality was undeniable: the pipes had corroded, leaching toxic levels of lead into the water supply.

It was a catastrophe of ethics, not of chemistry. The data was there. The community was vocal. What failed was valuing the right thing.[452]

The lesson is simple but sobering. When cost efficiency replaces compassion, systems fail - not because they're badly designed, but because they've lost the values they are aligned to.

Case Study: Māori-Led Infrastructure, Aotearoa New Zealand - Ethics as Relationship

Across the Tasman, a different story unfolds. Māori-led infrastructure projects are showing what ethical engineering can look like when Care for Country (or Kaitiakitanga) leads the design.

One example is the Te Ahi Tāmaki Makaurau (Auckland's Integrated Transport Network) planning initiative. Instead of consulting Māori as a stakeholder group after decisions were made, the project embedded Māori leadership from the outset - shaping everything from route selection to ecological restoration.[453]

The result wasn't just a better social outcome; it was a more resilient system. Decisions considered generational impacts, cultural narratives, and ecological balance.

Values here weren't a checklist. It was a relationship - between people, place, and future. This is engineering through the lens of reciprocity: to build is to belong, to decide is to care.

Case Study: The Roman Aqueducts - Ethics of Maintenance

Even in antiquity, Roman engineers demonstrated an ethical sense of stewardship that feels modern. Their aqueducts - some still functional two millennia later - were maintained through a legal and civic framework that recognised water as a public good.

Engineers swore oaths, similar to today's professional codes, to keep the systems pure and operational. They designed inspection shafts and maintenance access points centuries before asset management existed.[320]

They didn't just build for their lifetime. They built for the city's soul. That's values set in stone - engineering that sees maintenance as morality.

Case Study: Google's Project Aristotle - The Values of Teams

Sometimes values aren't about the big decisions, but the small ones that shape how we work. When Google studied what made its best teams perform, the top factor wasn't IQ, data skill, or technical excellence. It was psychological safety.[454]

Teams that trusted one another - where people could admit mistakes, raise doubts, and take interpersonal risks - were far more innovative and resilient. Embedding values in culture, it turned out, was the best risk mitigation strategy.

In other words: good values is good engineering.

When Data Ends, Dialogue Begins

When the data runs out, we don't stop thinking - we start talking. Values live in dialogue: the willingness to hear discomfort, to question the status quo, to pause when everyone else is rushing ahead.

That's why value based ethical decision-making isn't a solo act. It's a team sport. It lives in review meetings, site conversations, stakeholder engagement, and moments of pause before we commit.

A board member of a large piece of infrastructure shared something that has stuck with me for my whole career: "We can build anything. The question is - should we?"

It's the question we must all keep asking.

Values don't replace risk matrices or decision trees. It humanises them. It reminds us that every calculation is a moral act - and every decision a story about who we choose to be.

Consider: what's one decision you've made recently where the data didn't tell the full story? Who else could have added the missing perspective - and what might have changed if they had?

The Courage of Values-Based Engineers

Values-based engineers don't always make the easiest calls. They make the right ones - even when it costs them comfort, time, or approval.

They understand that uncertainty is not a flaw in engineering; it's the field we work within. And when the data runs out, what remains isn't just intuition. It's courage.

Because ultimately, every drawing, bridge, algorithm, and tunnel carries an invisible signature: the moment when someone decided to care.

The 60/30/10 Review Framework

To help teams apply this mindset, I often use what I call the **60/30/10 Review**:

- **60% Data:** What do we know from facts, tests, and models?
- **30% Consultation:** What do people who've seen similar problems think? What wisdom or experience can we borrow? You can ask industry experts, mentors, those with lived experience, or people in different disciplines.
- **10% Intuition:** What does our collective gut say? What pattern do we sense that data hasn't yet shown? This is your engineering judgement.

When the first 90% (data + consultation) aligns - or when intuition clearly signals something's off - that's your cue to act.

This framework builds confidence and shared understanding, while legitimising intuition as a valid part of engineering judgment. Intuition actually isn't guesswork - it's pattern recognition built over thousands of decisions, lived experience and trust in yourself.

Case Study: Apollo Engineers

The Apollo engineers faced decisions where no historical precedent existed. They were sending humans into an environment no one had survived in before - and much of their data was extrapolated, not tested.[440]

One famous example: the choice to use a lunar orbit rendezvous strategy - separating the spacecraft and landing only a small portion on the Moon. At the time, it seemed risky, even reckless. The data was incomplete; simulations were limited.

But NASA engineers used what we would now call the 60/30/10 framework:

- **Data:** Enough aerodynamic modelling to suggest feasibility
- **Consultation:** Heated debate among mission scientists and pilots
- **Intuition:** Gut conviction that this solution could work

It did. And that decision - made with incomplete information - remains one of the boldest, most successful calls in human history.

Case Study: The Venice MOSE Barriers - When Judgment Faltered

The MOSE flood-barrier system in Venice was born from urgent necessity: rising sea levels, increasingly frequent storm surges, and a sinking city running out of time. Engineers were tasked with designing a movable barrier system capable of protecting a 1,600-year-old city from catastrophic flooding - something no modern infrastructure had attempted at this scale.[455]

MOSE became a lesson in what happens when judgment is compromised, even when technical expertise is strong. Over decades, the project suffered political interference, corruption scandals, stalled decisions, and shifting priorities. It became a case study in how values - not calculations - determine outcomes when data is uncertain.

The decision-making environment looked like this:

- **Data:** Engineers had early modelling showing that rising tides threatened Venice's long-term survival and that a movable barrier could work in principle.
- **Consultation:** Marine ecologists, hydrologists, and cultural heritage experts raised concerns about lagoon ecology, sediment flows, and long-term impacts - but these voices were sidelined as political urgency grew.
- **Intuition:** Instead of engineering intuition guiding the final judgment, political intuition did: choosing speed, symbolism, and visibility over caution, transparency, and ecological stewardship.

The result? The system now works - it successfully protected Venice during the 2019 and 2023 acqua alta events. But it arrived decades late, at more than double the original cost, and only after widespread public distrust and a €1 billion corruption scandal.

MOSE is a reminder that engineering judgment fails not when data is incomplete, but when values are compromised.

Case Study: Indigenous Fire Management in Australia: Judgment Beyond Data

For tens of thousands of years, Aboriginal communities across Australia have practiced cool burning - low-intensity, patchy fires that reduce fuel load, encourage new growth, protect habitats, and prevent dangerous megafires. It is one of the most sophisticated, adaptive environmental management systems on Earth.

When European settlers arrived, they banned the practice. The result was immediate and devastating: fuel accumulated, landscapes changed, and catastrophic bushfires grew in frequency and intensity. Only in recent decades has Western science begun to recognise the depth of Indigenous knowledge - empirical, ecological, and cultural - embedded in these practices.

Cool burning is a model of engineering judgment without modern instruments:

- **Data:** Generations of observational evidence about vegetation patterns, wind, species behaviour, and seasonal cycles - recorded through oral tradition rather than sensors.
- **Consultation:** Community collaboration: Elders, rangers, and land custodians interpreting conditions, reading Country, and deciding together how to apply fire.
- **Intuition:** Cultural intuition - the embodied sense of when land is ready, how much heat it can take, and how fire can be used to heal, not harm.

Today, Indigenous-led programs like the North Australian Fire Management Initiative[456] have reduced carbon emissions, protected biodiversity, and prevented large-scale destructive wildfires. Scientific studies confirm that these early dry-season burns reduce late-season megafire risk by up to 50-60% in northern savannas.[457]

Indigenous fire management shows that some of the world's best risk systems were invented long before risk frameworks - and that learning beats control every time.

Case Study: Tokyo's G-Cans Flood System: Anticipation Made Tangible

- Tokyo's Metropolitan Area Outer Underground Discharge Channel - commonly known as G-Cans - is one of the most extraordinary flood protection systems on the planet. Built beneath Saitama Prefecture, it consists of giant subterranean "cathedrals" that store and redirect stormwater during typhoons, protecting millions of residents.
- What makes G-Cans remarkable is not scale, but foresight: it was built before climate change accelerated, before modelling predicted record rainfall, and before catastrophic floods struck the region.

- In the 1980s and 1990s, Japan's engineers made a bold call: prepare for a future you can't fully see yet.
- Their judgment looked like this:
  - **Data:** Hydrological trends showing increasing typhoon intensity, river overflow patterns, and early climate projections - incomplete but directionally clear.
  - **Consultation:** Urban planners, local governments, emergency response teams, and community groups shaped design priorities: protect life, protect infrastructure, minimise surface disruption.
  - **Intuition:** National intuition built on Japan's disaster history - the understanding that anticipation saves lives and that resilience must be built before catastrophe, not after.

The result?

- A 6.3 km underground system
- Storage capacity of 670,000 m^3
- Pressure-adjusting turbines comparable in size to jet engines
- A system capable of diverting an entire river's surge flow during peak rainfall

Since becoming operational, G-Cans has protected Tokyo from billions of dollars in flood damage and has become a global model for anticipatory risk engineering.[458]

This demonstrates resilient systems aren't the ones that avoid risk, but the ones that expect it - and are ready to adapt.

The Decision Matrix - Data + Values + Judgment = Decision

Engineering decisions have always been framed as rational. This is more of our hidden DNA: that if you gather enough data, analyse it well, and apply the right equations, the "best" decision will emerge.

But what happens when the data runs out? When the models contradict each other, or when the variables multiply faster than the answers? That's where most of the real work of engineering begins - in the space where facts meet judgment and where trust steps up.

I've come to think of this space as a kind of decision triangle - three forces in constant tension: Data, Values, and Judgment. When all three are present, decisions hold. When one dominates, they collapse.

"Every time we take a leap into the unknown, we're trusting that the ground will be there when we land." - Rachel Botsman Trust Expert

In engineering, there is absolutely a huge amount of data and science that is well researched, documented, and practiced in our standards and by experts in the industry. However, there are all the daily decisions we need to make - the ones that don't fit neatly into the manual because we are working with people, the natural environment, and complex systems. Every design decision in some ways is a small act of faith - faith in your data, your team, your process, and your judgment.

Botsman calls these "trust leaps": moments when we move forward without all the information, relying on relationships, competence, and shared intent to bridge the gap. They're not reckless leaps - they're structured acts of confidence.

If we didn't trust the people in our engineering teams, our projects wouldn't happen.

Engineers are daily building trust scaffolds: risk registers, peer reviews, simulations, and testing protocols.

The next time you're sitting in a design review or a project meeting, ask yourself:

- What leap are we being asked to take here?
- What confidence do we need to make it safely?
- Who do we trust - and to do what?

Progress doesn't come from having more and more perfect data; it comes from making trust visible in the way we decide, design, and deliver.

When Values Win in Decision Making

In my mid 20s, I co-founded a business with two partners. It started as a shared vision - three women building something that mattered.

But as the company grew and we prepared to take on investment, we faced one of those uncomfortable decisions that tested both friendship and principle.

One co-founder had been working part-time on the business while raising a young baby. Another had gone all-in - full-time, unpaid, for 18 months. I was somewhere in between, transitioning from part-time to full-time.

When we negotiated the investment deal, our new investor wanted 25% of the company. That left 75% to split between us - and the question was, how?

On one hand and logically, the split could reflect effort to date. The co-founder who'd worked full-time and unpaid wanted slightly more - 26 or 27%. That made sense. She'd sacrificed the most time and effort on the business. However this would mean that for majority decisions, she could have gone with our 25% investor to make the final call.

Something shifted in our discussions. We started talking not just about equity, but about equity in spirit. What we valued most was equality - the sense that we were building something together, not tallying hours but sharing purpose.

So we decided to split the ownership equally - 25% each. Instead of adjusting ownership, we agreed that when the business succeeded, we'd pay the difference in back salary to acknowledge the work already done.

It was one of the hardest conversations I've ever had, especially as we were all close friends and have different views. But it was also one of the most clarifying. It showed me something I now see in engineering all the time: when data and logic can't decide, values must.

That decision could easily have gone another way - toward hierarchy, resentment, or legal wrangling. Instead, it went toward trust. We left the conversation aligned, equal, and more committed than before.

That's what happens when values become the compass in uncertainty. They don't just tell you what's right; they show you who you are.

1. Data - The Ground Truth

Data gives us the scaffolding. It tells us what's measurable, what's observable, what's repeatable. Without it, we drift into speculation. But when we rely on it too much, we risk building systems that optimise numbers, not lives.

Consider a transport project where a traffic model shows that removing pedestrian crossings would increase vehicle throughput by 12%. From a data perspective, it is efficient and effective for cars. However, looking more closely at the site, you can now see the crossing connects a retirement village to a park. Efficiency for one group means exclusion for another. The data was right - but the decision would be wrong.

That's the first piece of the Decision Matrix: data can describe, but it can't decide.

2. Values - The Compass

Values step in where data stops. They remind us what matters most - who we are designing for, and why.

When we add values into the equation, the decision shifts from "What works?" to "What's right?"

Imagine a regional water supply system, where the community asked for design changes that didn't make much technical difference to performance but made a huge difference to trust. The engineers include them, even though it means a few more weeks of work.

The result? Higher community satisfaction, fewer complaints, and a stronger relationship between the council and the public.

Values don't make decisions easier - they make them clearer. They help us choose not just the most efficient path, but the one aligned with our ethics, our humanity, and our long-term vision.

3. Judgment - The Bridge

And then there's judgment - the often invisible force that links data and values. It's the ability to hold complexity, weigh competing realities, and act with confidence - even when you can't be certain.

Judgment grows with experience, but it thrives in cultures that allow people to talk about uncertainty. When engineers can openly say, "I don't know yet," it doesn't weaken trust - it strengthens it, when you are surrounded in the right culture with the right people.

Good judgment isn't just an individual skill - it's a collective practice. It's built in design reviews where multiple disciplines share their lenses. It's refined in risk workshops where psychological safety allows people to surface concerns. And it's reinforced when leaders model curiosity instead of control.

When Data Is Thin, Values Carry More Weight

The Decision Matrix combines data, values, and judgment into a single system for making confident, ethical choices when certainty isn't possible. It becomes most powerful when one leg - usually data - is weak. In those moments, values and judgment must compensate.

Take the example of a flood-prone city debating whether to relocate a community or invest in new protective infrastructure. The hydrological models can predict probabilities, but they can't predict the trauma of displacement, the erosion of community identity, or the long-term mental health impacts. Those decisions live in the space of values and judgment. The wrong move would be to say, "We don't have enough data - so let's wait." The right move is to say, "We have enough to understand the risks. Now let's decide together, guided by care, equity, and legacy."

That's how data, values, and judgment work together:

- **Data** shows what is.
- **Values** reveal what matters.
- **Judgment** decides what to do next.

It's a dynamic, iterative process - not a formula. But it's how engineers make confident, ethical decisions in an uncertain world.

A Simple Tool: The Decision Matrix in Practice

Here's how teams can use three simple tools to apply this thinking practically - in project meetings, design reviews, or strategic decisions. It's not just about managing risk; it's about designing with integrity.

1. **The 60/30/10 Review**
    - 60% Data: Decide when you have enough information to move forward.
    - 30% Consultation: Involve others. Ask "what might we be missing?"
    - 10% Intuition: Trust professional and lived experience to guide the final step. It's not about replacing rigour - it's about balancing it.

2. **Pre-Mortem Workshops**
    - Before committing to a decision, imagine it's six months later and the decision has failed.
    - Ask: What went wrong? What did we miss? This flips the risk mindset from defensive to proactive, and invites everyone to think critically without assigning blame.

3. **Decision Journals**
    - Record the reasoning behind major project decisions, not just the outcomes.
    - Include what data you had, what values you used, and what trade-offs you accepted. When revisited later, these journals become powerful learning tools for future teams - engineering not just structures, but wisdom.

Here is how this could look on some project examples you may have faced:

Decision Factor	Data Evidence	Values Consideration	Judgment Call/ Team Reflection
Safety Upgrade Cost	Quantified risk reduction of 15%	Does this reflect our value of care and inclusion?	Is the additional cost justified by public trust and peace of mind?
Design Speed Change	Modelled efficiency gain of 10%	Does this align with our value of community-first and wellbeing?	Will this improve or harm daily user experience?
Material Choice	Lower upfront cost, higher maintenance	Does this respect legacy and whole-of-life thinking?	How might future generations inherit this choice?
Engagement Process	Minimal compliance achieved	Are we listening deeply, or just informing?	What's the risk of lost trust if we don't involve them more?

When the data runs out, what will guide you - fear of being wrong, or courage to do what's right?

Decision Recovery - How We Learn, Adapt, and Rebuild After Mistakes

Every engineer remembers the moment they realised something didn't go as planned. A number that just didn't align. A detail that looked fine on paper but failed in real life. A meeting when everyone nodded, but no one really spoke.

To think of failure as a purely negative outcome is to miss its greatest gift. Failure is feedback. Data. A signpost. A teacher.

James Dyson, the inventor behind Dyson vacuums and more, faced it head-on. We know that over four years, he built **5,126 prototypes** that failed - but the 5,127th worked. Dyson wrote: "Learning by failure is a remarkably good way of gaining knowledge. Failure is to be welcomed rather than avoided."[74]

For Dyson, it wasn't just perseverance. It was structuring his work so that each "failure" taught a precise lesson: one variable changed at a time, one hypothesis tested, one question answered.

As engineers, our most important tool isn't always a finite element model or cost-benefit analysis. Sometimes it's the willingness to test, to measure what doesn't work, and to build again.

Case Study: The Tacoma Narrows Bridge

In 1940, the Tacoma Narrows Bridge in Washington State collapsed spectacularly in high winds - only a few months after it opened. The famous footage of the bridge oscillating wildly before breaking apart is still shown in engineering schools today.

It was a devastating failure, and it changed structural engineering forever. At the time, engineers didn't fully understand aeroelastic flutter - how wind could interact with a bridge's flexible structure to create self-reinforcing vibrations.

After the collapse, research into wind-structure interaction surged. Wind tunnels, aerodynamic modelling, and flexible design principles became standard practice.[459]

The lesson? The bridge failed, but the profession evolved. The world's safest modern bridges - from the Golden Gate to Denmark's Great Belt - are built on the data born of that failure.

That's the paradox of progress: every bridge that stands strong today does so because one bridge once fell.

Case Study: Fukushima Daiichi - The Cost of Complacency

In 2011, a 9.0-magnitude earthquake and tsunami struck Japan, devastating coastal regions and crippling the Fukushima Daiichi nuclear power plant. Backup generators were flooded, cooling systems failed, and radioactive material was released.

It was a disaster on multiple levels - technical, environmental, and social. But in the aftermath, the analysis revealed something deeper: systemic complacency.

The plant's seawalls were built for tsunamis up to 5.7 metres high. The waves that hit were 13 metres. Historical data had warned this could happen but decisions were deferred, risks rationalised.[460]

Since then, Japan's nuclear safety standards have been overhauled, introducing defence in depth systems - redundant layers of safety designed for "black swan" events. The world learned that resilience isn't built by predicting everything - it's built by preparing for the unimaginable.

Reframing Recovery: From Blame to Learning

In many organisations, failure triggers one of two responses: blame or learning. Blame shuts down conversation. It hides failure. It repeats mistakes. Learning opens questions. It logs what went wrong and evolves what comes next.

After every failure, there are two paths. The first is blame - who's responsible, who's at fault. The second is learning - what the system taught us, what patterns we missed, what needs to evolve.

The first satisfies emotion. The second serves the future.

Blame ends the story. Learning extends it.

In organisations with strong cultures of judgment and care, post-incident reviews aren't about punishment - they're about pattern recognition. They ask:

- What signals did we ignore?
- Where did communication fail?
- How can we embed learning into the next project?

Here's a simple structure engineers and teams can use:

Stage	Action	Purpose
Acknowledge	Name the issue factually, without reacting	Builds shared awareness and transparency
Analyse	Distinguish between technical and cultural root causes	Turns failure into data, not disaster
Adapt	Capture lessons as design inputs, procedural updates	Ensures learning becomes institutionalised
Apply	Use improved process or design in next project cycle	Converts insight into action
Amplify	Share findings across teams; celebrate transparency	Builds a culture where recovery becomes the norm

Engineering is not just about constructing. It's about caring for what we create. And when things go off plan - even just a little - recovery is not optional. It's essential. Because every asset we hand over will one day require repair, adaptation, or even letting go.

Ask yourself: what's a small failure in your past that changed how you work forever? How can you make your next Lessons Learned review less about liability, and more about learning?

The Courage to Repair

Engineering is hopeful: we build because the world can be better. But our legacy isn't just measured in what stands tomorrow - it's in how we respond when things don't hold up as planned.

The Romans called it cura. Indigenous Australians call it Care for Country. We may simply call it stewardship. And when future generations walk over our structures, drive on our roads, live in our cities, I hope they say: they didn't always get it right, but they always cared enough to try again.

The Art of Deciding

In the end, decision-making is never about having all the answers. It's about having enough integrity, courage, and humility to act - and then to adapt when reality teaches us something new.

When I think back on the moments that have defined my own career, very few of them were about perfect information. They were about people. They were about trust - trusting the team, trusting the process, trusting myself to make the best decision possible with what we knew, and to keep learning when we didn't.

That's what this chapter has really been about. Not erasing uncertainty, but reframing it. Not fearing risk, but designing with it. Not punishing mistakes, but using them to make us wiser.

Uncertainty is not the enemy of engineering - it's the medium we work in. Like water for a sailor or air for a pilot, it's what surrounds us. We can't remove it, but we can learn to navigate it with skill, ethics, and care.

The Rhythm of Decision-Making

Every good decision follows a rhythm:

- **Curiosity** opens the question.
- **Values** shape the options.
- **Judgment** makes the call.
- **Reflection** carries the learning forward.

This is what keeps engineering human. When data runs out, dialogue begins. When plans fail, purpose emerges. And when something breaks, repair becomes a form of care.

The New Definition of De-Risking

De-risking used to mean eliminating the unknown. Today, it means embracing it - early, honestly, and together.

It means shifting from control to curiosity, from blame to learning, from isolation to collaboration. It means holding every decision not as a verdict, but as a hypothesis - one that reality will test, and that we will refine.

This is what the next generation of engineers must learn: that judgment and empathy belong together. That technical excellence and values based courage are not separate skills. That risk is not something to be managed away - it's something to be managed with.

And maybe that's what the future of engineering leadership really is: not the absence of error, but the presence of care.

The Takeaway

When risk is defined narrowly, decisions optimise for what can be measured, insured, or defended on paper. Technical failure is avoided, but other forms of risk are inherently transferred, to communities, ecosystems, workforces, and future generations.

De-risking by design expands the lens. It recognises that some of the greatest risks in engineering are not structural or financial, but relational and systemic: loss of trust, social licence, adaptability, and legitimacy. When these risks are ignored, they eventually return, often as cost overruns, delays, legal disputes, or reputational damage.

The next chapter turns to the economic reality behind these choices. It examines the business case for the future, showing how long-term thinking, care, and regenerative design are not just ethical imperatives, but strategic ones. What looks like "doing the right thing" is often also the smartest decision, when value is measured over the life of a system rather than the life of a project.

Chapter 10: The Business Case for the Future

For decades, the business case in engineering has been framed around speed, certainty, and short-term return. Projects are justified through cost-benefit ratios, delivery timelines, and narrowly defined performance metrics. If the numbers add up, the decision is considered sound.

But many of the most expensive failures in engineering did not fail on paper. They failed in trust, adaptability, and long-term relevance. Costs were deferred rather than avoided, often to future generations. Risks were shifted rather than resolved. What looked efficient at the outset became a burden over time.

This chapter reframes the business case for the future. It explores how designing for people, place, and generations is not a departure from commercial logic, but an evolution of it. When value is assessed across the full life of a system, care, resilience, and legitimacy are not soft considerations. They become strategic advantages, and essential ingredients for sustainable success.

Why a Business Case? Why Now?

We are living in a moment when engineering is being asked to do more than deliver projects on time and on budget. Around the world, communities, governments, investors, and young engineers are raising the same expectation: build differently. Build with care for people, for Country, and for the generations who will inherit what we design. Build with an understanding that the systems we create shape society as much as they solve problems. Build with the courage to challenge the inherited DNA of our profession.

For leaders, this expectation is no longer philosophical - it is commercial. Trust, legitimacy, cultural safety, climate resilience, and community relationships now sit alongside cost, time, and safety as critical determinants of project success.

When these elements collapse, they create the most expensive failures in engineering: legal battles, redesigns, political intervention, project cancellations, reputational damage, and loss of social licence. Almost every high-profile engineering failure of the past decade has had one thing in common - not a technical error, but a values error. A decision narrowly optimised on paper, yet misaligned with people, place, or context in reality.

This is why leaders ask: where is the measurable business case for regenerative engineering? Not because they doubt the importance of doing the right thing, but because they are accountable for outcomes. They need confidence that values-driven decisions improve performance, reduce risk, and support the long-term viability of their organisations. They want metrics that can stand in front of boards, CFOs, procurement panels, and ministers. They want tools they can implement tomorrow, not concepts they appreciate philosophically.

In actual fact: the business case already exists. It is just not yet expressed in the language engineering organisations are accustomed to. The evidence shows that early community engagement reduces delays and rework. Values alignment improves team cohesion and innovation. Indigenous partnership deepens legitimacy and de-risks complex projects. Circular materials reduce lifecycle cost. Nature-based solutions outperform grey infrastructure in resilience and whole-of-life value. Trust, in measurable terms, accelerates approvals and reduces opposition. Purpose attracts and retains talent - which might be the most urgent competitive advantage of all.

We are in a new engineering economy - one where the costliest risks are no longer technical, but social, environmental, and reputational as well. The old business case was built for a world of predictable climate, stable communities, and narrow expectations. That world no longer exists.

This chapter provides leaders, boards, procurement teams, and engineers with a structure they can use immediately: a clear, quantifiable business case for designing with intention, values, and care. Not as an add-on, but as a core engineering competency. Not as a moral imperative alone, but as a strategic and financial one. Because the future will reward organisations who design for connection, regeneration, and long-term value - and it will expose those who continue to see engineering as separate from the world it shapes.

This is why a business case matters. And this is why it matters *now*.

The Limits of the Traditional Engineering Business Case

For decades, engineering business cases have been built around a narrow set of assumptions: optimise CAPEX and OPEX, manage engineering risk, compress schedules, and deliver the asset as specified. These are important metrics - but on their own, they can create a dangerously incomplete picture of project value. When we optimise only for cost, time, and technical compliance, we generate false economies: designs that appear efficient on spreadsheets but fail in the real world.

Traditional business cases systematically underestimate costs and under-measure benefits because they exclude the factors that most often derail projects. They rarely account for the financial and reputational impact of community opposition, cultural heritage breaches, misaligned values, or loss of trust. They overlook the costs of delays, litigation, redesign, political intervention, and media fallout - even though these routinely far exceed the initial CAPEX savings that "value engineering" was meant to achieve.[58]

They also fail to assess climate vulnerability or future operating conditions and sometimes forget to factor in decommissioning and remediation which would happen generations into the future, leaving assets exposed to risks outside their original design envelope. And they treat whole-of-life outcomes as secondary rather than central, focusing on short-term budget constraints at the expense of long-term performance, maintenance, resilience, and social acceptance.[58, 441]

When the model is narrow, so are the decisions. And when decisions are narrow, the true cost of engineering becomes invisible: the cost of rework, the cost of repairing trust, the cost of community resistance, the cost of ecological decline, and the cost of systems that are obsolete the moment they are commissioned.

In short: the traditional business case optimises the wrong parts of the system. It measures what is easy, not what is important and not from a legacy view. And in a world that demands resilience, legitimacy, and long-term value, that is no longer enough.

What the Data Shows: The Measurable Benefits of Regenerative Engineering

Across industries and continents, the data is telling a consistent story: when organisations design with values, community, and Country in mind, project performance improves. Regenerative engineering is not an aspiration - it is an evidence-based pathway to better commercial, technical, and social outcomes. What follows is a high-level snapshot of the measurable benefits, which we explore in detail throughout this chapter.

Early stakeholder engagement reduces objections and delays

Multiple infrastructure studies have found that early stakeholder engagement is associated with significantly fewer objections, approvals delays, and legal challenges compared with projects that engage late or reactively. Studies in infrastructure delivery show that when communities are engaged early - not after decisions are made - project teams face fewer redesign cycles, smoother approvals, and lower legal costs. Early engagement uncovers risks long before they become expensive problems, making upstream conversation one of the most commercially effective tools available to engineering organisations.[58, 461]

Nature-based solutions can outperform grey infrastructure on whole-of-life value

In many contexts, nature-based solutions have been shown to outperform conventional grey infrastructure on lifecycle cost, adaptability, and climate resilience. Restored waterways, wetlands, dunes, and vegetation buffers absorb storm events better, adapt over time, and cost less to maintain than concrete channels or hard barriers. They deliver multiple benefits simultaneously: climate resilience, cooling, biodiversity uplift, water quality improvements, and community amenity - none of which are delivered by traditional single-purpose assets.[390]

Circular and low-carbon materials reduce emissions and lifecycle costs

Circular material strategies can reduce emissions and whole-of-life costs, particularly when reuse and low-energy processing replace virgin material extraction. Recycled aggregates, reclaimed concrete, modular components, and closed-loop materials now meet engineering performance standards while lowering procurement costs, waste, and carbon output. When lifecycle cost is included, circular solutions often outperform virgin materials because they reduce maintenance, extraction costs, and future liabilities.[462]

Indigenous-led or co-designed projects achieve higher community acceptance and lower long-term maintenance burdens

Evidence from Indigenous-led and co-designed projects shows strong associations with improved community acceptance and, in many cases, more durable long-term outcomes. Projects shaped with Traditional Owners experience fewer objections, faster approvals, and stronger social licence because cultural, environmental, and community risks are identified early. Indigenous knowledge also improves ecological health and land stewardship, reducing the maintenance and remediation costs that arise when designs disregard Country's needs.[456]

High-trust engineering organisations outperform low-trust ones across productivity, retention, and innovation

Organisational research consistently shows strong correlations between high-trust cultures and improved productivity, retention, innovation, and safety performance. Teams that feel psychologically safe question assumptions earlier, surface risks sooner, and collaborate more effectively. Trust reduces friction, accelerates problem-solving, and lowers turnover - a significant cost saving in a sector facing global talent shortages.[454] For early-career engineers especially, values-aligned workplaces are a key driver of attraction and long-term retention.

These findings make one thing clear: Regenerative engineering is not a cost - it is a performance strategy. The organisations that embrace it reduce risk, strengthen trust, increase value, and build assets that last. Those who ignore it face the rising costs of rework, resistance, and irrelevance.

Cost Savings Through Early Engagement

Early engagement is one of the most commercially powerful tools available to engineers - yet it is often treated as optional, "soft", or something to do once the technical decisions are already made. But the data is clear, and the evidence from countless projects confirms it: when engineers engage early with the people, places, and systems their work affects, they save money, reduce delays, and avoid unnecessary rework.

Rework is one of the biggest hidden costs in engineering. It shows up as redesigns, scope changes, community objections, shifting approvals, political pressure, and reputational fallout. Most of this rework is not caused by technical error - it is caused by misalignment. Misalignment between what the design assumes and what the community, Country, environment, or broader system actually needs.[58]

Early engagement reduces this misalignment before it becomes expensive.

Reduced Design Changes and Avoided Rework

When engineers seek input early - when the brief is still fluid, when assumptions are still being shaped - potential issues surface long before they become baked into drawings or contracts. This prevents the most costly form of rework: late-stage revision.

Take the footpath retaining wall from my early-career days. A design that seemed perfectly sensible on paper led to a high-value land resumption once it met reality. It wasn't a technical failure. It was a failure of alignment. One conversation with the homeowner at the beginning would have saved thousands of dollars, weeks of redesign, and damage to trust.

Quantitatively, early engagement can lead to:

- 1-3 fewer redesign cycles per project
- 10-25% reduction in overall design hours
- Fewer variations during construction, where costs escalate exponentially

These are not marginal efficiencies - they compound across teams, portfolios, and financial years.[463, 464]

Faster Approvals and Fewer Legal Appeals

Early engagement reduces project friction.[465] When communities and stakeholders feel heard from the beginning, they are less likely to lodge objections, raise appeals, or block approvals later. Conversely, when engagement is late or perfunctory, resistance hardens.

Industry studies show that:

- Projects with early and meaningful engagement tend to face fewer objections and smoother approvals, because concerns are surfaced and resolved before they harden into formal opposition
- Approval cycles can shorten materially when key stakeholders are engaged early because fewer issues are escalated late, and fewer redesigns are triggered during assessment
- Legal challenges and environmental appeals reduce significantly

Every avoided objection represents avoided cost: legal fees, consultant reports, redesign labour, and - most expensive of all - time.

Time Saved During Delivery

We all know delays cost money. Even just admin costs can get astronomically high for even the smallest project. They compound into contractor claims, inflation exposure, political scrutiny, and reputational damage. Early engagement reduces delays because risks are surfaced early and expectations are aligned from the start.

Some examples include:

- Identifying cultural heritage sites before design starts rather than during construction
- Aligning road alignments with actual community movement patterns (as with Gap Creek Road)
- Reducing the need for reactive mitigation measures

Time saved is one of the most compelling metrics for leaders:

- A week saved during approvals can translate into substantial cost avoidance - often through reduced holding costs, consultant time, and schedule exposure - depending on project scale and risk profile

- A month saved during construction can translate into major cost avoidance on large projects through reduced preliminaries, claims exposure, escalation, and program risk - though the magnitude depends heavily on project size and contract structure

And this isn't aspirational - it is the lived experience of organisations that engage early.

Avoided Legal and Rectification Costs

Legal disputes, compensation claims, and forced redesigns are the hidden line items that erode project margins. Early engagement reduces the likelihood of:

- Land acquisition disputes
- Community compensation claims
- Appeals and objections
- Sudden scope changes forced by political or community pressure

These events are rare but devastating - and they almost always arise from inadequate early consultation.

I saw this firsthand on an intersection upgrade. Resume houses and offer $20,0000 more than market value for the property, or try to push the landowners down on price and go to court for anywhere from $300,000 to $700,000 in costs. With 14 properties to resume on the project, we opted for the first option.

If we quantified the cost avoided from even a single prevented dispute, the ROI of early engagement becomes unquestionable.

Reputational Uplift and Social Licence

Reputation is not soft. It is a hard financial asset. When organisations engage early, respectfully, and authentically:

- Trust increases
- Brand value strengthens
- Political relationships stabilise
- Future projects encounter less resistance

This creates a multiplier effect - organisations known for care and collaboration spend less time "selling" future projects and more time delivering them. The commercial value of legitimacy is difficult to overstate. Trust accelerates everything.

The Bottom Line

Early engagement is not a "nice to have." It is a cost-saving mechanism, a risk-reduction strategy, and a value generator. It reduces rework, accelerates approvals, avoids legal and political entanglements, strengthens community trust, and delivers projects that are better aligned with real needs - all of which translate directly into measurable financial performance.

In short, every hour spent engaging early saves dozens of hours - and thousands, sometimes millions of dollars - later.

One of the greatest missed opportunities in engineering is the gap between how assets are designed and how they are lived with. Traditional business cases focus heavily on CAPEX - the cost to design and build - even though for many long-life assets, the majority of total cost is incurred during operations, maintenance, and renewal-often in the order of 70-80% over the life of the asset. For CFOs, asset owners, and long-term operators, the true financial picture emerges after ribbon-cutting: in maintenance, renewal, social acceptance, failure response, and resilience to changing conditions.

Regenerative engineering expands the design horizon. It shifts the question from "What is the cheapest way to build this now?" to "What is the most valuable way to own, operate, and sustain this over time?" That shift reveals financial, ecological, and social benefits that traditional models routinely overlook.

Whole-of-Life Value: Seeing the Full Cost Curve

When we design only for initial cost, we optimise the wrong part of the equation. Whole-of-life analyses consistently show that regenerative and values-aligned decisions reduce total expenditure by:

- Lowering long-term maintenance needs
- Avoiding costly reactive repairs

- Extending asset life through better alignment with environmental conditions
- Reducing asset vulnerability to climate stressors
- Eliminating future retrofit costs required to meet evolving standards

A slightly higher CAPEX invested in alignment, resilience, or circular materials often results in significant OPEX savings across decades of use.[58, 462, 441]

For asset owners and operators, whole-of-life optimisation is not philosophical; it is financially responsible stewardship.

Maintenance Avoidance: Designing Out Problems Before They Occur

A regenerative approach reduces failure and degradation at the source. Design decisions rooted in connection to place, hydrology, microclimate, and community behaviour tend to produce assets that require less intervention over time. Consider:

- Drainage systems aligned with natural water paths resist blockage and erosion
- Materials suited to the local climate require fewer replacements
- Landscapes designed with Country regenerate naturally rather than needing constant upkeep
- Community-led design reduces vandalism, unsafe behaviours, and misuse

Every avoided maintenance cycle returns value directly to the organisation. The savings are cumulative, compounding, and attractive to CFOs managing multibillion-dollar asset portfolios.

Reduced Failure Likelihood: Technical and Social Resilience

Failures rarely occur solely because of technical miscalculations. They happen when technical design is misaligned with human behaviour, environmental realities, or cultural context. Regenerative engineering reduces both technical failures (e.g., water ingress, heat stress, ecological breakdown) and social failures (e.g., public rejection, unsafe use, reputational damage).

This reduction in failure likelihood delivers measurable benefits:

- Fewer insurance claims
- Lower liability exposure
- Fewer emergency repairs
- Reduced political scrutiny
- Lower asset downtime
- Improved safety outcomes

When alignment improves, fragility decreases. Resilient assets cost less to own.

Climate Adaptation Savings: Building for the World We're Actually In

Climate impacts - heat, flooding, storms, sea-level rise - are now a material cost factor. Traditional business cases often treat these risks as peripheral or "externalities," but asset owners bear the financial consequences.

Regenerative engineering:

- Incorporates future climate modelling into design decisions
- Integrates nature-based buffers that adapt over time
- Reduces heat stress on materials, improving durability
- Supports water retention and infiltration, decreasing flooding damage
- Ensures assets remain functional and safe under extreme conditions

Every avoided climate-related failure represents avoided cost: remediation, downtime, insurance, political fallout, and emergency works. Resilience is not a premium - it is a cost-saving strategy.

Biodiversity and Ecosystem Co-Benefits: Benefits We Can Now Monetise

Ecosystem services - cooling, water filtration, carbon storage, habitat creation, flood mitigation - can now be valued using widely accepted economic frameworks. This is a seismic shift for engineering, because it allows regenerative design to be expressed in financial terms that CFOs recognise.

Nature-based and Indigenous-informed approaches often produce:

- Cooler public spaces (reduced energy demand)
- Improved stormwater quality (lower treatment costs)
- Stronger soil and landscape stability (reduced erosion and maintenance)
- Increased biodiversity (improved ecosystem resilience)
- Carbon sequestration (meeting emissions targets)

These co-benefits are no longer intangible. They are quantifiable and increasingly required by regulators, insurers, and investors.

Regenerative engineering reveals value that traditional cost models simply fail to see.

Aligning Regeneration With Asset Management Reality

For asset owners and CFOs, the message is clear:

- Regenerative design lowers whole-of-life cost.
- It reduces failure and maintenance risks.
- It increases resilience in the face of climate volatility.
- It strengthens public trust, reducing social and political friction.
- It creates assets that last longer, perform better, and cost less to operate.

This is not a shift away from engineering rigour - it is an expansion of it. It recognises that the real measure of engineering excellence is not how an asset looks at handover, but how it performs over years, decades, and generations.

In the long run, regenerative engineering is not only the right thing to do - it is the financially superior approach for any organisation responsible for the health, safety, and sustainability of its assets.

Risk Reduction: Technical, Social, and Reputational

Engineering risk has traditionally been defined in narrow terms: structural failure, system malfunction, safety hazards, or non-compliance with standards. These risks matter - they always will. But the past decade of high-profile failures has shown that the most damaging engineering failures are not caused by technical miscalculation. They are caused by values misalignment.

When care, context, community, and Country are missing from the brief, catastrophic blind spots emerge. Projects stumble not because engineers don't know how to build, but because the system around the project no longer supports what they built, why they built it, or for whom it was built. Regenerative engineering reduces risk by widening our field of view - by expanding the boundary conditions of design beyond the technical to include the human, environmental, and social systems an asset must operate within.

This section breaks risk into three categories - technical, social, and reputational - to show how a broader values lens reduces the full spectrum of risks leaders must manage.

Technical Risk: Reducing Catastrophic Blind Spots

Technical risk is rarely just technical. When failures occur, they are almost always the result of narrow framing: designing for compliance rather than context, meeting standards rather than meeting needs, optimising for cost rather than consequences.

Recent events make this painfully clear:

- **Optus 000 outage** - A technically functional system that failed a fundamental duty because designers didn't account for cascading vulnerabilities in the ecosystem.[18, 19, 20]

- **Grenfell Tower** - A technically certified cladding system that proved lethal when disconnected from a holistic understanding of fire behaviour, materials science, and human safety.[31, 32]

- **Rozelle Interchange** - A complex system meeting specifications on paper but misaligned with real-world traffic patterns, human behaviour, and community impact.[26, 27]

These failures were not simply engineering errors - they were value errors. They came from decisions that were too narrow, too isolated, too focused on short-term performance rather than long-term consequence.

Regenerative engineering reduces technical risk by:

- Broadening the criteria used to define "success"
- Integrating ecological, social, and behavioural data into early design
- Questioning the inherited assumptions of the brief
- Designing for resilience, not minimal compliance
- Involving diverse perspectives that catch what a technical-only lens can miss

When values expand, blind spots shrink.

Social Risk: The Cost of Misalignment with People and Place

Social risk is one of the most underestimated - and most expensive - risks in engineering. It emerges when communities, Traditional Owners, or other stakeholders feel excluded, harmed, or undermined by project decisions. And unlike technical issues, social issues cannot be fixed with a design amendment alone.

Social risk shows up as:

- Community opposition
- Loss of legitimacy
- Erosion of trust
- Delayed approvals
- Increased regulatory scrutiny
- Political pressure and media attention

These risks translate directly into financial impacts:

- Months or years added to approval cycles
- Redesign costs
- Legal and consultation expenses

- Escalated project budgets
- Reduced ability to deliver future projects

When engineers engage early, listen deeply, and design with care for Country and community, social risk drops dramatically. Projects become collaborative rather than adversarial. Stakeholders become partners rather than opponents. This doesn't just save time - it protects reputational capital, which is far more valuable than any individual project line item.

Reputational Risk: The Hidden Risk That Drives All Others

Reputational risk used to be treated as intangible, but it is now one of the most financially significant risks in engineering and infrastructure. A loss of trust can collapse project pipelines, trigger political intervention, and erode shareholder confidence.

Reputational risk appears as:

- Media fallout
- Political intervention or inquiries
- Shareholder or investor concern
- Loss of social licence
- Long-term brand damage
- Employee resignation

Once reputation is compromised, everything costs more: engagement, approvals, recruitment, retention, compliance, and delivery.

- Regenerative engineering reduces reputational risk by:
- Embedding community and Country early
- Creating transparency in decision-making
- Prioritising values that align with public expectations
- Designing assets that perform socially as well as technically
- Demonstrating care in every stage of delivery

Reputation improves when projects improve people's lives. It degrades when engineering retreats into a narrow, technical-only frame.

What Really Causes Engineering Projects to Fail

When we trace project failures to their root cause, a pattern emerges. Many high-profile project failures are driven less by technical miscalculation than by misalignment between engineering decisions, social expectations, and organisational values.

When care is absent, risk multiplies. When care is present, risk is managed - holistically, intentionally, and sustainably.

Regenerative engineering reframes care not as a "soft skill," but as a risk strategy. It ensures we design with awareness of the full system - the people, the land, the culture, the context, the climate, the future - that our projects operate within.

This isn't idealism. This is the most reliable way to reduce the costliest risks in modern engineering.

Regulatory and Policy Alignment

The regulatory environment for infrastructure and engineering is undergoing a huge shift. Where compliance once focused narrowly on technical standards, safety, and environmental impact assessments, today's regulatory frameworks increasingly demand transparency, accountability, and long-term stewardship.

Regenerative engineering - with its emphasis on values, community, ecological health, and future generations - aligns directly with these emerging requirements. In many cases, it positions organisations ahead of compliance rather than scrambling to meet it.

Leaders who embrace regenerative principles now aren't doing "extra work." They are preparing their organisations for the policy landscape of the next decade.

Climate Disclosure Requirements

Mandatory climate disclosure is reshaping engineering practice. Regulations and reporting standards - including the Task Force on Climate-Related Financial Disclosures (TCFD)[466] and its successors - require organisations to assess and disclose:

- Climate-related physical risks
- Transition risks

- Adaptation strategies
- Resilience measures
- Scenario modelling

Regenerative engineering inherently integrates climate adaptation and resilience into design decisions. It reduces exposure to climate-induced failures, positions assets for long-term viability, and creates the exact documentation and foresight required under these frameworks.

ESG Reporting and Investor Expectations

Environmental, Social, and Governance (ESG) criteria are no longer niche investor interests; they now guide major capital flows. Infrastructure and engineering organisations are assessed not just on financial performance but on:

- Carbon reduction
- Biodiversity impact
- Social licence
- Cultural heritage protection
- Stakeholder engagement
- Governance transparency

Regenerative engineering strengthens every dimension of ESG performance. It provides tangible actions, measurable indicators, and evidence-based practices that improve ESG scores and reduce investor concern. In a capital environment increasingly shaped by ESG screens, this is a competitive advantage.

Modern Slavery Legislation

Modern slavery laws in Australia and many other jurisdictions require organisations to report on modern slavery risks in their operations and supply chains, and to describe the actions they are taking to assess and address those risks - shifting procurement from price-only decisions toward transparency and due diligence. Regenerative engineering supports compliance by:

- Prioritising local, circular, and transparent supply chains

- Reducing reliance on high-risk materials
- Embedding ethical criteria into procurement frameworks
- Strengthening governance and accountability

This reduces legal exposure while aligning engineering decisions with global expectations for ethical production and labour rights.

First Nations Engagement Frameworks

Government agencies and project approvals increasingly require early and meaningful engagement with Traditional Owners. Many agencies and approval pathways now require, or strongly expect:

- Indigenous-led consultation
- Cultural heritage assessments
- Co-design of key project components
- Recognition of Country as a stakeholder

Regenerative engineering not only meets these requirements but enhances them by embedding Indigenous knowledge early in the process. This strengthens legitimacy, reduces delays, and aligns organisations with the direction of national policy commitments.

Local Government Procurement Criteria

Local governments across Australia and globally are reshaping procurement to prioritise:

- Social value
- Environmental outcomes
- Circularity
- Community benefits
- Cultural inclusion
- Nature-positive design

Regenerative engineering offers a direct pathway to meeting - and often exceeding - these criteria. It transforms procurement from a compliance exercise into a strategic advantage.

Infrastructure Australia's Sustainability and Resilience Metrics

Infrastructure Australia's assessment framework embeds sustainability and resilience considerations, including expectations around long-term outcomes like resilience, liveability and wellbeing, and broader social and environmental impacts like[461]:

- Climate resilience
- Social cohesion
- Environmental regeneration
- Cultural outcomes
- Long-term adaptability
- Stakeholder relationships

These expectations strongly overlap with regenerative engineering - especially where they push decision-makers to value long-term resilience, community outcomes, and whole-of-life performance, not just upfront cost. The message is clear: national infrastructure policy is evolving toward the outcomes regenerative engineering already delivers.

Regulation Is Catching Up With Regeneration

The systems that govern engineering - policy, investment, procurement, approvals - are being redesigned to prioritise long-term value, resilience, inclusion, and ecological health. Therefore, regenerative engineering should be seen not as a moral luxury but as a compliance multiplier: a design philosophy that inherently satisfies a widening web of regulatory expectation.

For leaders, the question is not whether they can afford to shift toward regenerative practice. It is whether they can afford not to.

Reduced Rework and Efficiency Gains

Engineers love efficiency. We optimise systems, streamline processes, and eliminate waste wherever we find it. Yet some of the greatest inefficiencies in engineering don't come from technical miscalculations - they come from assumptions that go untested too long, decisions made without full context, and problems discovered downstream instead of upstream.

Regenerative engineering reduces these inefficiencies by widening our field of view early. It ensures the right conversations happen at the right time, so rework is avoided, friction decreases, and teams move with greater clarity and confidence.

Early Questioning Reduces Waste and Rework

Some of the most expensive mistakes in engineering occur because a simple question wasn't asked early enough.

Take the footpath retaining wall from my early-career days: a technically sound design that triggered an unnecessary land resumption once it met the real world. Or the Gap Creek Road alignment, where the community's lived experience revealed design opportunities that engineers could not see from drawings alone.

In both cases, the rework was avoidable - if the right conversations had happened upstream.

Across the industry, the data is clear:

- A significant share of rework traces back to early decisions made without full context. Early engagement is one of the most reliable ways to reduce it.
- Early engagement reduces design iterations by 1-3 cycles.
- Upstream alignment saves thousands of engineering hours across a project.

Rework is not a technical failure. It is a system failure - a failure of alignment.

Clarity of Intention Reduces Friction

When teams understand why a project exists - not just what it is meant to deliver - design decisions become faster, cleaner, and more coordinated.

Clarity of intention:

- Reduces internal debate and confusion
- Speeds up decision-making

- Decreases the number of "non-value-adding" design cycles
- Aligns technical, environmental, and community considerations
- Prevents expensive last-minute scope shifts

Projects slow down when no one is clear on the purpose. They accelerate when everyone is aligned on intention.

Design Efficiencies From Broader Perspectives

Bringing in community, Country, behavioural insights, ecological knowledge, and diverse engineering voices doesn't make projects slower - it makes them smarter.

Broader perspectives:

- Spot unintended consequences early
- Reveal simpler, more elegant solutions
- Identify issues before they escalate
- Reduce reliance on costly mitigation measures
- Avoid downstream conflicts that force redesign

This isn't "extra work." It's efficiency work. Every perspective added early removes a problem later.

Time Saved by Addressing Issues Upstream

The downstream cost of upstream silence is enormous. Every issue moved one stage earlier in a project can save weeks or months of time.

Addressing issues upstream reduces:

- Approval delays
- Redesign cycles
- Contractor claims
- Costly construction phase surprises
- Corrective work that breaks schedules and budgets

In complex systems, small misalignments early become large disruptions later. Upstream clarity is the single most effective way to prevent that.

The 60% Rule: Make Decisions With Enough Information, Not All of It

Engineers often wait for certainty before acting. But certainty arrives too late in complex, dynamic systems. A regenerative engineering mindset uses the 60% rule. This means that you make decisions when you have about 60% of the information - enough to act, not enough to be stuck.

Waiting for 100% of the information slows teams down, invites unnecessary modelling, and causes decision bottlenecks. Moving at 60%:

- Accelerates progress
- Encourages early testing and iteration
- Reduces wasted design effort
- Improves adaptability
- Surfaces risks earlier
- Supports collaboration

The 60% rule is not reckless - it is responsive. It acknowledges that engineering is not just about calculations, but about judgement, context, and iteration.

Efficiency Is a Regenerative Outcome

Reduced rework, faster decisions, smoother collaboration, fewer redesigns - these are not side effects. They are the practical efficiency gains that regenerative engineering produces by aligning people, purpose, and place early in the process.

When engineers design with intention, context, and care:

- Systems flow better
- Teams perform better
- Assets cost less to deliver
- Organisations move faster with less friction

Efficiency isn't lost when we broaden our perspective - it's what we gain.

Trust and Legitimacy as Performance Drivers

In engineering, trust has often been considered intangible - something we hope for, not something we design for. Yet trust and legitimacy are powerful performance drivers. They influence every part of project delivery: how communities respond, how regulators assess proposals, how quickly approvals progress, and how teams collaborate under pressure.

When trust is strong, the entire engineering system flows more smoothly. When trust collapses, the cost is immense.

Legitimacy is more than compliance. It is the sense that a project is aligned with community values, respectful of Country, technically sound, and ethically delivered. It is earned through care, transparency, and genuine engagement - and it has measurable commercial impact.

Legitimacy Smooths Stakeholder Interactions

When stakeholders view a project team as respectful, collaborative, and transparent, interactions become easier and more productive. Meetings move from adversarial to problem-solving. Issues are surfaced early rather than withheld until late. Complexity becomes a shared challenge, not a battleground.

Trust reduces friction and reduced friction reduces cost.

Legitimacy Accelerates Approvals

Regulators move faster when they have confidence in the people behind the submission. Projects that demonstrate early engagement, cultural awareness, and environmental responsibility face fewer objections and less scrutiny. They move through assessments more smoothly because risk feels better managed.

Approvals delayed by one month can cost millions. Legitimacy shortens those timelines.

Legitimacy Strengthens Social Licence

Social licence is no longer optional - it is a currency. Without it, even technically excellent projects can stall, attract protest, or become politically unviable. With it, projects gain momentum.

Regenerative engineering builds social licence by showing care for community, Country, and long-term outcomes. When communities feel included and respected, they shift from resisting to supporting. This is not public relations; it is values alignment.

Legitimacy Reduces Political Interference

When a project lacks trust, political actors step in to pacify community concerns, respond to media pressure, or correct perceived wrongdoing. Political intervention triggers redesigns, pauses, budget reallocations, and sometimes full project resets.

Conversely, projects seen as legitimate attract less political risk. They create fewer headlines and fewer surprises, and deliver stability to ministers, executives, and the public. Political peace is commercial value.

Legitimacy Stabilises Budgets

The most expensive project blowouts are rarely caused by poor estimating. They are caused by community resistance, rushed rework, cultural or environmental breaches, or political pauses - all downstream consequences of lost legitimacy.

Legitimate projects:

- Experience fewer variations
- Maintain clearer and more predictable cashflow
- Avoid escalated contractor claims
- Remain aligned with their intended scope

Legitimacy Improves Safety Outcomes

Teams that trust one another speak up earlier, challenge assumptions safely, and escalate concerns without fear. Communities that trust the project are more likely to collaborate around safety issues rather than work against the team.

Where trust is strong:

- Near misses decline
- Critical risks are identified earlier
- Behavioural safety improves
- Psychological safety increases
- Teams perform better under pressure

Legitimacy Builds Long-Term Organisational Resilience

Trust compounds across years, projects, and generations of staff. Organisations known for doing the right thing attract talent, retain leaders, build strong partnerships, and recover faster from setbacks.

Legitimacy is a form of organisational capital. It is slow to build, quick to lose, and profoundly valuable.

Where Moral Courage Meets Technical Excellence

Regenerative engineering shows that technical excellence and moral courage are not competing forces - they are co-requisites. The same behaviours that build trust (listening deeply, engaging early, respecting Country, widening the design brief, caring for people) also produce better engineering outcomes.

Talent Attraction and Retention

Engineering organisations around the world are facing an unprecedented challenge: attracting and retaining the talent they need to deliver the next generation of infrastructure, energy, digital systems, and adaptive urban environments. The shortage is not only technical - it is cultural. While companies continue to recruit based on skills, younger engineers are choosing workplaces based on values.

A new generation of engineers wants to work where their contributions matter, where their voice is heard, and where the organisation's actions align with its stated principles. They want purpose, agency, and integrity. They want to know that their work contributes to a future they believe in - not one they have to justify to themselves.

Regenerative engineering speaks directly to this desire.

Younger Engineers Want Purpose and Values Alignment

Multiple workforce studies indicate that many early-career engineers place strong value on purpose, ethics, and alignment between organisational values and personal beliefs.[467] Across universities, graduate programs, and early-career teams, the pattern is clear: Young engineers choose organisations that align with their values.

They want to work in places that:

- Care about people and community
- Respect Country and culture
- Consider long-term impact, not just short-term delivery
- Engage with complexity openly
- Are psychologically safe and trusting
- Challenge harmful norms in the industry
- Create solutions that support life, not extract from it

Purpose is not a "nice-to-have" add-on to a graduate recruitment brochure - it is a strategic workforce lever.

Regenerative Organisations Retain High Performers Longer

High-performing engineers - not just the new graduates - stay longer in organisations where their work feels meaningful and their judgement is valued. Regenerative and conscious engineering cultures:

- Give teams more ownership and autonomy
- Encourage systems thinking and creativity
- Offer clarity of intention, reducing internal conflict
- Foster collaboration instead of competition
- Create a sense of belonging and shared responsibility

These workplaces attract people who want to contribute to something bigger than a project, and keep them long enough for their expertise to compound.

Retention is not only a cultural win; it is a financial one. The cost of turnover in engineering roles - recruitment, onboarding, lost productivity, project delays - is enormous. Values alignment reduces that cost dramatically.

Toxic Cultures Are Expensive

Toxicity shows up as burnout, disengagement, blame, silence, fear, and turnover. It erodes trust, undermines psychological safety, and creates environments where engineers are unwilling to speak up - even when they see something wrong.[454] This is dangerous and expensive.

Toxic culture costs organisations through:

- High staff turnover
- Increased sick leave and burnout
- Reduced innovation
- More incidents and near misses
- Lower productivity
- Slower project delivery
- Diminished reputation

The industry has too many examples where silence and pressure created the conditions for failure. Regenerative engineering - with its focus on care, connection, and intention - creates cultures that prevent these risks, rather than manage them after the fact.

Inclusive, Values-Based Cultures Enhance Productivity, Innovation, and Safety

Innovation does not come from technical specifications alone. It comes from the interactions between people: how freely they share ideas, how boldly they question assumptions, how safely they challenge each other, and how collaboratively they solve problems.

Inclusive cultures:

- Unlock innovation through diversity of thought
- Strengthen safety through psychological security
- Increase productivity through reduced friction and conflict

- Improve quality through richer design inputs
- Accelerate delivery through trust

When people feel safe, respected, and included, they think more clearly, speak more courageously, and design more intelligently. Regenerative engineering provides the values base that makes this possible.

Talent Is the Ultimate Competitive Advantage

In a future defined by complex systems, climate uncertainty, and massive technical transformation, the organisations that thrive will be those able to attract, grow, and retain exceptional engineers.

Regenerative engineering gives organisations:

- A purpose young people believe in
- A culture where high performers excel
- A workplace where people stay
- A brand that stands out in a competitive market
- A leadership narrative that inspires rather than intimidates

Boards and CEOs care about this because they know the reality: you cannot deliver the future if you cannot staff it.

Regenerative engineering is not only an ethical pathway - it is a talent strategy. It is how engineering organisations become places where people want to work, want to stay, and want to lead.

KPIs Leaders Can Adopt Tomorrow

The fastest way to shift an engineering organisation is to change what it measures. If we continue to track only cost, time, and safety, we will continue to optimise for narrow outcomes.

But when we measure trust, alignment, regeneration, and early clarity, behaviours shift. Teams focus differently. Decisions evolve. The organisation begins to value what truly drives long-term performance.

Below is a set of KPIs that leaders could adopt tomorrow - metrics that translate regenerative engineering into everyday practice across people, environment, and delivery. They are simple, actionable, and aligned with the direction of modern regulation and community expectation.

People & Community KPIs

1. Stakeholder Engagement Score

A measure of the quality, timing, and depth of engagement with community, Traditional Owners, and stakeholders.

Why it matters: Higher scores correlate with fewer objections, reduced delays, and clearer project alignment.

2. Community Benefit Indicators

Tracks how projects enhance local wellbeing, access, amenity, mobility, or cultural outcomes.

Why it matters: Demonstrates social value and strengthens legitimacy.

3. Cultural Safety Assessment

Evaluates whether project processes, communication, and decision-making are culturally respectful and inclusive.

Why it matters: Improves relationships, reduces reputational risk, and strengthens Indigenous partnerships.

Environment & Regeneration KPIs

4. Net-Positive Dashboard

Measures whether projects deliver more ecological value than they remove - through biodiversity uplift, carbon sequestration, shading, cooling, or water quality improvements.

Why it matters: Shows alignment with climate and ESG commitments.

5. Ecosystem Services Uplift

Quantifies improvements to ecosystem functions (e.g., stormwater filtration, soil stability, habitat creation).

Why it matters: Captures financial value that was previously invisible.

6. Circularity Percentage

Tracks the proportion of materials that are recycled, reclaimed, reused, or designed for disassembly.

Why it matters: Reduces lifecycle cost and regulatory risk while improving sustainability metrics.

Process & Delivery KPIs

7. Upstream Decision Clarity Index

Assesses whether project briefs, design intentions, and values are defined early enough to prevent rework.

Why it matters: Clarity upstream creates efficiency downstream.

8. Number of Rework Incidents

Counts design or construction rework events and their root causes.

Why it matters: Reveals misalignment early and drives continuous improvement.

9. Approval Cycle Time

Measures time from submission to approval, tracking delays and their sources.

Why it matters: Strong indicator of legitimacy, trust, and alignment.

10. Escalation Events

Tracks how often issues require executive or political attention.

Why it matters: Fewer escalations = stronger trust and lower reputational risk.

11. Community Objection Rate

Number and type of objections per project, including their stage and resolution.

Why it matters: A proxy for trust, cultural safety, and design alignment.

12. Whole-of-Life Cost Variance

Tracks divergence between forecast and actual lifecycle costs.

Why it matters: Shows whether regenerative design is delivering the expected financial benefits.

Why These KPIs Are Important

These metrics do more than track performance - they reshape behaviour. They tell teams that meaning, community, culture, ecology, and clarity matter as much as cost and schedule. They give leaders the ability to spot misalignment early, respond to risk holistically, and reward decisions that create long-term value.

Most importantly, these KPIs make regenerative engineering operational. They turn

- Values into measurable outcomes.
- Intention into accountability.
- Care into performance.

The Case for Value-Based, Regenerative Engineering

Doing the right thing for our future is also the lowest-risk, lowest-waste, highest-performance path forward.

For decades, engineering has been framed as a discipline of optimisation - the pursuit of efficiency, precision, and control. But the variables we optimised for were too narrow. We maximised short-term cost savings at the expense of long-term value. We streamlined delivery while overlooking the true drivers of risk. We engineered for compliance rather than consequence.

Regenerative engineering turns that equation around. It expands what we optimise for. It recognises that resilience, trust, community, Country, and ecology are not soft considerations; they are structural components of performance. They determine whether projects proceed, whether assets endure, whether organisations maintain legitimacy, and whether engineering continues to serve society in the way it was always intended.

Regenerative engineering is not a departure from engineering rigour - it is an evolution of it. It gives leaders a framework that aligns technical excellence with human and environmental outcomes. It meets the expectations of communities, regulators, investors, and future generations. And it reduces the risks that cost organisations the most: rework, rejection, reputational damage, climate vulnerability, and talent loss.

When we design with intention and values at the centre, we create:

- Fewer delays
- Fewer redesigns
- Fewer disputes
- Fewer failures
- More resilient assets
- More efficient processes
- More trust
- More alignment
- More long-term value

Projects grounded in care and connection cost less to deliver and less to operate. They stand up better under pressure. They perform more reliably across decades. They earn the trust required to move quickly and collaboratively. They strengthen the organisations behind them, making them more resilient to change, disruption, and uncertainty.

This is not an alternative to traditional engineering. It is traditional engineering done well - with the full system in view.

As the world becomes more complex, more interconnected, and more unpredictable, the organisations that thrive will be those who embrace this broader definition of value. They will recognise that engineering is not only about solving technical problems, but about shaping the conditions for life to flourish.

Regenerative engineering is the most economically rational form of engineering available today because it reduces risk, increases resilience, and delivers benefits that compound over time. It is how we honour the privilege and responsibility of shaping the world - not just for today, but for decades and generations to come.

Doing the right thing is good engineering. And good engineering is good business.

The Takeaway

When the business case is expanded beyond short-term cost and compliance, a different picture of value emerges. Investments in care, resilience, trust, and long-term adaptability are no longer hard to justify. They become the foundations of systems that endure, organisations that retain legitimacy, and projects that remain fit for purpose long after delivery.

What becomes clear is that the future of engineering does not require choosing between doing what is right and doing what is viable. The two are increasingly inseparable. In a world of rising complexity and scrutiny, the most robust decisions are those grounded in integrity, foresight, and responsibility.

This shift is not just about projects or economics. It is about the profession itself. The final chapter turns to the role of the engineer, exploring what it means to reclaim engineering's deeper purpose, and what becomes possible when technical excellence is paired with care, courage, and stewardship.

Chapter 11: The Return of the True Engineer

This book has traced the future of engineering through values, systems, decisions, and legacy. But beneath all of that lies a more fundamental question: who is the engineer we are becoming, and who have we always been?

This final chapter is about identity. It reflects on the deeper purpose of engineering, beyond delivery schedules and risk registers, and asks what it means to practise this profession with imagination, care, and courage in a time of profound change.

The future of engineering does not require us to abandon rigour or reinvention. It asks us to remember. To reconnect with the social, creative, and ethical roots of our work, and to reclaim engineering as a profession that serves life, shapes possibility, and leaves the world better than we found it.

Remembering Who We Are as Engineers

I was walking through the city one early morning, long before the rush of commuters filled the streets. As I passed through the familiar rhythm of the city I grew up in, I was reminded, as I often am, of the invisible fingerprints of engineers everywhere.

The curve of the many bridges. The hum of a water pump deep beneath the pavement where the creek used to be. The way the morning sun refracted through the patterned glass of a building. The ferry terminals that were redesigned and rebuilt after the 2011 floods. The trees planted along the river banks. The hidden Indigenous engineering that has taken care of this Country in ways that we might never know or fully understand.

Everywhere I looked was the story of someone's ingenuity. Someone who once stood on a construction site, drawing in hand, trying to solve a problem that mattered. Someone who used mathematics, material science, and imagination to make everyone's life better.

On this particular day, it struck me that the world we walk through every day is, in many ways, a living museum of engineering - an expression of our creativity, courage, innovation, and care for people and society. And yet, in the rush of delivery schedules, procurement models, and risk registers, it sometimes feels as if we've forgotten who we really are.

We didn't become engineers to tick boxes or charge billable hours. We became engineers because we wanted to make things work - not just functionally, but beautifully - and to make a difference. Because we actually want to solve big challenges, wicked problems, and invent brilliant things that make people's lives safer, easier, and more connected - things that make a positive difference in the world and make it a better place.

That's the part of us that I feel has been somewhat forgotten. It's not gone - just buried under layers of systems and values that were built for another era.

The coming decade will ask more of engineers than any that came before. We are stepping into an age defined by extremes - a changing climate, technological acceleration, cyber vulnerability, global supply strain, and a cultural shift in what it means to work, belong, and create.

It might seem daunting. But I believe it's actually an invitation - a chance to reconnect with the very best parts of our DNA as engineers. It is the most exciting time to be an engineer, as we are the ones best positioned to actually solve these huge challenges.

Engineers have always been pioneers. We have always adapted to new tools, new ideas, and new frontiers. From the aqueduct builders of Rome to the First Nations custodians who designed living systems of care and regeneration across millennia, engineering has always been about invention, discovery, and innovation for society, people, and planet.

When we remember that, we remember our power.

This next decade isn't just about surviving disruption - it's about shaping it. It's about rediscovering the social, creative, and ethical roots of what we do. It's about designing systems that serve life, not just demand more from it. And it's about serving people and being pioneering, connected to humanity, grounded, imaginative, relentless in pursuit of "better", courageous, and unafraid to fail.

We must put these things first and then use technology as the solution - not the other way around. The next decade belongs to engineers who can do exactly that - who can balance the logic of the machine with the humanity of care. Who can design for the storms ahead while also leaving beauty and belonging in their wake.

Engineering isn't just about making things work. It's about making the world work better. And to do that, we don't need to become something new. We just need to remember who we already are.

The Future Engineer

As we look ahead, it's clear that the world the next generation of engineers will inherit is profoundly different from the one many of us trained for. Climate volatility, aging infrastructure, global supply-chain fragility, demographic shifts, labour shortages, new materials, new technologies, and rising community expectations are no longer abstract trends - they are the daily operating environment. Complexity and uncertainty is no longer the exception; it is the baseline.

This final chapter of the book turns us toward that future. It explores the pressures already reshaping our profession: rebuilding in an era of extremes, delivering projects in a world of unpredictable supply, designing for ecological regeneration rather than mere mitigation, leading teams through cultural and generational change, and navigating a talent landscape where retention, belonging, and meaning matter as much as salary and title.

The "future engineer" is not defined by tools or titles, but by mindset. They will need to be deeply connected humans able to navigate climate shocks calmly, engage communities with care, collaborate across disciplines with humility, and lead with values that endure even when circumstances do not. They will be asked to solve problems no one has solved before - and to steward systems long after their design life has technically expired. They will be required to think like integrators, negotiators, ecologists, storytellers, and guardians of public trust.

This chapter is an invitation to imagine that engineer - not as an idealised figure, but as a very real response to the world we are entering. The sections that follow unpack the major forces shaping our future and the capabilities we must cultivate to meet them. If the past taught us how to build, the future will ask us how to endure, regenerate, and lead with intention.

The future engineer is not merely someone who designs for the world - they are someone who helps the world continue to work, adapt, and thrive.

Climate Extremes and Resilience as Baseline

If there's one thing the last decade has shown us, it's that climate change is no longer an abstract scenario. It's a design condition.

In the past, resilience was a section in a report - a table at the back that listed "considerations" for floods, heat, or wind. Today, it's becoming the baseline for engineering itself. But even with that shift, I still see teams struggling to embed it deeply into how they think, not just what they design.

Resilience isn't something you can simply calculate. It's relational. It's about how systems adapt, absorb, and regenerate - not just survive. And that requires engineers to return to an older wisdom: designing with nature, not against it and creating adaptability in design.

When I first moved to the Sunshine Coast, I spent time working out of the Peregian Digital Hub - a place buzzing with innovators, technologists, and people who think beyond the obvious. It was there I learned about an emerging collaboration between local technologists, startups, and Queensland Fire and Emergency Services to explore new ways of anticipating bushfire risk using drones and AI.

After the devastating 2019 Peregian bushfires, the community didn't simply ask how to respond better - they asked how to detect danger earlier. The trials that followed used thermal camera-equipped drones and AI mapping to identify hotspots and flare-ups long before they became visible from the ground. Instead of waiting for smoke, these tools allowed fire crews to see heat signatures early, giving them crucial minutes - sometimes hours - to act.[468]

This wasn't technology for its own sake; it was anticipation over reaction. A partnership between engineers, firefighters, technologists, and local innovators all united by one intention: protect people and place before crisis strikes.

It was a simple but powerful reminder that the most effective resilience isn't built on fear of failure, but on curiosity, collaboration, and the courage to see risk differently.

There are many innovative ways people are seeking to solve these global issues.

In 2023, facing a world where wildfires grow larger, faster, and more unpredictable, XPRIZE launched one of its boldest challenges yet: to end destructive wildfires. With an $11 million prize pool and a four-year timeline, the competition calls on innovators globally to shift wildfire management from reaction to anticipation.[469]

The challenge is divided into two tracks. One aims to revolutionise detection by using satellites, remote sensing, AI and advanced sensors to spot incipient fire across vast lands with unprecedented speed. The other demands a fully autonomous response system - robots, drones, or hybrid systems that can locate, assess, and suppress a fire within 10 minutes, even across remote or hazardous terrain.[469]

As of 2025, a roster of semifinalist and finalist teams from around the world - including Australia - are building technologies that could fundamentally change how we co-exist with fire. These systems are designed not only to protect lives and property, but also to safeguard ecosystems, cultural landscapes, and future resilience.[469, 470]

For engineers committed to regenerative design, XPRIZE Wildfire offers a powerful reminder: the most effective resilience doesn't wait for the crisis. It anticipates it - with curiosity, care, and collaboration.

Around the world, we're seeing more examples of this kind of thinking - of engineers returning to that spirit of curiosity and partnership with the natural world. In the Netherlands, the Room for the River program did something extraordinary: instead of fighting water with walls, engineers gave the rivers back their floodplains.[301, 390] They redesigned entire urban systems so that water could flow freely during floods, absorbing pressure rather than resisting it. What was once seen as "surrendering control" turned out to be the smartest, safest, and most sustainable choice.

In the United States, cities like Miami and New Orleans are adopting this mindset too - shifting from "build higher walls" to "build with wetlands".[471] Some of the most promising research now focuses on nature-based infrastructure - mangrove restoration, blue-green corridors, and coastal buffers that heal the environment while protecting people.[472]

But what strikes me most about these examples isn't the innovation - it's the mindset. Each began when someone asked a different kind of engineering question: "What would nature do here?"

This question doesn't mean we give up technical rigor; it means we broaden it. It means recognising that resilience isn't a single system's property, but a fundamental foundation to start from. It's the relationship between all systems: water, soil, structure, and people.

I remember once starting a bikeway project where we calculated and modelled a "one in 100 year" storm event. On my first site visit, a local approached me asking what I was doing. I said a bikeway was proposed for this location. He said that there had just recently been a flood in the area and showed me where the water had come up to. It was higher than our model. That local knowledge changed how we viewed the projects and meant we had to go back and review our modelling.

We can no longer afford to design in isolation. Climate extremes will test our structures, but they will also test our values - especially our ability to listen, to collaborate, and to think in systems. We must move beyond the traditional project outcomes of "The road is going from A to B and here is the alignment" to instead look more broadly at resilient systems.

When I think about the engineers of the next decade, I don't imagine them just in hard hats and site boots. I see them standing in wetlands, talking with ecologists and Indigenous custodians, and listening to how nature organises itself. I see them sketching flow diagrams - part of a system that looks more like rivers, rather than a standalone model or drawing.

Resilience begins with respect for the forces we can't control and the systems that sustain us. And when engineers design with that respect, resilience stops being an add-on. It becomes the baseline for everything we build.

AI, Automation, and Complexity

If climate extremes are reshaping the physical world engineers design for, then AI and automation are reshaping the cognitive world we design from. For the first time in history, engineers are working alongside systems that learn, adapt, predict - and occasionally surprise us. And this shift is not theoretical; it's already transforming how we design, decide, and deliver.

For decades, engineering was built around linearity: define the inputs, run the calculations, check against the standards, and produce a predictable outcome. But the systems we are now building - autonomous sensors, adaptive control systems, predictive maintenance engines, AI-assisted design tools - don't behave linearly. They behave like ecosystems. They interact, evolve, and respond to changing conditions.

This means the role of the engineer is also changing. Not from designer to bystander - but from controller to conductor.

Instead of waiting for bridges, tunnels, and water systems to degrade before intervening, engineers are now training models to detect stress signatures long before they become visible to the human eye. Algorithms are learning from vibration patterns, temperature fluctuations, and micro-fracture acoustics - detecting the signals we were never trained to hear.

This is already happening across New South Wales, where the Transport for NSW team asked: "What if we could see every defect across the network before someone had to report it?"

The answer became Asset AI - a collaborative project between Transport for NSW, 32 local councils, researchers, and technology partners. The premise was deceptively simple: equip everyday council vehicles - waste trucks, utes, ranger cars - with small, windscreen-mounted cameras that film the road network as they complete their normal rounds. Those video streams feed into a machine-learning platform trained to detect more than a dozen asset issues: potholes, cracks, edge breaks, ponding, faded line-marking, damaged signage, uneven footpaths, encroaching vegetation, and more.[473]

The idea was not to replace human inspectors but to change the rhythm of maintenance. Traditionally, councils conduct road audits every three to five years, walking or driving long stretches manually. Issues appear between inspections, and are often only discovered after a complaint, collapse, or near miss. Asset AI flips this model: every vehicle becomes a rolling sensor, creating continuous, geospatially tagged insight across thousands of kilometres of network.

Early pilots showed how powerful this shift could be. In Shellharbour City Council, the system analysed more than 6,000 km of footage in its first phase, identifying thousands of defects that would have taken months to discover manually.[474]

The AI didn't just flag problems - it prioritised them based on severity, helping maintenance teams schedule repairs efficiently and transparently. Shoalhaven City Council reported similar outcomes: waste trucks, doing their routine weekly routes, generated real-time asset data at a scale previously unimaginable for a regional council.

Asset AI exists because the partnership itself was engineered differently. Rather than a top-down contract, TfNSW created a cross-sector collaboration involving councils, data scientists, software engineers, and policy researchers. Councils shaped the feature set. Engineers trained the algorithms. Local crews tested accuracy in the field. Each partner contributed expertise that the others lacked.

It became a living demonstration of "global capability meets local wisdom."

Most importantly, the program signalled a shift from reactive maintenance to anticipatory design. Instead of waiting for failures, councils now have the ability to:

- Detect early-stage cracking long before potholes emerge
- Identify patterns of deterioration linked to heavy vehicles or drainage issues
- See systemic problems across the network, not just isolated events
- Communicate more transparently with communities about repair priorities

Asset AI hasn't eliminated uncertainty - but it has changed the posture of engineering from "Tell us when something breaks" to "Let's see the early signals before impact occurs." And in the process, it has become a model for how data, collaboration, and curiosity can reshape the future of civil infrastructure.

With this capability comes complexity. AI systems amplify both our strengths and our blind spots. We know a poorly trained model can reinforce bias. A misunderstood dataset can mislead decision-making. An automated process can create failure at scale if not stewarded carefully.

AI does not remove the need for engineering judgment - it increases it.

One of the clearest examples of this shift is in the energy sector. As Australia transitions to a renewable grid, operators now rely on AI to stabilise systems with high penetration of solar and wind, forecasting fluctuations and automatically adjusting supply flows.[475] These systems are too dynamic for manual operation alone - too many variables, too fast.

Engineers now design not just for physical reliability, but for informational reliability:

- Are the datasets accurate and ethical?
- Do we understand what the models are optimising for?
- Do we know how decisions are being made inside the "black box"?
- And could a system still explain itself in a moment of crisis?

Globally, we're seeing powerful examples of AI and automation enhancing engineering while revealing new responsibilities. In Singapore's Changi Airport, autonomous cleaning robots navigate terminal floors, adjusting paths based on pedestrian flow - a micro-example of systems that respond to human behaviour rather than force humans to work around machines.[476] In Tokyo, AI-assisted metro scheduling analyses millions of passenger movements daily, adjusting frequencies with remarkable accuracy to reduce crowding and energy use.[477] In Australia's mining industry, autonomous haul trucks and drills now operate 24/7 in the Pilbara, monitored by control centres thousands of kilometres away.[478] These systems reduce human exposure to risk - but introduce new layers of cyber, ethical, and systemic vulnerability.

What ties these examples together is not the technology itself, but the mindset required to work with it: curiosity, humility, systems thinking, and ethical clarity.

What strikes me most is how often AI reveals the same lesson nature teaches:

- Systems thrive when they are adaptable, not rigid.
- When they learn continuously, not episodically.
- When they are designed for uncertainty rather than illusionary control.

The next generation of engineers will not simply "use AI tools". They will need to interpret, challenge, guide, and govern them. They will need to understand when to trust the model - and when to trust their intuition and engineering judgement. They will need fluency not just in code, but in complexity, ethics, emergent behaviour, and cross-disciplinary dialogue.

And perhaps most importantly, they will need to stay deeply human in a world of accelerating automation.

AI can analyse patterns, optimise systems, and predict failures, but it cannot yet do one thing engineers must do every day: decide what is worth building, protecting, or changing. That remains a human responsibility - and a human privilege.

Cyber Is Engineering

When I was working on a major infrastructure project, I believed the greatest risks lay in concrete, steel, and earthwork. Little did I realise that the more formidable threat could actually come from code, connections, and the "invisible pipes" of a network.

In September 2025, Optus experienced a serious emergency call outage that prevented some customers from reaching Triple-Zero (000) emergency services across parts of South Australia, Western Australia, the Northern Territory, and adjacent areas after a routine firewall upgrade caused a critical system failure.

The outage lasted around 14 hours, and approximately 600 calls failed. Tragically, it is alleged at least three deaths have been linked in enquiries to that outage.[19]

What stands out isn't just the scale of the failure, but the layered human and system behaviours behind it:

- Despite early warning signs (including customer reports of failed emergency calls), internal escalation protocols were slow and the issue wasn't fully recognised for many hours, delaying corrective action and notification to authorities.
- Internal escalation protocols were not followed in time; welfare checks began late; regulatory communication lagged.[17]
- The system's design premise assumed physical robustness, not the vulnerability of digital upgrade controls.

For infrastructure engineers, this is a resounding wake-up call: physical resilience alone is no longer enough. Every asset, site, and network now carries a cyber-physical dimension where failure isn't measured only in repair cost, but in lost trust, lives, and legitimacy.

While from the outside it looks like the roads of the city run with the physical pavement, engineers will know that there are control centres full of cameras behind the roads, with the ability to change certain traffic lights to help emergency vehicles through. These IT systems are more obvious in our modern tunnels and Australia's are some of the most sophisticated in the world. This is a fantastic system, but also opens us up for cyber attack threats on our critical infrastructure.

We often focus on loads, stresses, floods, and earthquakes. In the cyber age, we must design for informational loads - surge traffic, malicious interference, system drift, human error. A collapsed bridge makes headlines; a network failure that leaves people unable to call emergency services may leave no visible damage, but the moral injury is just as great.

This example forces us to map our engineering identity into the digital realm:

- When we design a system, we must assume it will one day be attacked, misused, or degraded.
- When we handover, our operational language must include code-versioning, digital logs, real-time alerting, and human intervention processes.
- Our judgement must ask: "What if this upgrade goes wrong? How quickly can we detect it? How many lives are we putting at risk if we assume everything will just work?"

In many ways, the cyber challenge is what engineering was born for - but in the past, we may have treated it as optional. Now it's central.

We will need to become bilingual engineers - fluent in both steel and software, oversight and orchestration, system failure and human consequence. We must understand and embed cybersecurity not as an adjunct but as an innate design requirement - part of the very DNA of trusted infrastructure.

Because whether you build a bridge, a tunnel or a telecom network - the job is the same: protect people, connect places, extend possibilities. And in the 21st century, that includes protecting the digital lifelines as much as the physical ones.

Supply Chains, Cost Pressures, and Productivity

There was a time when efficiency was the measure of excellence in engineering. We streamlined our processes, refined our designs, and built systems that could deliver faster, cheaper, leaner. It was progress - or so we thought.

But in the last few years, the world has shown us that efficiency alone is no longer enough. A global pandemic, wars, resource shortages, and climate disruption have made something unavoidable: our systems were optimised for predictability, not for volatility. They were designed to run smoothly, not to bend without breaking.

During COVID, when ships stopped sailing and factories paused, projects across the world stalled. Steel prices soared, delivery times doubled, and design schedules unravelled. For the first time in a generation, engineers everywhere felt the fragility of the systems we'd built - a system tuned for speed, not endurance.

We had built a global supply chain like a high-performance engine: every component finely calibrated, every cost shaved down to the cent. But one missing bolt - or one ship going off course - can halt the entire machine.

In March 2021, the world watched a single container ship - the Ever Given - wedge itself sideways in the Suez Canal. For six days, 12% of global trade stood still. An estimated US$9.6 billion of goods per day were held up. Over 400 ships queued at either end of the canal.[479] Factories in Europe slowed due to missing parts; exporters in Asia couldn't deliver; oil prices briefly surged. A single vessel, knocked off course by high winds, became a global economic choke point - proof that a system optimised for efficiency has no slack when disruption occurs.

Then in 2024, the container ship Dali struck Baltimore's Francis Scott Key Bridge, collapsing a major U.S. transport artery within seconds. The Port of Baltimore - one of America's busiest for cars, farm machinery, and bulk goods - was paralysed. Billions of dollars of cargo were diverted. Thousands of jobs were disrupted. Supply chains across the eastern United States had to rapidly reroute. Investigations found the ship suffered multiple electrical blackouts and a loss of propulsion shortly before the collision. Although power was restored, there was not enough time to regain propulsion before impact - and the consequences cascaded across logistics, manufacturing, and local economies.[480]

Both incidents revealed the same vulnerability: When a system is too optimised, it has no capacity to absorb shock. One ship can fail - but it's the system that collapses.

We had confused efficiency with resilience. And we learned the hard way that they are not the same thing.

Talent, Culture, and Retention

It is clear the challenges of the next decade and the future are not technological - they are human.

In Australia, the warning signs are already here. Engineers Australia estimates that the nation is short by at least 50,000 engineers, with demand outpacing supply across every major discipline - civil, electrical, mechanical, environmental, and software. Around 8.5% of Australian university graduates hold engineering qualifications - a relatively low share compared with many other OECD countries, according to Engineers Australia.[481] Migration once filled the gaps and recent data shows skilled engineering migration has returned to, and in some measures exceeded, pre-pandemic levels, easing some pressure while demand remains high.

The demographic challenge is stark. Engineers Australia has warned of significant retirement-driven outflows over the next decade and beyond, creating a capability gap that cannot be filled quickly without stronger domestic pipelines and retention.[482] This means that over the next decade, tens of thousands of senior engineers will retire, creating a leadership and capability vacuum that can't be filled quickly. The defence and infrastructure sectors are already reporting difficulty securing senior project engineers, estimators, commercial specialists, and technical leaders.

At the same time, retention is emerging as a critical threat. Engineers Australia has found that a large share of people with engineering qualifications are not employed in engineering occupations, and that retention and progression, particularly for women, are strongly shaped by workplace culture, flexibility, and visible pathways to leadership. The profession is losing talent not because they lack capability - but because they lack belonging.

This is the real cliff: a shrinking senior cohort, a stretched mid-career layer, and a generation of emerging engineers who will inherit unprecedented responsibility earlier and under far more complexity than any generation before them.

The next decade will show if engineering in Australia becomes a thriving, future-shaping profession - or an overstretched system trying to do too much with too little.

And the only way across the talent cliff is culture: how we attract, develop, support, and lead the people who will build our future.

The systems our new engineers enter were designed for a different time - one of hierarchy, compliance, and linear career paths. But the world they are inheriting is complex, interconnected, and fast-moving.

If we want to retain and empower the next generation of engineers, we must first re-engineer the culture they're stepping into.

Building the Next Generation of Engineers

I didn't grow up knowing what engineering was. No one in my family was an engineer, and it certainly wasn't a word that came up at the dinner table. What I did know was that I loved creating things. For years, I thought I'd become a visual artist because creativity felt like home.

Everything changed because of one person - my physics teacher, Mr Adamson, at All Hallows' School in Brisbane. I adored physics, mostly because of how he taught it: full of curiosity, possibility, and connection to the real world.

One day, he said something that completely surprised me: "Have you ever thought about engineering?"

I remember laughing on the inside - genuinely confused. In my mind, engineering was people sitting in a room doing calculations all day. It felt rigid, serious, unattainable. I didn't think I was "smart enough". I didn't think it was for someone like me.

But that one question planted a seed. It shifted something. I put engineering down on my university preferences - a little curious, a little unsure, and not as my first preference.

When I got accepted straight in, I decided to give it a chance. It challenged me, absolutely. But it also unlocked a new way of seeing the world: as something you could shape.

After a few years in the industry, I kept hearing the same conversation everywhere I went: diversity in engineering, the low numbers of women, the lack of representation, the pipeline problem. Report after report. Panel after panel. PowerPoints full of statistics that made my stomach sink.

I saw it too. I was regularly the only woman in a meeting or on site. And I remember thinking: how many more times are we going to talk about this before we actually do something? When will it actually change?

So when the opportunity came up in late 2011 to run an engineering outreach event for high school students, I didn't hesitate. That one event became the beginning of Power of Engineering. What started as a single program for Year 9 and 10 girls in Brisbane grew into more than 140 events across Australia, plus Engineering in a Box, reaching over 15,000 students, with a focus on young women and regional and remote communities.

But the real joy - the thing that keeps me going - is that we are now meeting young women who did our programs in high school, then studied engineering, and are now working as engineers in the real world. We're watching the next generation arrive - confident, creative, values-led, and ready to solve problems we can't yet see.

This work has taught me so much about attraction, belonging, and culture. About what brings people into engineering - and what pushes them away. And that's why this following section matters. Because the future of our profession isn't just about the projects we build. It's about the people we nurture, support, and keep. My hope is that the ideas in this section help you play a part in creating that future. The future engineer is already on their way - our job now is to build an industry worthy of them.

Attracting the Next Generation to Engineering

When we created the Power of Engineering program, we saw a remarkable shift. Large numbers of people moved from saying, "No, I've never considered engineering," to "Yes, I can see this as a career for me."

Wanting to understand why it was working so well, the team - led by my co-founder, Jillian Kenny - undertook research into the program's impact. Later, we found the Invergowrie Foundation independently explored the same question.[483] Both identified the same four core elements that made a difference:

1. **Participation and collaboration:** Everyone took part in hands-on activities and site tours, often with assigned roles, so that each person played an active part. Those roles were never gendered, creating space for genuine collaboration and shared problem-solving.
2. **Language:** Research consistently shows that when engineering is described only in terms of "building bridges", "analysing data", or "writing code", many people silently opt out. But when we talk about engineering as creative, adaptable, and organised work,

more people can see themselves in it. This inclusive language is embedded throughout the program.

3. **Role models:** The research tells us that both men and women can be powerful role models, but what matters most is that they feel attainable. I often share with young people that I failed maths at university - something still uncomfortable to admit - but it matters, because it shows you don't need perfection to belong in engineering.

4. **Real-world relevance:** Through workshops and site tours, participants could see how engineering connects directly to everyday life - whether that was through solving problems like access to clean water, designing cities that actually work for people, or tackling challenges they could recognise from their own communities.

It's this combination that helped people not just understand engineering, but imagine themselves within it.

Reimagining Engineering at University

When I worked at Swinburne University, I joined a team, led by visionary engineering educator Llewellyn Mann, asking a radical question: if we could design an engineering degree from scratch - for the next century, not the last - what would it look like?

We started with a blank slate and interviews from hundreds of engineering professionals on what they thought engineers of the future needed to know. We learnt that the future engineer must be as creative and connected as they are technical. The skills the industry craved were human skills.

To get into the course, there was no ATAR cut-off. Instead, the team came up with an assessment centre with interviews and practical challenges to find students who thought differently, who were curious, who cared. For those who hadn't studied advanced maths or physics, we offered bridging courses. We built a system that believed in potential - not perfection.

The incredible academic team set out a brand new curriculum of micro-credentials. One fifth were professional skills like communication and the people skills industry wanted.

From day one, students were like consultants and worked on real-world, industry-led projects. They weren't memorising equations in isolation. They were applying them to future-focused problems -

questions like "What is the future of paint?" or "How can we design non mechanical air conditioning systems for remote locations in Australia?"

Their assessment was never an exam. It was a pitch to industry panels, collaborating across disciplines, and presenting their ideas to executives, communities, and councils - just as they would in the real world.

And the results were extraordinary. Students who might have been overlooked in traditional programs thrived. Their creativity flourished. Their confidence grew. They weren't just learning engineering - they were becoming engineers connected to purpose, the real world, and projects that mattered from day one.

James Dyson saw the same reality and built it into the Dyson Institute of Engineering and Technology. His model is simple but revolutionary: engineers learning by engineering. Students spend three days a week working on live Dyson projects and two days studying theory. They live and learn together on campus, and their tuition is covered.There's no summer break, no gap between theory and practice. The rhythm mirrors real life.

Dyson understands that to cultivate inventive, ethical engineers, you must immerse them in a culture where curiosity, courage, and care are the norm. His institute is more than an educational model - it's a living prototype of what an engineering culture could be: creative, practical, human, and purposeful.

Retaining Engineers for the Future

Here we could talk about talent pipelines, succession planning, skills shortages, or graduate recruitment strategies. But we need to talk about something far more fundamental: purpose.

Engineers leave not because they can't do the work, but because the work stops feeling connected to something that matters. They burn out. They feel invisible. They can't see the impact of what they're doing. They lose their why.

And when purpose disappears, people do too.

We often respond to this with more training, more recruitment campaigns, more leadership programs - all valuable, but none of them address something more significant: engineers stay when they feel their work has meaning. They stay when their values are alive in the work. They stay when they can see how their efforts change something for people, community, or Country.

At a few points in my career, I've stayed in organisations with horrible cultures that were psychologically unsafe. Of course, I don't recommend this or suggest anyone does it. We absolutely need safe workplaces and great cultures, but it is startling that I stayed because I was so motivated by my purpose and the positive impact my projects were having. That is how strong the pull of purpose is.

Purpose is not a nice-to-have. It is a simple and effective retention strategy.

When organisations reconnect engineers to why they're here - to service, legacy, problem-solving, impact - something remarkable happens: engagement rises, innovation increases, collaboration strengthens, and people want to stay.

And this is not abstract. You can feel the difference on a project with purpose. The energy lifts, conversations deepen, ideas sharpen. The work becomes more than the output; it becomes part of who people are.

The future of engineering will be shaped not by how many people we can hire, but by how many we can keep - and that depends on their connection to purpose.

This is how we retain people in the profession: not through perks, not through pressure, but through purpose.

When I think about the engineers of the past - the ones who built the great bridges, tunnels, and water systems - I wonder what their why and their purpose was. They knew why they were building. Their work was grounded in service to people and legacy.

Somewhere along the way, we lost that thread.

We became efficient. Productive. Brilliant at doing what we were asked - but slower to ask why.

Our systems rewarded compliance, not curiosity. They celebrated output, not insight. The profession became measured in delivery, not direction.

We started valuing the what and how of engineering - and forgot to protect the why.

Most engineers, when I ask why they chose this path or why they are still here, tell the same kind of story: "I wanted to make a difference. I wanted to build things that help people."

That's how I know our why isn't lost. It's just buried.

Reclaiming it isn't nostalgic - it's vital, because our why is what gives meaning to every calculation, every drawing, every late-night decision. It's the heartbeat behind the blueprint.

The Leadership We Need

Reconnecting to purpose is not just a personal act - it's a leadership imperative.

The world doesn't just need more engineers. It needs more engineering leaders - people who can translate between systems, stories, and stakeholders. People who can hold complexity without collapsing into control. People who can connect teams back to the meaning behind the work.

For too long, leadership in engineering has been confused with management - climbing ladders, signing drawings, overseeing delivery. But leadership in the 21st century looks different. It's about alignment, not authority. It's about leading with questions, not certainty. It's about creating workplaces where people feel safe enough to be curious, brave enough to challenge assumptions, and grounded enough to stay.

Tomorrow's engineering leaders will navigate systems of extraordinary complexity. They'll have to bridge worlds - between climate scientists and construction crews, AI modellers and community Elders, policy-makers and Traditional Owners.

To do that, they'll need more than technical excellence. They'll need human fluency, which is often lacking in engineering programs today.

These skills must be learned intentionally and cultivated - not assumed to appear with seniority. In engineering, I have seen time and time again technical people who were promoted because they are technically excellent, but not great with people. Often they don't want to lead and want to stay on the engineering tools. They take on leadership roles because they feel they are expected to. They struggle leading, and then are expected to train the next group of leaders when they haven't mastered the skill themselves. The cycle plays out over and over again, creating poor leaders and average cultures.

This problem in our industry is why I co-founded WeAspire - to help train, support, and coach people in technical professions like construction and infrastructure to excel in leadership roles.

It is the responsibility of leaders in our profession to create the cultures that people want to stay in. To protect psychological safety. To connect their teams to purpose and values. To hold space for reflection, failure, learning, and growth. To help engineers rediscover the "why" that got them here in the first place.

But we cannot expect leaders to do this without support.

The work we've done through WeAspire - training more than 10,000 leaders across the construction and infrastructure industries - has shown me the same pattern again and again: we train engineers intensely for technical excellence, and then give them almost no training for leadership.

We spend four years at university learning how to design bridges, calculate loads, and run models - and little time, sometime no time, learning how to lead people. Yet the first leadership role is often the hardest of an engineer's entire career. You're expected to manage people for the first time while still being "on the tools". You're juggling urgent fires and long-term strategy simultaneously. And you're doing it with no roadmap, no training, and often no psychological safety.

We typically invest in leadership development far too late - at the senior or executive level, long after habits have formed and burnout has set in. But the moment that matters most is earlier: the leap from technical expert to first-time leader.

When we teach engineers how to lead - not just deliver - something extraordinary happens. People who never saw themselves as leaders discover their strengths and lead with them, rather than try to be someone they are not. Teams become safer and more cohesive. Cultures stabilise. And retention grows, not because we've asked people to stay, but because we've given them something worth staying for.

Imagine what could be possible if every engineer got that message early in their career - along with the tools, confidence, and support to lead well. Imagine the culture we could build. Imagine the profession we could become.

If we want a future engineering workforce - diverse, resilient, purpose-driven - we must develop the leaders who will protect it, inspire it, and hold it. Leadership is not a bonus skill. It is the foundation for retention, culture, and legacy.

And when we get it right, we don't just keep people in engineering - we help them become the kind of engineers the future needs.

Creating the Culture that Retains

Unfortunately, I can tell you many, many examples of bad cultures and toxic workplaces that I've seen throughout my career. Everything from being sexually assaulted to bullying, negative comments and bosses who undermined me.

But I've also experienced great cultures too. I think if there is one thing that would make the biggest difference to the culture that we need for the future of engineering, it is creating psychologically safe spaces.

I often think back to a meeting I chaired on the huge tunnel project where I had 20 stakeholders in the room - the client, design engineers, contractors, and future operators. The topic was contentious, but I felt the discussion was going well.

That is, until one of the senior men leaned back in his chair and said casually, "The gentlemen around the room will make the decision."

It stopped me cold. I was caught off guard.

This was someone who, outside that room, had been one of the biggest advocates for diversity. He'd even supported several of the initiatives I'd been involved in.

But in that moment, his words pulled the air out of the room. I felt like my leadership had just evaporated. I felt tiny. Like I didn't belong. Like my contribution didn't matter.

I shut down completely. It was hard for me to continue to chair the meeting.

Afterwards, I gathered the courage to approach him privately. I said, "Hey, did you know you said that in the meeting?"

He looked genuinely shocked. He hadn't realised. There was no malice. It was habit - a relic of the old culture we all inherited.

I was proud of myself for raising it. But when I asked a more senior female colleague for advice, she shrugged and said, "That's the industry. Get used to it."

It was heartbreaking, because the subtext was clear: silence is survival.

That's what happens when psychological safety is missing. When small words carry heavy weights. When unspoken hierarchies define who gets heard.

It taught me something important: that courage is contagious, but so is silence.

And then I saw the opposite.

A few years later, I met an incredible leader - the CEO of a large company in Brisbane. After a company-wide town hall, he went up to one of the most junior staff members and said, "Hey, how do you think that went?"

She gave him honest feedback - some praise, some constructive comments - and he thanked her sincerely. She later told him she was stunned. She'd never expected her voice to matter that much.

He replied, "Of course it matters. It's everyone's organisation."

And he kept asking for her feedback - not once, but regularly.

That's what real leadership looks like. It doesn't claim to have all the answers. It creates a space where the answers can surface.

I've also seen leaders model this brilliantly in more structured ways. When I was on Arup's Diversity and Inclusion Committee, our chair was exceptional at creating a balanced, reflective space. He would always wait before speaking, and would listen to everyone else's perspective first. He knew that his voice carried weight, so he deliberately used it last. It was an act of generosity, and it made the conversation richer every single time.

I learned to do the same when running strategy sessions for Power of Engineering. My co-founder and I would meet beforehand, brainstorm our thoughts, and come up with ideas for the year ahead. But when we got to the group workshop, we would hold those ideas back. We'd start instead by asking everyone else: "What do you think? What should we focus on?"

The ideas that emerged were always better than what we'd imagined on our own. The best designs are born in cultures where it's safe to question, to challenge, and to care. When people hesitate to speak, risks multiply. When people feel seen and respected, judgment sharpens and creativity flourishes. People are open about failure and everyone learns and makes things that are better.

That's why the best engineering organisations in the world - from Toyota to NASA - invest as much in how people communicate as in what they design. They've learned that you can have all the technical excellence in the world, but without trust and psychological safety, you can't have resilience.

The Future Awaits

The 21st century is teaching us that the systems we create are not linear machines - they are living organisms. They breathe, adapt, and depend on relationships between parts we can no longer treat as separate.

When one system falters, another feels the strain. A decision made in a design meeting can ripple into social, ecological, and cultural change decades later. We are no longer just designing things. We are designing conditions, and in those conditions, humanity is the stabilising force.

There are five leadership capacities that will define the next generation of engineers:

1. **Judgment** - the courage to make decisions when the data is incomplete, and the stakes are human. Being deeply connected to values and purpose.
2. **Courage** - to question the brief, speak truth to power, and admit uncertainty.
3. **Empathy** - to design with, not just for, the people who live the outcomes.
4. **Communication** - to translate complexity into connection and make engineering understandable, relatable, and trusted.
5. **Vision** - to lift our eyes beyond compliance and see the legacy we are building.

These are not "soft skills." They are structural skills - the ones that hold systems, projects, and people together when everything else changes.

When I look at the next generation of engineers - the students I've taught, the graduates I've mentored - I'm filled with hope. They're already redefining what engineering can be: collaborative, creative, inclusive, regenerative.

They remind me that engineering is, at its core, a social act rather than a solely technical one. Every bridge, every hospital, every tunnel - they're all built for people. And when we remember that, our designs change.

If we build a culture that's curious, kind, and courageous, engineers will not only stay - they'll lead. They'll build a world where every structure, system, and story carries the imprint of care.

The future of engineering isn't just something we design. It's something we become.

The next decade and the ones that follow will test us in ways we cannot yet imagine. As the world around us accelerates, we can come back to the framework of Values, Principles, and Trust. It slows us down and anchors us in what matters. When uncertainty grows, it gives us a compass. When systems fail, it gives us a way to rebuild - not just technically, but ethically, relationally, and regeneratively.

- **Values anchor us.** They remind us who we are when everything else is shifting. They are our roots in a fast-moving world.
- **Principles guide our behaviour.** They turn intention into action, ethics into everyday design.
- **Trust binds the system.** It's the invisible architecture - the thing that holds projects, teams, and communities together when logic alone can't.

These three elements form the circuitry of modern engineering leadership. They are how we stay human inside systems that are becoming more-than-human.

If you trace the threads of every story in this book - from climate resilience to AI ethics, from cyber security to circular design, from culture to care - one pattern becomes unmistakable: as complexity rises, so must our humanity. The future of engineering is human.

There are no isolated frontiers anymore - only living systems, woven together by the choices we make, the values we hold, and the way we choose to lead.

The World We Could Build

Imagine a world where the structures, systems, and cities we create don't just serve us today, but actively regenerate life for tomorrow. A world where engineering doesn't hide in the background - it leads with intention. Where every bridge strengthens a community, every road connects rather than divides, every building holds the story of the Country it stands on, and every project leaves the world a little better than when we found it.

In this world, we pause before we solve. We challenge our briefs not out of defiance, but out of devotion - to people, to place, to possibility. We expand what we optimise for.

Instead of efficiency standing alone as the hero of progress, it takes its place alongside equity, connection, biodiversity, and cultural integrity. Engineers begin to measure success not just by traffic volumes or cost savings, but by shade created, trust strengthened, carbon restored, and communities brought closer together. Value expands - and so does our imagination.

Picture young engineers entering the profession and recognising themselves immediately - not in the stereotypes of calculators and compliance, but in a lineage of designers, creators, and custodians. They learn to see systems as living, dynamic, interconnected. They learn that creativity is a prerequisite for progress, that care is a design principle, and that standing up for better outcomes is not a risk - it's a responsibility of the role. And the leaders who guide them don't only ask for technical excellence; they ask for wisdom, courage, and compassion too.

Companies shift as well. Instead of choosing between shareholder value and societal good, they create models that deliver both. Procurement becomes a tool for partnership, not a race to the bottom. Risk expands into a larger definition - one that includes social trust and future generations. Imagine boardrooms where the first question isn't "How fast?" or "How much?", but "Who is this for?" and "What will this mean for the world long after we are gone?" Decisions begin to change. Outcomes begin to change. Legitimacy grows.

And beneath it all, imagine a renewed relationship with Country. Projects are shaped not just by regulations, but by respect - for the land's history, its limits, its rhythms, and its gifts. Engineering becomes a practice of reciprocity: we take only what we can regenerate, and we give back more than we ever imagined possible. Country becomes a collaborator, not a constraint. We build with land, not on top of it.

This world is not fantasy. The seeds of it already exist - in the pockets of innovation, the leaders choosing courage over convenience, the communities insisting on better, and the engineers questioning the inherited rules. When these seeds grow together, when values shift alongside standards, when intention becomes as important as calculation, we don't just design projects - we design futures.

And perhaps the greatest realisation of all is this: we've built the world once before. The systems, cities, and networks we rely on today were imagined and brought to life by engineers who believed in what was possible then.

We can build it again - this time, with intention, compassion, and care.

The Takeaway

The future of engineering isn't waiting for us to discover it. It's waiting for us to design it. And the world we could build is already within reach, ready for those brave enough to begin.

The reinvention we need isn't about adding something new - it's about remembering something that has always been there. It's like Glinda the Good Witch tells Dorothy in *The Wizard of Oz*: "You had it in you all along." I think this is true for the future of engineering. We have everything we need inside of us; we just need to tap into it.

We often imagine transformation as something grand - but it rarely starts that way. Instead, what if each of us made just a 1% shift in our practice as engineers and in our leadership? One value embodied, one conversation started, one design made gentler for both people and planet. Those small, deliberate choices, multiplied across a profession, become the great reinvention of our time.

I can't even begin to imagine how different our world might look with this shift. The true engineer is not lost. They are simply waiting to be remembered - in the way we listen, lead, and build.

And if we lead with courage, creativity, and care, design with humility, and build with love, then the future won't just be engineered. It will be healed.

This isn't just about the future of engineering. It's about the future of everything engineering touches.

We are the custodians of the future. The stewards of systems that shape lives. And if we can can build with care for Country, care for each other, and care for the generations who will inherit our work, then our legacy will not be measured in what we made, but in what we made possible.

That is the true reinvention of engineering. And it begins with us.

We are the custodians of the future. Let's engineer it well.

Acknowledgements

This book was never written alone.

This book exists because of a community.

While my name sits on the cover, every idea in these pages has been shaped, challenged, softened, strengthened, and made possible by the people who walked alongside me throughout my career and my life. Engineering may be taught as an individual discipline, but everything meaningful I have learned about it has come through relationships and through people.

To my mentors - thank you for seeing something in me before I could see it myself. Catherine Tanna, Belinda Gibson, Peter Bailey, and many more - your leadership, courage, and generosity showed me what it looks like to hold power with integrity and to lead without losing humanity. You helped me understand that engineering leadership is not about control, but about care, responsibility, and courage. I carry your lessons into every room I enter.

To the managers, leaders, and colleagues who trusted me early, backed me when I was young, and gave me space to grow - thank you. Some of you believed in me when I didn't yet have the confidence to believe in myself. Others challenged me, stretched my thinking, and asked more of me than I thought I could give. You taught me that good engineering cultures don't just deliver projects; they develop people.

To the people named throughout this manuscript - thank you for the moments, conversations, and experiences that shaped the questions at the heart of this book. From educators and engineers to leaders and collaborators, each of you played a role in helping me see engineering more clearly for what it is: a deeply human, values-driven profession, whether we acknowledge it or not.

To the First Nations culture around the world, and particularly the First Nations of Australia - the oldest continually living culture on the planet and the first engineers of the country, on the lands to which I was born. Thank you for taking care of this land that I call home for thousands of years, for understanding the land in ways I never will. This always was and always will be your land, and I will continue to learn from you. This book was proudly written on your lands, on Gubbi Gubbi country (Sunshine Coast) where I live, and also on the lands I travelled through, including Jagera and Turrbal (Brisbane); Gai-mariagal of the Eora Nation (Sydney) and Gumbaynggirr, Bundjalung, Wiradyuri,

Ngunnawal, Wonnarua (New South Wales); Dhudhuroa, Jaithmathang, Taungurung, Waywurru, Gunaikurnai (Victoria); Kaurna (Adelaide); and Muwinina, Plairhekenillerplue, Pydairrerme, Tommeginne (Tasmania). May we learn to take care of the land, water, and sky as you always have.

To the Power of Engineering community - this book would not exist without you. To the advisory group - thank you for your wisdom, your challenge, and your belief that engineering can and must do better. To the volunteers who gave their time, energy, and heart to the program - you are the quiet force behind every young person who left seeing themselves differently. And to the many people who helped deliver, shape, research, and support the program along the way - this work is yours as much as mine.

To my coach, Paris Cutler - thank you for asking the questions I couldn't avoid, and for holding me to the work when it felt easier to stay silent. You helped me find the courage to write this book not as a technical exercise, but as an act of leadership. Your belief that this story mattered helped me believe it too.

To my husband, the incredible Michael Furey - thank you for being my constant, my sounding board, and my safe place. You carried more than your share so I could write, think, and reflect. This book holds your support on every page. And to my children, Winston and Jemima - this book is for you, even though you may not realise it yet. Becoming your parent changed everything about how I see the world and my role in shaping it. You made legacy real for me - not as an abstract idea, but as a responsibility. When I think about the future engineers this book is written for, I think about the world you will inherit. You taught me that engineering decisions are never just technical; they are hugely personal. You are the reason I care so fiercely about what we leave behind.

Finally, to the wider engineering community past, present, and future - thank you for the work you do, often unseen and often uncelebrated. This book is an offering to you. It is shaped by you. And it exists because of a shared belief that engineering can be a regenerative force for good. This book was never written alone. It was built, like all meaningful things, by many hands, many voices, and a community willing to care.

With gratitude,

Felicity

References

1. Taylor, Frederick Winslow. *The Principles of Scientific Management*. New York: Harper & Brothers, 1911.

2. Kanigel, Robert. *The One Best Way: Frederick Winslow Taylor and the Enigma of Efficiency*. New York: Viking, 1997.

3. Criado Perez, Caroline. *Invisible Women: Data Bias in a World Designed for Men*. New York: Abrams Press, 2019.

4. Bose, Devajyoti, Mary Lesch, and Joonho Ryu. "Female Occupant Injury Risk in Frontal Crashes. A Review of Crash Test Dummy Limitations." *Traffic Injury Prevention* 12, no. 5 (2011): 434–41. https://doi.org/10.1080/15389588.2011.600098.

5. Bohman, Katarina, Anna Jakobsson, and Maria Stockman. "Seat Belt Load Limiting for Female Drivers: Effects of Seating Position and Anthropometry." *Traffic Injury Prevention* 14, no. 6 (2013): 615–22. https://doi.org/10.1080/15389588.2012.748162.

6. Kingma, Boris R. M., and Wouter D. van Marken Lichtenbelt. "Energy Consumption in Buildings and Female Thermal Demand." *Nature Climate Change* 5, no. 12 (2015): 1054–56. https://doi.org/10.1038/nclimate2741.

7. Institute of Medicine. *Exploring the Biological Contributions to Human Health: Does Sex Matter?* Washington, DC: National Academies Press, 2001. https://doi.org/10.17226/10028.

8. Klein, Sabra L., and Katie L. Flanagan. "Sex Differences in Immune Responses." *Nature Reviews Immunology* 16, no. 10 (2016): 626–38. https://doi.org/10.1038/nri.2016.90.

9. Knapik, Joseph J., Katy L. Reynolds, and Everett Harman. "Soldier Load Carriage: Historical, Physiological, Biomechanical, and Medical Aspects." *Military Medicine* 169, no. 1 (2004): 45–56. https://doi.org/10.7205/MILMED.169.1.45

10. Austroads. *Guide to Road Design Part 6: Roadside Design, Safety and Barriers (AGRD06-22), Edition 3.2*. Sydney: Austroads, 2022.

11. Infrastructure Australia. *Australian Infrastructure Audit 2019: An Assessment of Australia's Future Infrastructure Needs*. Canberra: Infrastructure Australia, 2019. https://www.infrastructureaustralia.gov.au/publications/australian-infrastructure-audit-2019.

12. Pfautsch, Sebastian. *Benchmarking Summer Heat: Across Blacktowns Diverse Urban Landscapes*. Western Sydney University, 2025. https://researchers.westernsydney.edu.au/files/282754150/2025_Blacktown_Heat_Mapping_FIN_003_.pdf

13. Williamson, Bess. *Accessible America: A History of Disability and Design*. New York: NYU Press, 2019.

14. Downey, Gregory J. *Closed Captioning: Subtitling, Stenography, and the Digital Convergence of Text with Television*. Baltimore: Johns Hopkins University Press, 2008.

15. U.S. Environmental Protection Agency. *Notice of Violation of the Clean Air Act to Volkswagen AG, Audi AG, and Volkswagen Group of America, Inc*. Washington, DC, September 18, 2015.

16. U.S. Department of Justice. "Volkswagen AG Agrees to Plead Guilty and Pay $4.3 Billion in Criminal and Civil Penalties; Six Volkswagen Executives and Employees Indicted in Connection with Conspiracy to Cheat U.S. Emissions Tests." Press release, January 11, 2017.

17. Australian Communications and Media Authority (ACMA). "ACMA Statement on Optus Triple Zero Investigation." September 22, 2025.

18. Australian Broadcasting Corporation (ABC News). "How the Optus Triple-0 Outage Unfolded — Timeline and Failures." *ABC News*, September 23, 2025.

19. Reuters. "Australia's Optus Emergency Services Outage Review Flags Urgent Protocol Gaps." *Reuters*, December 18, 2025.

20. Australian Communications Consumer Action Network (ACCAN). "Fundamental Change Urgent After Fatal Optus Outage." Media release, September 21, 2025.

21. Australian Broadcasting Corporation (ABC). "Qantas: The Inside Story." *Four Corners*, September 5, 2022. https://www.abc.net.au/news/programs/four-corners/2022-09-05/qantas-the-inside-story/101401390.

22. Australian Competition and Consumer Commission (ACCC). *Airline Competition in Australia.* Canberra: ACCC, 2024.

23. Australian Transport Safety Bureau. *Foreign Object Debris Event Involving Airbus A380, VH-OQI: Final Report (AO-2024-006).* Canberra: ATSB, 2024.

24. Australian Transport Safety Bureau (ATSB). *Safety Investigation Report AO-2024-006: Maintenance Error Involving an Airbus A380, Qantas Airways.* Canberra: ATSB, February 2024.

25. Australian Broadcasting Corporation (ABC News). "Why Qantas Is Flying One of the World's Oldest Domestic Jet Fleets." *ABC News*, September 6, 2023.

26. The Guardian. "'All the Lanes Work, but the Community Doesn't': Sydney Interchange Deemed 'Unreasonable and Unacceptable'." *The Guardian*, July 15, 2024.

27. News.com.au. "Urban Planning Expert Says Rozelle Interchange Working as Planned Despite Chaos and Community Outrage." *News.com.au*, December 7, 2023.

28. Australian Broadcasting Corporation (ABC News). "Public Inquiry Finds Sydney Interchange Prioritised Toll Traffic Over Communities." *ABC News*, July 15, 2024.

29. The Guardian. "'A Berlin Wall of Barriers': Residents Say Sydney Interchange Has Cut Their Suburbs in Two." *The Guardian*, April 10, 2024.

30. New South Wales Parliament, Legislative Council. Submission no. 19 (Mr Nathan English) to the inquiry *Impact of the Rozelle Interchange*, March 21, 2024.

31. Grenfell Tower Inquiry. *Grenfell Tower Inquiry: Final Report (Phase 2).* London: UK Government, September 2024.

32. Grenfell Tower Inquiry. *Grenfell Tower Inquiry: Phase 1 Report.* London: UK Government, October 2019.

33. The Guardian. "Grenfell Tower Fire Was Result of 'Decades of Failure', Final Inquiry Report Finds." *The Guardian*, September 4, 2024.

34. The Guardian. "Grenfell Fire: How Dangerous Cladding Turned a Kitchen Blaze into a National Tragedy." *The Guardian*, June 14, 2018.

35. The Independent. "Grenfell Tower Refurbishment Switched to Cheaper Cladding to Save Money, Inquiry Hears." *The Independent*, December 6, 2018.

36. Booth, Robert, and Sarah Johnson. "Grenfell Tower Refurbishment Used Cheaper Cladding to Save Money, Inquiry Told." *The Guardian*, December 6, 2018. https://www.theguardian.com/uk-news/2018/dec/06/grenfell-tower-refurbishment-cheaper-cladding-save-money-inquiry.

37. IFSEC Global. "Grenfell Tower Inquiry: How the 'Stay Put' Policy Failed." *IFSEC Global*, October 30, 2019. https://www.ifsecglobal.com/fire-safety/grenfell-tower-inquiry-how-the-stay-put-policy-failed/.

38. London Fire Brigade. *Grenfell Tower Inquiry Phase 1: London Fire Brigade Response and Action Plan.* London: London Fire Brigade, 2020. https://www.london-fire.gov.uk/about-us/our-strategy-and-performance/good-governance/grenfell-tower-inquiry/.

39. London Fire Brigade. *Fire Survival Guidance Policy and Procedures (Public Summary).* London: London Fire Brigade, 2021. https://www.london-fire.gov.uk/safety/the-greater-london-fire-and-rescue-authority/guidance-and-policies/.

40. UK Home Office. *Grenfell Tower Inquiry Phase 1 Recommendations: Government Response.* London: Home Office, January 2020. https://www.gov.uk/government/publications/grenfell-tower-inquiry-phase-1-recommendations-government-response.

41. National Fire Chiefs Council. *Simultaneous Evacuation Guidance for High-Rise Residential Buildings.* London: NFCC, January 2021. https://www.nationalfirechiefs.org.uk/write/MediaUploads/NFCC_Simultaneous_Evacuation_Guidance_January_2021.pdf.

42. UK Parliament. *The Fire Safety (England) Regulations 2022* (SI 2022/547). London: The Stationery Office, 2022. https://www.legislation.gov.uk/uksi/2022/547/contents/made.

43. Greater London Authority. *Evacuation Alert Systems for High-Rise Residential Buildings.* London: Greater London Authority, 2021. https://www.london.gov.uk/questions/2021/3537.

44. UK Government. *Ban on Combustible Materials in and on the External Walls of Buildings.* London: Ministry of Housing, Communities and Local Government, 2018; updated 2022. https://www.gov.uk/government/publications/ban-on-combustible-materials.

45. UK Government. *Building Safety Programme: Remediation of Unsafe Cladding.* London: Department for Levelling Up, Housing and Communities, 2023. https://www.gov.uk/guidance/remediation-of-unsafe-cladding.

46. Ward Hadaway. *Procurement in a Nutshell – Grenfell Tower Report: Key Procurement Lessons.* September 19, 2024. Accessed 1 December 2025. https://www.wardhadaway.com/insights/procurement-in-a-nutshell/procurement-in-a-nutshell-grenfell-tower-report-key-procurement-lessons/

47. Bucciarelli, Louis L. *Designing Engineers.* Cambridge, MA: MIT Press, 1994.

48. Petroski, Henry. *To Engineer Is Human: The Role of Failure in Successful Design.* New York: Vintage Books, 1992.

49. Reason, James. *Human Error.* Cambridge: Cambridge University Press, 1990.

50. Aven, Terje. *Risk Analysis: Assessing Uncertainties Beyond Expected Values and Probabilities.* Chichester, UK: Wiley, 2008.

51. Winner, Langdon. "Do Artifacts Have Politics?" *Daedalus* 109, no. 1 (1980): 121–36.

52. Busch, Lawrence. *Standards: Recipes for Reality.* Cambridge, MA: MIT Press, 2011.

53. Vaughan, Diane. *The Challenger Launch Decision: Risky Technology, Culture, and Deviance at NASA.* Chicago: University of Chicago Press, 1996.

54. Schön, Donald A. *The Reflective Practitioner: How Professionals Think in Action.* New York: Basic Books, 1983.

55. Whorf, Benjamin Lee. *Language, Thought, and Reality: Selected Writings of Benjamin Lee Whorf.* Edited by John B. Carroll. Cambridge, MA: MIT Press, 1956.

56. Chen, M. Keith. "The Effect of Language on Economic Behavior: Evidence from Savings Rates, Health Behaviors, and Retirement Assets." *American Economic Review* 103, no. 2 (2013): 690–731. https://doi.org/10.1257/aer.103.2.690.

57. Winawer, Jonathan, Nathan Witthoft, Michael C. Frank, Lisa Wu, Alex R. Wade, and Lera Boroditsky. "Russian Blues Reveal Effects of Language on Color Discrimination." *Proceedings of the National Academy of Sciences* 104, no. 19 (2007): 7780–85. https://doi.org/10.1073/pnas.0701644104.

58. Flyvbjerg, Bent. *Survival of the Unfittest: Why the Worst Infrastructure Gets Built — and What We Can Do About It.* Oxford: Oxford University Press, 2014.

59. Frangopol, Dan M., M. J. Kallen, and Jan M. van Noortwijk. "Probabilistic Models for Life-Cycle Performance of Deteriorating Structures: Review and Future Directions." *Progress in Structural Engineering and Materials* 6, no. 4 (2004): 197–212.

60. Sen, Amartya. *Development as Freedom.* New York: Alfred A. Knopf, 1999.

61. Perrow, Charles. *Normal Accidents: Living with High-Risk Technologies.* New York: Basic Books, 1984.

62. Power, Michael. *The Risk Management of Everything: Rethinking the Politics of Uncertainty.* London: Demos, 2004.

63. Flyvbjerg, Bent. *How Big Things Get Done: The Surprising Factors That Determine the Fate of Every Project.* New York: Currency, 2023.

64. Gilligan, Carol. *In a Different Voice: Psychological Theory and Women's Development.* Cambridge, MA: Harvard University Press, 1982.

65. Scott, James C. *Seeing Like a State: How Certain Schemes to Improve the Human Condition Have Failed.* New Haven, CT: Yale University Press, 1998.

66. Bayles, David, and Ted Orland. *Art & Fear: Observations on the Perils (and Rewards) of Artmaking.* Santa Cruz, CA: Image Continuum Press, 1993.

67. Queensland Government. *Paradise Dam: Project Overview and Construction Summary.* Brisbane: Queensland Department of Natural Resources and Mines, 2005.

68. Sunwater. *Paradise Dam Safety Review and Temporary Full Supply Level Reduction.* Brisbane: Sunwater, September 2019.

69. Queensland Government. *Paradise Dam Independent Technical Review: Final Report.* Brisbane: Department of Regional Development, Manufacturing and Water, 2021.

70. Queensland Government. *Paradise Dam Improvement Project: Business Case Summary.* Brisbane: Queensland Government, 2023.

71. Massachusetts Department of Transportation. *Longfellow Bridge Rehabilitation Project: Project History and Cost Overview.* Boston: MassDOT, 2015. https://www.mass.gov/doc/longfellow-bridge-project-history-and-cost-overview/download

72. Parvin, Alastair. "Architecture for the People by the People." TED Talk, filmed 2013, published February 2014. https://www.ted.com/talks/alastair_parvin_architecture_for_the_people_by_the_people.

73. WikiHouse Foundation. *WikiHouse: An Open Source Building System.* London: WikiHouse Foundation, ongoing technical documentation (first released c. 2011–2014). https://www.wikihouse.cc.

74. Dyson, James. *Invention: A Life.* London: Penguin Random House UK, 2021.

75. Bradfield, J. J. C. *The Official Opening of the Sydney Harbour Bridge, March 19, 1932.* Sydney: Government Printer, 1932.

76. Australian Heritage Council. *Sydney Harbour Bridge: National Heritage List Citation.* Canberra: Australian Government Department of Climate Change, Energy, the Environment and Water, 2007. https://www.dcceew.gov.au/parks-heritage/heritage/places/national-heritage-list/sydney-harbour-bridge.

77. Standards Australia. *AS 5100:2017 Bridge Design, Parts 1–9.* Sydney: Standards Australia, 2017.

78. Transport for New South Wales. *Sydney Harbour Bridge Asset Management and Maintenance Program.* Sydney: Transport for NSW, 2022.

79. Transport for New South Wales. *Sydney Harbour Bridge Asset Management and Maintenance Program.* Sydney: NSW Government, 2022.

80. Crowe, Timothy D. *Crime Prevention Through Environmental Design.* 3rd ed. Burlington, MA: Butterworth-Heinemann, 2013.

81. Austroads Ltd. *Guide to Road Design Part 3: Geometric Design.* Sydney: Austroads Ltd, 2021, publication no. AGRD03-16, ed. 3.4.

82. Arup. "Designing the Night-Time City." Arup, n.d. Accessed January 2026. https://www.arup.com/perspectives/designing-the-night-time-city.

83. Arup. "Randwick Health & Innovation Precinct." Arup Projects, n.d. Accessed January 2026. https://www.arup.com/projects/randwick-health-and-innovation-precinct.

84. Brand, Peter, and Julio D. Dávila. "Mobility Innovation at the Urban Margins: Medellín's Metrocables." *World Development* 41 (2013): 1–14. https://doi.org/10.1016/j.worlddev.2012.05.031.

85. Echeverri, Alejandro, and Giancarlo Mazzanti. "Medellín's Social Urbanism: Transforming the City from the Margins." *Harvard Design Magazine*, no. 35 (2012).

86. Inter-American Development Bank. *Medellín: Transformation Through Social Urbanism*. Washington, DC: Inter-American Development Bank, 2015.

87. City of Vienna. *Gender Mainstreaming in Urban Planning and Urban Development*. Vienna: Municipal Department for Gender Mainstreaming, 2013.

88. Franzen, Brigitte, and Eva Kail. "Gender-Sensitive Urban Planning: The Case of Vienna." *Built Environment* 37, no. 3 (2011): 307–21.

89. European Institute for Gender Equality. *Gender Equality and Urban Planning: The Case of Vienna*. Vilnius: EIGE, 2016.

90. Scheurer, Jan, and Peter Newman. "Vauban: A European Model Bridging the Green and Brown Agendas." *Journal of Urban Affairs* 31, no. 5 (2009): 651–76.

91. Buehler, Ralph, and John Pucher. "Sustainable Transport in Freiburg: Lessons from Germany's Green City." *Transport Reviews* 31, no. 1 (2011): 43–70.

92. City of Freiburg. *Vauban District: Sustainable Urban Development*. Freiburg im Breisgau: City of Freiburg, 2008. https://www.freiburg.de/pb/site/Freiburg/get/params_E-1025911322/207264/vauban_sustainable_development.pdf

93. European Commission. *Sustainable Urban Development in Freiburg-Vauban*. Brussels: European Commission, Directorate-General for Regional Policy, 2012. https://ec.europa.eu/regional_policy/sources/docgener/presenta/urban2009/vauban_case_study.pdf

94. Nightingale Housing. *The Nightingale Model*. Melbourne: Nightingale Housing, accessed March 2025. https://www.nightingalehousing.org.

95. Australian Broadcasting Corporation (ABC News). "Nightingale Housing: The Radical Model Challenging Australia's Property Market." *ABC News*, June 8, 2019.

96. Ulrich, Roger S., et al. "A Review of the Research Literature on Evidence-Based Healthcare Design." *HERD: Health Environments Research & Design Journal* 1, no. 3 (2008): 61–125.

97. Jencks, Charles, and Maggie Keswick Jencks. *Maggie's Centres: The Architecture of Hope*. London: Frances Lincoln, 2010.

98. Ulrich, Roger S. "View Through a Window May Influence Recovery from Surgery." *Science* 224, no. 4647 (1984): 420–21.

99. City of Karlskoga. *Gender Equality Assessment of Snow Removal Operations*. Karlskoga: Municipality of Karlskoga, 2011.

100. Gender Equality Council. *Gender Equality in Transport: Policy and Practice in Sweden*. Stockholm: Swedish Association of Local Authorities and Regions (SALAR), 2012.

101. European Institute for Gender Equality. *Gender Equality in Transport: Good Practices and Policies*. Vilnius: EIGE, 2019.

102. Lyons, Oren, and John Mohawk. "The Iroquois Confederacy." In *Exiled in the Land of the Free: Democracy, Indian Nations, and the U.S. Constitution*, edited by Oren Lyons et al., 28–48. Santa Fe, NM: Clear Light Publishers, 1992.

103. Frederick, Shane, George Loewenstein, and Ted O'Donoghue. "Time Discounting and Time Preference: A Critical Review." *Journal of Economic Literature* 40, no. 2 (2002): 351–401.

104. Hershfield, Hal E., Daniel G. Goldstein, William F. Sharpe, Jesse Fox, Leo Yeykelis, Laura L. Carstensen, and Jeremy N. Bailenson. "Increasing Saving Behavior Through Age-Progressed Renderings of the Future Self." *Journal of Marketing Research* 48, no. S1 (2011): S23–S37. https://doi.org/10.1509/jmkr.48.SPL.S23.

105. Thaler, Richard H., and Cass R. Sunstein. *Nudge: Improving Decisions About Health, Wealth, and Happiness.* New Haven, CT: Yale University Press, 2008.

106. Raworth, Kate. *Doughnut Economics: Seven Ways to Think Like a 21st-Century Economist.* London: Random House Business, 2017.

107. Haushofer, Johannes, and Ernst Fehr. "On the Psychology of Poverty." *Science* 344, no. 6186 (2014): 862–67. https://doi.org/10.1126/science.1232491.

108. Norman, Donald A. *Emotional Design: Why We Love (or Hate) Everyday Things.* New York: Basic Books, 2004.

109. Porges, Stephen W. *The Polyvagal Theory: Neurophysiological Foundations of Emotions, Attachment, Communication, and Self-Regulation.* New York: W. W. Norton & Company, 2011.

110. World Economic Forum. "Why Urban Living Is Bad for Your Mental Health." *World Economic Forum*, October 14, 2019.

111. Evans, Gary W. "The Built Environment and Mental Health." *Journal of Urban Health* 80, no. 4 (2003): 536–55.

112. McEwen, Bruce S. "Protective and Damaging Effects of Stress Mediators." *New England Journal of Medicine* 338, no. 3 (1998): 171–79.

113. Meerlo, Peter, et al. "Restricted and Disrupted Sleep: Effects on Autonomic Function, Neuroendocrine Stress Systems and Stress Responsivity." *Sleep Medicine Reviews* 12, no. 3 (2008): 197–210.

114. New South Wales Parliament. *Inquiry into the Design, Delivery and Effectiveness of Built Environments in Addressing Loneliness.* Sydney: Parliament of New South Wales, 2021.

115. Maas, Jolanda, et al. "Green Space, Urbanity, and Health: How Strong Is the Relation?" *Journal of Epidemiology & Community Health* 60, no. 7 (2006): 587–92.

116. Vartanian, Oshin, et al. "Architectural Design and the Brain: Effects of Ceiling Height and Perceived Enclosure on Beauty Judgments and Approach-Avoidance Decisions." *Journal of Environmental Psychology* 41 (2015): 10–18.

117. Bringslimark, Tina, et al. "The Psychological Benefits of Indoor Plants: A Critical Review of the Experimental Literature." *Journal of Environmental Psychology* 29, no. 4 (2009): 422–33.

118. Nieuwenhuis, Marlon, et al. "The Relative Benefits of Green Versus Lean Office Space: Three Field Experiments." *Journal of Experimental Psychology: Applied* 20, no. 3 (2014): 199–214.

119. Kellert, Stephen R., Judith H. Heerwagen, and Martin L. Mador, eds. *Biophilic Design: The Theory, Science, and Practice of Bringing Buildings to Life.* Hoboken, NJ: Wiley, 2008.

120. Browning, William D., Catherine O. Ryan, and Joseph O. Clancy. *14 Patterns of Biophilic Design.* New York: Terrapin Bright Green, 2014.

121. Rosa, Hartmut. *Social Acceleration: A New Theory of Modernity.* New York: Columbia University Press, 2013.

122. McEwen, Bruce S. *The End of Stress As We Know It.* Washington, DC: Joseph Henry Press, 2002.

123. Stevens, Richard G., et al. "Meeting Report: The Role of Environmental Lighting and Circadian Disruption in Cancer and Other Diseases." *Environmental Health Perspectives* 115, no. 9 (2007): 1357–62.

124. Falchi, Fabio, et al. "The New World Atlas of Artificial Night Sky Brightness." *Science Advances* 2, no. 6 (2016): e1600377.

125. Sennett, Richard. *The Corrosion of Character: The Personal Consequences of Work in the New Capitalism.* New York: W. W. Norton & Company, 1998.

126. Maslach, Christina, and Michael P. Leiter. *The Truth About Burnout: How Organizations Cause Personal Stress and What to Do About It.* San Francisco: Jossey-Bass, 1997.

127. Power, Michael. *The Audit Society: Rituals of Verification.* Oxford: Oxford University Press, 1997.

128. Hollnagel, Erik, David D. Woods, and Nancy Leveson, eds. *Resilience Engineering: Concepts and Precepts*. Aldershot, UK: Ashgate, 2006.

129. Boston Consulting Group. *Decoding Global Talent 2024: AI, Skills, and the New Workforce*. Boston: Boston Consulting Group, 2024.

130. Dill, Kathi. "Burnout Is the Top Workplace Challenge for Employees in 2024." *HR Brew*, January 23, 2024.

131. Soga, Masashi, Kevin J. Gaston, and Yuichi Yamaura. "Gardening Is Beneficial for Health: A Meta-Analysis." *Ecology and Society* 22, no. 3 (2017): 1–16.

132. Carrus, Giuseppe, Giuseppe Scopelliti, Massimiliano Lafortezza, Giovanni Colangelo, Francesco Ferrini, Francesco Salbitano, Gianluca Sanesi, and Giuseppe Semenzato. "Go Greener, Feel Better? The Positive Effects of Biodiversity on the Well-Being of Individuals Visiting Urban and Peri-Urban Green Areas." *Landscape and Urban Planning* 134 (2015): 221–228.

133. White, Mathew P., et al. "Spending at Least 120 Minutes a Week in Nature Is Associated with Good Health and Wellbeing." *Scientific Reports* 9 (2019): 7730.

134. Foucault, Michel. *Discipline and Punish: The Birth of the Prison*. Translated by Alan Sheridan. New York: Vintage Books, 1977.

135. McEwen, Bruce S., and Peter J. Gianaros. "Central Role of the Brain in Stress and Adaptation: Links to Socioeconomic Status, Health, and Disease." *Annals of the New York Academy of Sciences* 1186, no. 1 (2010): 190–222.

136. Petty, James. "The London Spikes Controversy: Homelessness, Urban Securitisation and the Question of 'Hostile Architecture.'" *International Journal for Crime, Justice and Social Democracy* 5, no. 1 (2016): 67–81.

137. Rosenberger, Robert. "Governing the City through Design: Hostile Architecture and the Ethics of Public Space." *Design Issues* 33, no. 2 (2017): 52–64.

138. Holt-Lunstad, Julianne, Timothy B. Smith, and J. Bradley Layton. "Social Relationships and Mortality Risk: A Meta-Analytic Review." *PLoS Medicine* 7, no. 7 (2010): e1000316.

139. Evans, Gary W., and Elyse Kantrowitz. "Socioeconomic Status and Health: The Potential Role of Environmental Risk Exposure." *Annual Review of Public Health* 23 (2002): 303–31.

140. Grandjean, Philippe, and Esben Budtz-Jørgensen. "Immunotoxicity of Perfluorinated Alkylates: Calculation of Benchmark Doses Based on Serum Concentrations in Children." *Environmental Health* 12, no. 1 (2013): 35.

141. Whyatt, Robin M., et al. "Perfluoroalkyl Substances and Cardiometabolic Outcomes in Young Adults." *Environmental Health Perspectives* 126, no. 5 (2018): 057003.

142. Lind, Lars, et al. "Circulating Levels of Perfluoroalkyl Substances Are Associated with Metabolic Syndrome." *Science of the Total Environment* 514 (2015): 1–7.

143. Yale School of the Environment. "Yale Experts Explain PFAS 'Forever Chemicals'." *Yale University*, 2 December, 2025. https://sustainability.yale.edu/explainers/yale-experts-explain-pfas-forever-chemicals.

144. Moss, Michael. *Salt Sugar Fat: How the Food Giants Hooked Us*. New York: Random House, 2013.

145. Monteiro, Carlos A., et al. "Ultra-Processed Foods: What They Are and How to Identify Them." *Public Health Nutrition* 22, no. 5 (2019): 936–41.

146. Australian Broadcasting Corporation (ABC News). "Australians Are Eating More Ultra-Processed Foods amid Rising Cost of Living." *ABC News (Australia)*, December 15, 2023. https://www.abc.net.au/news/dependency-ultra-processed-foods-history-health-concerns-experts/103196858 (accessed January 23, 2026).

147. Baker, Phillip, et al. "Ultra-Processed Food Consumption and Dietary Intake in Australia." *Public Health Nutrition* 21, no. 1 (2018): 234–47.

148. Lane, Melissa M., et al. "Ultra-Processed Food Consumption and Mental Health: A Systematic Review and Meta-Analysis." *BMJ Nutrition, Prevention & Health* 4, no. 1 (2021): 36–45.

149. Eyal, Nir. *Hooked: How to Build Habit-Forming Products.* New York: Portfolio/Penguin, 2014.

150. Mark, Gloria, Shamsi Iqbal, Mary Czerwinski, and Paul Johns. "Capturing the Mood: Measuring the Effects of Interruptions on Mood and Performance in the Workplace." *Human-Computer Interaction* 29, no. 3–4 (2014): 1–28.

151. Thomée, Sara, et al. "Mobile Phone Use and Stress, Sleep Disturbances, and Symptoms of Depression among Young Adults." *BMC Public Health* 11 (2011): 66.

152. Jahncke, Helena, Scott Hygge, Niklas Halin, Annika Green, and Kajsa Dimberg. "Open-Plan Office Noise: Cognitive Performance and Restoration." *Journal of Environmental Psychology* 31, no. 4 (2011): 373–82.

153. Evans, Gary W., and Dana Johnson. "Stress and Open-Office Noise." *Journal of Applied Psychology* 85, no. 5 (2000): 779–83.

154. Sundstrom, Eric, et al. "Office Noise, Satisfaction, and Performance." *Environment and Behavior* 26, no. 2 (1994): 195–222.

155. Hollnagel, Erik. *Safety-I and Safety-II: The Past and Future of Safety Management.* Farnham, UK: Ashgate, 2014.

156. Wickens, Christopher D., Justin G. Hollands, Simon Banbury, and Raja Parasuraman. *Engineering Psychology and Human Performance.* 4th ed. New York: Routledge, 2015.

157. Cho, Yong-Gyu, and Charles A. Czeisler. "Exposure to Artificial Light at Night and Circadian Disruption: Health Consequences." *Sleep Medicine Clinics* 14, no. 4 (2019): 387–95.

158. Stevens, Richard G., George C. Brainard, David E. Blask, Scott W. Lockley, and Michael J. Motta. "Adverse Health Effects of Nighttime Lighting: Comments on American Medical Association Policy Statement." *American Journal of Preventive Medicine* 45, no. 3 (2013): 343–46.

159. Cho, Young Min, et al. "Outdoor Artificial Light at Night Is Associated with Obesity and Diabetes." *Diabetologia* 58, no. 7 (2015): 1479–88.

160. Fonken, Laura K., and Randy J. Nelson. "The Effects of Light at Night on Circadian Clocks and Metabolism." *Endocrine Reviews* 35, no. 4 (2014): 648–70.

161. Hancock, Peter A., and Joel S. Warm. "A Dynamic Model of Stress and Sustained Attention." *Human Factors* 31, no. 5 (1989): 519–37.

162. Coutts, Lauren V., and Kylie D. J. Jones. "Self-Tracking, Health Anxiety, and the Quantified Self." *Health Sociology Review* 28, no. 3 (2019): 316–31.

163. Wargocki, Pawel, and David P. Wyon. "The Effects of Moderately Raised Classroom Temperatures and Classroom Ventilation Rate on the Performance of Schoolwork by Children." *HVAC&R Research* 13, no. 2 (2007): 193–220.

164. World Health Organization. *WHO Guidelines for Indoor Air Quality: Selected Pollutants.* Copenhagen: WHO Regional Office for Europe, 2010.

165. Twohig-Bennett, Caroline, and Andy Jones. "The Health Benefits of the Great Outdoors: A Systematic Review and Meta-Analysis of Greenspace Exposure and Health Outcomes." *Environmental Research* 166 (2018): 628–37.

166. Hajat, Shakoor, and Thomas Kosatsky. "Heat-Related Mortality: A Review and Exploration of Heterogeneity." *Journal of Epidemiology & Community Health* 64, no. 9 (2010): 753–60.

167. Burr, Christopher, Markus Christl, and Nikola Biller-Andorno. "Cognitive Overload and Digital Decision-Making." *AI & Society* 35 (2020): 321–34.

168. Swart, Tara. *The Source: The Secrets of the Universe, the Science of the Brain.* London: Vermilion, 2019.

169. Zeki, Semir, Oliver Y. Chén, and Romy Frackowiak. "The Neurobiology of Love." *NeuroImage* 17, no. 1 (2002): 119–29.

170. Arnheim, Rudolf. *Art and Visual Perception: A Psychology of the Creative Eye.* Berkeley: University of California Press, 1974.

171. Zeki, Semir. *Inner Vision: An Exploration of Art and the Brain.* Oxford: Oxford University Press, 1999.

172. Dissanayake, Ellen. *Homo Aestheticus: Where Art Comes From and Why.* Seattle: University of Washington Press, 1992.

173. Brown, Steven, et al. "Universals in the World's Musics." *Behavioral and Brain Sciences* 29, no. 1 (2006): 1–54.

174. Kang, Jaehee, and Steven Heine. "Cultural Psychology of Self-Decoration: Why Humans Adorn the Body." *Journal of Personality and Social Psychology* 103, no. 2 (2012): 318–35.

175. Coburn, Alex, Oshin Vartanian, and Anjan Chatterjee. "Buildings, Beauty, and the Brain: A Neuroscience of Architectural Experience." *Journal of Cognitive Neuroscience* 29, no. 9 (2017): 1521–31.

176. Lewicka, Maria. "Place Attachment, Place Identity, and Place Memory: Restoring the Forgotten City Past." *Journal of Environmental Psychology* 28, no. 3 (2008): 209–231.

177. Mouratidis, Kostas, et al. "Walking through Beautiful Urban Environments Improves Emotional Wellbeing and Reduces Stress." *Landscape and Urban Planning* 243 (2024): 104965.

178. Nasar, Jack L. *The Evaluative Image of the City.* Thousand Oaks, CA: Sage Publications, 1998.

179. Kuo, Frances E., and William C. Sullivan. "Environment and Crime in the Inner City: Does Vegetation Reduce Crime?" *Environment and Behavior* 33, no. 3 (2001): 343–67.

180. Fraser, Adam. *The Third Space: Using Life's Little Transitions to Find Balance and Happiness.* Milton, QLD: Wiley, 2012.

181. Assure Programs Australia. "How to Be Mentally Healthy Every Day." *Assure Programs Australia*, n.d. https://assureprograms.com.au/about/news-and-updates/how-to-be-mentally-healthy-every-day/ (accessed 2 December 2025).

182. Sweller, John. *Cognitive Load Theory.* New York: Springer, 2011.

183. Sweller, John, Jeroen J. G. van Merriënboer, and Fred G. W. C. Paas. "Cognitive Architecture and Instructional Design." *Educational Psychology Review* 10, no. 3 (1998): 251–96.

184. Edmondson, Amy C. *The Fearless Organization: Creating Psychological Safety in the Workplace for Learning, Innovation, and Growth.* Hoboken, NJ: Wiley, 2018.

185. Edmondson, Amy C. "Psychological Safety and Learning Behavior in Work Teams." *Administrative Science Quarterly* 44, no. 2 (1999): 350–83.

186. Kaplan, Stephen, and Rachel Kaplan. *The Experience of Nature: A Psychological Perspective.* Cambridge: Cambridge University Press, 1989.

187. Yegül, Fikret K. *Baths and Bathing in Classical Antiquity.* Cambridge, MA: MIT Press, 1992.

188. Carruthers, Mary. *The Craft of Thought: Meditation, Rhetoric, and the Making of Images, 400–1200.* Cambridge: Cambridge University Press, 1998.

189. International Organization for Standardization (ISO). *ISO 9241-210: Human-Centred Design for Interactive Systems.* Geneva: ISO, 2019.

190. Goldhagen, Sarah Williams. *Welcome to Your World: How the Built Environment Shapes Our Lives.* New York: HarperCollins, 2017.

191. Salingaros, Nikos A. *A Theory of Architecture.* Solingen: Umbau-Verlag, 2006.

192. Maggie's. "About Maggie's: Our History, Values and Impact." London: Maggie's Centres, n.d. https://www.maggies.org (accessed 1 December 2025).

193. JR East Group. *Tokyo Station City: Universal Design and Passenger Wayfinding.* Tokyo: East Japan Railway Company, 2013.

194. Calori, Chris, and David Vanden-Eynden. *Signage and Wayfinding Design: A Complete Guide to Creating Environmental Graphic Design Systems.* 2nd ed. Hoboken, NJ: Wiley, 2015.

195. Bjarke Ingels Group (BIG), Superflex, and Topotek 1. *Superkilen: A Participatory Urban Space.* Copenhagen: BIG, 2012.

196. Gehl, Jan, and Ditte Svarre. *How to Study Public Life.* Washington, DC: Island Press, 2013.

197. Wellington Waterfront Ltd. *Wellington Waterfront Framework and Design Principles.* Wellington: Wellington Waterfront Ltd, 2001.

198. Project for Public Spaces. "Wellington Waterfront: Reconnecting a City to Its Harbour." *Project for Public Spaces*, n.d. https://www.pps.org/article/wellington-waterfront-reconnecting-a-city-to-its-harbour (accessed 1 December 2025).

199. British Council for Offices (BCO). *Designing for Neurodiversity in the Workplace.* London: British Council for Offices, 2022.

200. Austin, Robert D., Gary P. Pisano, and Lareina Yee. "Neurodiversity as a Competitive Advantage." *Harvard Business Review*, May–June 2017.

201. EDGE Technologies. *The Edge, Amsterdam: Portfolio Project Profile.* Amsterdam: EDGE Technologies. https://edge.tech/portfolio/the-edge (accessed 23 November 2025).

202. Montessori, Maria. *The Montessori Method.* New York: Frederick A. Stokes Company, 1912.

203. Lillard, Angeline Stoll. *Montessori: The Science Behind the Genius.* 3rd ed. New York: Oxford University Press, 2017.

204. Lillard, Angeline S., et al. "Montessori Preschool Elevates and Equalizes Child Outcomes: A Longitudinal Study." *Frontiers in Psychology* 8 (2017): 1783. https://doi.org/10.3389/fpsyg.2017.01783.

205. Verbeek, Hilde, et al. "Small-Scale, Homelike Dementia Care: Effects on Quality of Life, Behaviour, and Use of Psychotropic Drugs." *Journal of the American Medical Directors Association* 11, no. 3 (2010): 165–72.

206. de Bruin, Simone R., et al. "Green Care Farms as Innovative Nursing Homes, Promoting Activities and Social Interaction for People with Dementia." *Dementia* 9, no. 3 (2010): 365–78.

207. The Atlantic. "The Dutch Village Where Everyone Has Dementia." *The Atlantic*, November 14, 2014. https://www.theatlantic.com/health/archive/2014/11/the-dutch-village-where-everyone-has-dementia/382195/

208. Friends of the High Line. *The High Line: History and Impact.* New York: Friends of the High Line, n.d. https://www.thehighline.org/history/ (accessed 1 December 2025).

209. Corner, James, Diller Scofidio + Renfro, and Piet Oudolf. *The High Line: Foreseen / Unforeseen.* London: Phaidon Press, 2015.

210. Bratman, Gregory N., J. Paul Hamilton, and Gretchen C. Daily. "The Impacts of Nature Experience on Human Cognitive Function and Mental Health." *Annals of the New York Academy of Sciences* 1249, no. 1 (2012): 118–36.

211. Korea Forest Welfare Institute (FoWI). *Forest Welfare Services in Korea: Healing Forests and Forest Therapy Programs.* Daejeon: Korea Forest Welfare Institute, 2019. https://www.fowi.or.kr/sustainabilityReport/2022/2022_%ED%95%9C%EA%B5%AD%EC%82%B0%EB%A6%BC%EB%B3%B5%EC%A7%80%EC%A7%84%ED%9D%A5%EC%9B%90_%EC%86%8C%EC%B1%85%EC%9E%90_%EC%98%81%EB%AC%B8.pdf (accessed January 23, 2026).

212. Park, Bum-Jin, et al. "Physiological Effects of Shinrin-Yoku (Taking in the Forest Atmosphere) in Young Japanese Male Subjects." *Public Health* 125, no. 2 (2011): 93–100.

213. Kotera, Yasuhiro, et al. "Effects of Shinrin-Yoku (Forest Bathing) and Nature Therapy on Mental Health: A Systematic Review and Meta-Analysis." *International Journal of Mental Health and Addiction* (online 2020; in print 2022).

214. Jewel Changi Airport Development. *Jewel Changi Airport: Design, Sustainability and Experience.* Singapore: Jewel Changi Airport Development, 2019. https://www.jewelchangiairport.com/en/the-jewel-story.html (accessed 1 December 2025).

215. Safdie Architects. *Jewel Changi Airport*. Somerville, MA: Safdie Architects, 2019. https://www.safdiearchitects.com/projects/jewel-changi-airport (accessed 1 December 2025). Safdie Architects.

216. Popovic, V., B. Kraal, and M. Chamorro-Koc. "Passenger Experience in Airports: A Review of Human-Centred Design Factors." *Journal of Air Transport Management* 67 (2018): 131–44.

217. Fagan, Garrett G. *Bathing in Public in the Roman World*. Ann Arbor: University of Michigan Press, 1999.

218. Wycherley, R. E. *How the Greeks Built Cities*. New York: W. W. Norton & Company, 1962.

219. Wright, Kenneth R., and Alfredo Valencia Zegarra. *Machu Picchu: A Civil Engineering Marvel*. Reston, VA: American Society of Civil Engineers, 2000.

220. Mannikka, Eleanor. *Angkor Wat: Time, Space, and Kingship*. Honolulu: University of Hawai'i Press, 1996.

221. Nitschke, Günter. *Japanese Gardens: Right Angle and Natural Form*. Cologne: Taschen, 1993.

222. Gammage, Bill. *The Biggest Estate on Earth: How Aborigines Made Australia*. Sydney: Allen & Unwin, 2011.

223. Pascoe, Bruce. *Dark Emu: Aboriginal Australia and the Birth of Agriculture*. Broome, WA: Magabala Books, 2014.

224. Coaldrake, William H. *Architecture and Authority in Japan*. London: Routledge, 1996.

225. Eberhard, John P. *Brain Landscape: The Coexistence of Neuroscience and Architecture*. New York: Oxford University Press, 2009.

226. Kalat, James W. *Biological Psychology*. 13th ed. Boston: Cengage Learning, 2018.

227. Cronon, William. "The Trouble with Wilderness; or, Getting Back to the Wrong Nature." *Environmental History* 1, no. 1 (1996): 7–28.

228. Odum, Eugene P., and Gary W. Barrett. *Fundamentals of Ecology*. 5th ed. Belmont, CA: Thomson Brooks/Cole, 2005.

229. Hobsbawm, Eric. *The Age of Revolution: 1789–1848*. New York: Vintage Books, 1996.

230. Mumford, Lewis. *Technics and Civilization*. Chicago: University of Chicago Press, 1934.

231. Welch, Jack, and John A. Byrne. *Jack: Straight from the Gut*. New York: Warner Books, 2001.

232. Gelles, David. *The Man Who Broke Capitalism: How Jack Welch Gutted the Heartland and Crushed the Soul of Corporate America—and How to Undo His Legacy*. New York: Simon & Schuster, 2022.

233. Useem, Michael. *The Leadership Moment: Nine True Stories of Triumph and Disaster and Their Lessons for Us All*. New York: Times Books, 1998.

234. Tickner, David, et al. "Bending the Curve of Global Freshwater Biodiversity Loss: An Emergency Recovery Plan." *Nature* 595, no. 7868 (2021): 388–95.

235. Associated Press. "The World's Rivers Faced the Driest Year in Three Decades in 2023, the UN Weather Agency Says." *AP News*, October 7, 2024. https://apnews.com/article/86183afa4d917fe9777f730159849c8f (accessed 10 January 2026).

236. Food and Agriculture Organization of the United Nations (FAO). *Status of the World's Soil Resources (SWSR): Main Report*. Rome: FAO, 2015.

237. United Nations Environment Programme (UNEP). *Global Land Outlook*. Nairobi: UNEP, 2022.

238. World Wide Fund for Nature (WWF). *Living Planet Report 2022: Building a Nature-Positive Society*. Gland, Switzerland: WWF, 2022.

239. Howard, Brian Clark, and Alejandra Borunda. "8 Mighty Rivers Run Dry from Overuse." *National Geographic*, March 8, 2021. https://www.nationalgeographic.com/environment/article/rivers-run-dry (accessed 3 December 2025).

240. Intergovernmental Panel on Climate Change (IPCC). *Sixth Assessment Report (AR6): Impacts, Adaptation and Vulnerability*. Geneva: IPCC, 2022.

241. Middleton, Nick, and David Thomas. *World Atlas of Desertification*. 3rd ed. Luxembourg: European Commission, 2017.

242. Torres, Arturo, et al. "Global Sand and Aggregate Extraction Exceeds Safe Operating Space." *Nature Sustainability* 4 (2021): 951–58. https://doi.org/10.1038/s41893-021-00812-0.

243. United Nations Environment Programme (UNEP). *Global Waste Management Outlook*. Nairobi: UNEP, 2015.

244. United Nations Environment Programme (UNEP). *From Pollution to Solution: A Global Assessment of Marine Litter and Plastic Pollution*. Nairobi: UNEP, 2021. https://www.unep.org/resources/report/pollution-solution-global-assessment-marine-litter-and-plastic-pollution.

245. Australian Institute of Aboriginal and Torres Strait Islander Studies (AIATSIS). *Aboriginal and Torres Strait Islander Peoples: Australia's First Nations*. Canberra: AIATSIS, 2021.

246. Berkes, Fikret. *Sacred Ecology: Traditional Ecological Knowledge and Resource Management*. 4th ed. New York: Routledge, 2018.

247. Engineers Australia. *Creating a Culture of Indigenous Engineering Knowledge*. Canberra: Engineers Australia, 2017.

248. UNESCO World Heritage Centre. *Budj Bim Cultural Landscape*. Paris: UNESCO, 2019. https://whc.unesco.org/en/list/1577/.

249. Haviland, John B. "Guugu Yimithirr Cardinal Directions." *Ethos* 6, no. 3 (1978): 197–227.

250. Levinson, Stephen C. *Space in Language and Cognition: Explorations in Cognitive Diversity*. Cambridge: Cambridge University Press, 2003.

251. Palmer, Bill, et al. "Linguistic Spatial Frames of Reference across the Languages of Australia." *Language Sciences* 33, no. 1 (2011): 1–14.

252. Majid, Asifa, Melissa Bowerman, Sotaro Kita, Daniel B. M. Haun, and Stephen C. Levinson. "Can Language Restructure Cognition? The Case for Space." *Trends in Cognitive Sciences* 8, no. 3 (2004): 108–14. https://doi.org/10.1016/j.tics.2004.01.003.

253. Edmonds-Wathen, Caroline. *Talking Mathematics: Aboriginal Perspectives on Learning Mathematics*. Adelaide: Australian Association of Mathematics Teachers, 2012.

254. Cooper, Tom, Chris Matthews, and Aboriginal and Torres Strait Islander Education Unit, Queensland University of Technology. *YuMi Deadly Maths: Culturally Responsive Mathematics from an Indigenous Perspective*. Brisbane: Queensland University of Technology, 2015.

255. Rose, Deborah Bird. *Nourishing Terrains: Australian Aboriginal Views of Landscape and Wilderness*. Canberra: Australian Heritage Commission, 1996.

256. Rose, Deborah Bird. *Reports from a Wild Country: Ethics for Decolonisation*. Sydney: University of New South Wales Press, 2004.

257. Kimmerer, Robin Wall. *Braiding Sweetgrass: Indigenous Wisdom, Scientific Knowledge, and the Teachings of Plants*. Minneapolis: Milkweed Editions, 2013.

258. McNiven, Ian J., and Bruno David. "Eel Traps and Shell Middens in Western Victoria: New Archaeological Evidence for Sustainable Aquaculture at Budj Bim." *Australian Archaeology* 79, no. 1 (2014): 1–14.

259. Bird-David, Nurit. "'Animism' Revisited: Personhood, Environment, and Relational Epistemology." *Current Anthropology* 40, no. S1 (1999): S67–S91.

260. Yunkaporta, Tyson. *Sand Talk: How Indigenous Thinking Can Save the World*. Melbourne: Text Publishing, 2019.

261. Graham, Mary. "Some Thoughts about the Philosophical Underpinnings of Aboriginal Worldviews." *Australian Humanities Review* 45 (2008): 181–94.

262. UNESCO World Heritage Centre. *Rice Terraces of the Philippine Cordilleras*. Paris: UNESCO, 1995.

263. Gómez-Pompa, Arturo, and Andrea Kaus. "Taming the Wilderness Myth." *BioScience* 42, no. 4 (1992): 271–79. https://doi.org/10.2307/1311675.

264. Beaumont, Peter. "Qanat Systems in Iran." *Journal of the Royal Asiatic Society* 3, no. 1 (1993): 1–20. https://www.jstor.org/stable/25182601.

265. Denevan, William M. "Stone Terraces in the Andes: A Reappraisal." *Annals of the Association of American Geographers* 78, no. 2 (1988): 249–62. https://www.jstor.org/stable/2562760.

266. Roberts, Mere, et al. "Kaitiakitanga: Māori Perspectives on Conservation." *Pacific Conservation Biology* 1, no. 1 (1995): 7–20. https://doi.org/10.1071/PC950007.

267. Mollison, Bill, and David Holmgren. *Permaculture One: A Perennial Agriculture for Human Settlements*. Melbourne: Tagari Publications, 2002.

268. Arup. *Regenerative Design: Reimagining the Built Environment*. London: Arup, 2023. https://www.arup.com/perspectives/regenerative-design.

269. United Nations Environment Programme (UNEP). *Making Peace with Nature: A Scientific Blueprint to Tackle the Climate, Biodiversity and Pollution Emergencies*. Nairobi: UNEP, 2021.

270. EcoLogic Studio. *PhotoSynthetica / BioSolar Leaf: Living Architecture for Urban Air Purification*. London: EcoLogic Studio, 2018.

271. Pattinson, William, Claudia Pasquero, and Marco Poletto. "Urban Microalgae Systems for Air Quality Improvement." In *Proceedings of the International Conference on Sustainable Design and Manufacturing*, 2019.

272. Bullitt Foundation. *The Bullitt Center: A Living Building Case Study*. Seattle: Bullitt Foundation, 2014.

273. International Living Future Institute. *Living Building Challenge 3.1*. Seattle: ILFI, 2014.

274. Bullitt Foundation. *Bullitt Center Financial Performance and Cost Analysis*. Seattle: Bullitt Foundation, 2014.

275. Bullitt Foundation. *The Bullitt Center: 250-Year Building Case Study*. Seattle: Bullitt Foundation, 2016.

276. International Living Future Institute. *Net Zero and Living Building Challenge Financial Study*. Washington, DC: International Living Future Institute, 2022. Accessed 3 December 2025. https://living-future.org/wp-content/uploads/2022/05/NZEB_LBC_-DC_Financial_Study.pdf

277. Public Broadcasting Service (PBS). "The World's Greenest Building." *PBS NewsHour*, April 22, 2013.

278. The Guardian. "Inside the World's Greenest Office Building." *The Guardian*, April 22, 2013.

279. Public Utilities Board (PUB) Singapore. *Active, Beautiful, Clean Waters Programme: Managing Stormwater for Liveable and Sustainable Cities*. Singapore: PUB, 2013.

280. Dreiseitl, Herbert, and Ramboll Studio Dreiseitl. *Water Is Life: The Projects of Studio Dreiseitl*. Basel: Birkhäuser, 2015.

281. Landezine International Landscape Award. "Bishan–Ang Mo Kio Park." Landezine, 2012.

282. Planning Institute of Australia (NSW). "Bishan–Ang Mo Kio Park: Working with Water." *Planning News* (NSW Chapter), 2013.

283. Melbourne Water. *Greening the Pipeline: Transforming Sewer Corridors into Green Infrastructure*. Melbourne: Melbourne Water, 2017.

284. Wyndham City Council. *Greening the Pipeline – Federation Trail Pilot Project*. Wyndham Vale, VIC: Wyndham City Council, 2018.

285. Queensland Government, Department of Environment and Science. *Moreton Bay Marine Park: Seagrass Protection and Environmentally Friendly Moorings*. Brisbane: Queensland Government, 2018.

286. Grech, Alana, Hugh P. Possingham, and Anthony J. Preen. "Marine Protected Areas and Dugong Conservation: Spatial Protection of Seagrass Habitats in Moreton Bay." *Biological Conservation* 142, no. 10 (2009): 2293–2303.

287. African Union Commission. *Great Green Wall for the Sahara and the Sahel Initiative: Implementation Status and Way Ahead to 2030.* Addis Ababa: African Union, 2020.

288. United Nations Convention to Combat Desertification (UNCCD). *The Great Green Wall: Implementation Status and Way Ahead to 2030.* Bonn: UNCCD, 2020.

289. International Union for Conservation of Nature (IUCN). *The Great Green Wall: Advancing Restoration in the Sahel.* Gland, Switzerland: IUCN, 2019.

290. Headley, Tom R., and Chris M. Tanner. "Constructed Floating Wetlands: Their Ecological Value, Design Characteristics, and Performance." *Water Science and Technology* 55, no. 11 (2007): 165–72.

291. Chicago Department of Transportation, Chicago Riverwalk Authority, and Urban Rivers. *Wild Mile Framework: A Vision for Ecological Restoration of the Chicago River* (Part 1). Chicago: City of Chicago, 2021. https://wateractionhubfrontdoor-d6dwaqhbgwebcfg2.z01.azurefd.net/media/files/2021/01/22/Wild_Mile_Framework_Final_Part1.pdf

292. Browne, Mark A., et al. "Living Seawalls: Enhancing Coastal Defence Structures to Support Biodiversity." *Journal of Applied Ecology* 52, no. 4 (2015): 1296–1306.

293. Bromenshenk, Jerry J., et al. "Monitoring Honey Bee Colonies for Environmental Stress Using Sensor Networks." *Environmental Monitoring and Assessment* 189, no. 8 (2017): Article 368.

294. World Economic Forum. *Harnessing the Internet of Things for Environmental Monitoring: Bees as Bio-Sensors.* Geneva: World Economic Forum, 2020.

295. Waitangi Tribunal. *The Report on the Waikato River Claim (Wai 212).* Wellington: Legislation Direct, 2010.

296. Te Arawhiti (Office for Māori Crown Relations). *Waikato River Co-Governance and Co-Management Framework.* Wellington: New Zealand Government, 2019.

297. New Zealand Department of Conservation. *Te Urewera: Governance, Restoration, and Partnership.* Wellington: Department of Conservation, 2018.

298. Jones, Mike, Antoni Gandia, and Sarah Bismarck. "Mycelium Composites: A Review of Engineering Characteristics and Growth Parameters." *Journal of Materials Science* 55, no. 30 (2020): 1–21.

299. Heisel, Felix, et al. "The MycoTree: Building with Biological Materials." *Architectural Design* 87, no. 4 (2017): 44–51.

300. Ecovative Design. *Mycelium-Based Materials for the Built Environment.* Green Island, NY: Ecovative Design, 2021.

301. Rijkswaterstaat (Ministry of Infrastructure and Water Management). *Room for the River Programme: A New Approach to Flood Risk Management.* The Hague: Government of the Netherlands, 2016.

302. Delta Programme Commissioner. *Delta Programme 2015: Working on the Delta.* The Hague: Ministry of Infrastructure and the Environment, 2015.

303. Organisation for Economic Co-operation and Development (OECD). *Water Governance in the Netherlands: Fit for the Future?* Paris: OECD, 2014.

304. National Oceanic and Atmospheric Administration (NOAA). *Oyster Reef Restoration: A Key Strategy for Coastal Resilience.* Washington, DC: NOAA, 2017.

305. La Peyre, Megan K., et al. "Oyster Reef Restoration in the Northern Gulf of Mexico: Extent, Methods and Outcomes." *Estuaries and Coasts* 37, no. 4 (2014): 813–28. https://doi.org/10.1007/s12237-013-9689-8.

306. Aga Khan Development Network. "Infrastructure Development: Energy Access and Renewable Solutions." *Aga Khan Development Network.* https://the.akdn/en/what-we-do/generating-economic-growth/infrastructure-development (accessed 29 November 2025)

307. Ritchie, Hannah, and Max Roser. "Energy Access." *Our World in Data.* Oxford: University of Oxford, updated regularly. https://ourworldindata.org/energy-access.

308. Takeuchi, Kazuhiko. "Rebuilding the Relationship between People and Nature: The Satoyama Initiative." *Ecological Research* 25, no. 5 (2010): 891–97.

309. United Nations University Institute for the Advanced Study of Sustainability (UNU-IAS). *Satoyama–Satoumi Ecosystems and Human Well-Being.* Tokyo: United Nations University, 2014.

310. Bjarke Ingels Group (BIG). *CopenHill / Amager Bakke.* Copenhagen: BIG, project description and performance overview. https://big.dk/#projects-amab.

311. Amager Ressourcecenter (ARC). *Amager Bakke: Waste-to-Energy with Urban Recreation.* Copenhagen: ARC, official project documentation.

312. Attia, Peter, with Bill Gifford. *Outlive: The Science and Art of Longevity.* New York: Harmony Books, 2023.

313. Roychand, Rajeev, Shannon Kilmartin-Lynch, Mohammad Saberian, Jie Li, and C. Q. Li. "Translating Lab Success to the Field: Evaluating Coffee Biochar-Enhanced Concrete in Real-World Construction." *Case Studies in Construction Materials* 22 (2025): e04233.

314. Harris, Joseph. *The Eiffel Tower: The History, the Stories, and the Legacy of Paris's Most Famous Landmark.* London: Amberley Publishing, 2018.

315. Sydney Institute of Marine Science (SIMS). *Living Seawalls: Increasing Biodiversity on Urban Seawalls.* Sydney: SIMS, 2020. https://www.sims.org.au/living-seawalls/.

316. Dellinger, Drew. *Love Letter to the Milky Way.* Berkeley, CA: Planetize the Movement Press, 2010.

317. Lehner, Mark. *The Complete Pyramids: Solving the Ancient Mysteries.* London: Thames & Hudson, 1997.

318. Thapar, Romila. *Aśoka and the Decline of the Mauryas.* 3rd ed. Oxford: Oxford University Press, 2012.

319. Lewis, David. *We, the Navigators: The Ancient Art of Landfinding in the Pacific.* 2nd ed. Honolulu: University of Hawai'i Press, 1994.

320. Lane, Frederic C. *Venice: A Maritime Republic.* Baltimore: Johns Hopkins University Press, 1973.

321. Maathai, Wangari. *The Green Belt Movement: Sharing the Approach and the Experience.* New York: Lantern Books, 2004.

322. Gehl, Jan. *Cities for People.* Washington, DC: Island Press, 2010.

323. Figueres, Christiana, and Tom Rivett-Carnac. *The Future We Choose: Surviving the Climate Crisis.* New York: Alfred A. Knopf, 2020.

324. Deemal-Hall, Eileen. *Aboriginal Cultural Values and Engineering Practice.* Brisbane: Indigenous Consulting Australia, 2018.

325. Hodge, A. Trevor. *Roman Aqueducts & Water Supply.* 2nd ed. London: Duckworth, 2002.

326. Wilson Jones, Mark. *Principles of Roman Architecture.* New Haven, CT: Yale University Press, 2000.

327. Dehejia, Vidya. *Early Buddhist Art of India.* New York: Abrams, 1997.

328. Needham, Joseph. *Science and Civilisation in China. Vol. 4, Physics and Physical Technology. Part III: Civil Engineering and Nautics.* Cambridge: Cambridge University Press, 1971.

329. Queensland Government, Cherbourg Aboriginal Shire Council. *Cherbourg: Community History and Heritage.* Cherbourg, QLD: Queensland Government, 2025. https://www.qld.gov.au/aboriginal-and-torres-strait-islander-peoples/communities/cherbourg/history (accessed 1 December 2025)

330. Find & Connect. *Cherbourg Aboriginal Settlement (c. 1904–1986).* National Centre for Australian Indigenous Studies, 2025.

331. Remplan / Australian Bureau of Statistics. *Cherbourg Labour Force Status Profile.* REMPLAN Community, 2025.

332. Memmott, Paul. *Aboriginal Housing: The State of the Art (or Non-State of the Art).* Canberra: Australian Housing and Urban Research Institute, 2007.

333. Lear, Linda J. *Beatrix Potter: A Life in Nature.* New York: St. Martin's Press, 2007.

334. Department of Water and Environmental Regulation (Western Australia). *Pilbara Groundwater Allocation and Management Framework*. Perth: Government of Western Australia, 2020.

335. Australian Human Rights Commission. *Free, Prior and Informed Consent: An Australian Perspective*. Sydney: AHRC, 2021.

336. Legacie. *Ngarluma Water Desalination Project – Project Information Sheet*. Legacie Water Infrastructure Developers, 2024. https://legacie.com.au/wp-content/uploads/2024/03/LEGA13643_Project-Information-Sheet.pdf

337. Davidson, Christopher M. *Dubai: The Vulnerability of Success*. New York: Columbia University Press, 2008.

338. Government of Dubai. *Dubai Economic Report 2022*. Dubai: Department of Economy and Tourism, 2022. https://www.dubaided.gov.ae/Publications/Dubai-Economic-Report-2022.pdf

339. Fernandes, G. W., et al. "Deep into the Mud: Ecological and Socio-Economic Impacts of the Catastrophic Tailings Dam Failure in Mariana, Brazil." *Perspectives in Ecology and Conservation* 15, no. 4 (2017): 145–51.

340. do Carmo, F. F., et al. "Fundão Tailings Dam Failures: The Environmental Tragedy of the Fundão Tailings Dam." *Perspectives in Ecology and Conservation* 15, no. 3 (2017): 156–62.

341. United States Environmental Protection Agency. *Love Canal Emergency Declaration Area: A Special Report to the Congress*. Washington, DC: U.S. Environmental Protection Agency, 1978.

342. Levine, Adeline Gordon. *Love Canal: Science, Politics, and People*. Lexington, MA: Lexington Books, 1982.

343. Michigan Civil Rights Commission. *The Flint Water Crisis: Systemic Racism Through the Lens of Flint*. Lansing, MI: Michigan Civil Rights Commission, 2017.

344. Hanna-Attisha, Mona, Jenny LaChance, Richard C. Sadler, and Allison Champney Schnepp. "Elevated Blood Lead Levels in Children Associated with the Flint Drinking Water Crisis." *American Journal of Public Health* 106, no. 2 (2016): 283–90.

345. Government of Samoa. *Cyclone Evan 2012: Post-Disaster Needs Assessment*. Apia: Government of Samoa, World Bank, Asian Development Bank, and United Nations, 2013.

346. Nunn, Patrick D. *Climate, Environment and Society in the Pacific during the Last Millennium*. Amsterdam: Elsevier, 2007.

347. Heggelund, Gørild. *Environment and Resettlement Politics in China: The Three Gorges Project*. Farnham, UK: Ashgate, 2004.

348. Wu, Jingjing, Jun Xia, and Changming Liu. "Impact of the Three Gorges Project on the Eco-Environment and Its Countermeasures." *Journal of Hydrology* 400, no. 3–4 (2011): 241–49.

349. Jackson, Kenneth T. *Crabgrass Frontier: The Suburbanization of the United States*. New York: Oxford University Press, 1985.

350. Marohn, Charles L. *Strong Towns: A Bottom-Up Revolution to Rebuild American Prosperity*. Hoboken, NJ: Wiley, 2019.

351. Ford, Henry. *My Life and Work*. Garden City, NY: Doubleday, Page & Company, 1922.

352. Harrison, Max Dulumunmun. *My Country My Life: Fight for Indigenous Culture and Sacred Land*. Sydney: Bantam Australia, 2004.

353. City of Stockholm. *The Hammarby Model: A Sustainable Urban Development Concept*. Stockholm: Stockholm City Planning Administration, 2007.

354. Pandis Iveroth, Sonja, and Anna Brandt. "The Development of a Sustainable Urban District: Hammarby Sjöstad in Stockholm." *Journal of Cleaner Production* 18, no. 13 (2010): 1306–14.

355. Brisbane City Council. *Oxley Creek Transformation: Reconnecting Creeks, Community and Country*. Brisbane: Brisbane City Council, 2019.

356. Engineers Without Borders Canada. *Failure Reports: Why Engineering Projects Fail and What We Can Learn.* Toronto: Engineers Without Borders Canada, 2011.

357. OzHarvest. *OzHarvest Impact Report 2022–23.* Sydney: OzHarvest, 2023.

358. Great Place to Work Australia. *Australia's Best Workplaces™ in Health & Wellbeing – OzHarvest.* Sydney: Great Place to Work Australia, 2022.

359. UK Government. *The Five Case Model: A Guide to the Strategic Business Case.* London: HM Treasury and Infrastructure and Projects Authority, 2020.

360. City of Melbourne. *Social Procurement Framework.* Melbourne: City of Melbourne, 2018.

361. Sydney Opera House Trust. *Conservation Management Plan.* Sydney: Sydney Opera House Trust, 2018.

362. Queensland Audit Office. *Report to Parliament No. 9: Transport Infrastructure Development and Delivery.* Brisbane: Queensland Audit Office, 2011.

363. Sumida City Office. *The Sumida Hokusai Museum: Urban Context, Design Intent and Community Integration.* Tokyo: Sumida City, 2016.

364. Fujimoto, Sou. "Architecture as Forest: The Sumida Hokusai Museum." *Japan Architect* 98 (2017): 44–53.

365. Scottish Parliament Public Audit Committee. *Edinburgh Trams Inquiry: Final Report.* Edinburgh: Scottish Parliament, 2016.

366. Audit Scotland. *Edinburgh Trams: Lessons from Transport Projects.* Edinburgh: Auditor General for Scotland, 2011.

367. Øresundsbro Konsortiet. *The Øresund Fixed Link: Environmental Concerns and Design Solutions.* Copenhagen–Malmö: Øresundsbro Konsortiet, 2000.

368. Stoltze, Michael, and Knud Falk. *Naturen pá Peberholm [Nature on Peberholm].* Copenhagen: Dansk Ornitologisk Forening, 2012.

369. State of Victoria. *Wind Environment and Pedestrian Comfort in High-Density Urban Precincts: Lessons from Docklands.* Melbourne: Victorian Planning Authority, 2019.

370. Audit Office of New South Wales. *Sydney Light Rail: Planning and Implementation.* Sydney: Audit Office of NSW, 2016.

371. New South Wales Government. *Inquiry into the Sydney Light Rail Project.* Sydney: NSW Legislative Council, Public Accountability Committee, 2018.

372. Transport Scotland. *The Borders Railway: Project Overview and Outcomes.* Edinburgh: Scottish Government, 2016.

373. Scottish Government. *Borders Railway Blueprint: Evaluation of Economic, Social and Transport Impacts.* Edinburgh: Scottish Government, 2017.

374. Crossrail Ltd. *Learning Legacy: Lessons from the Elizabeth Line Programme.* London: Crossrail Ltd, 2018.

375. Davies, Glyn, Mark Dodgson, and Michael Hobday. *Projects, Micro-Organizations and Dynamic Capabilities: Lessons from Crossrail.* London: Saïd Business School, University of Oxford, 2016.

376. U.S. House Committee on Transportation and Infrastructure. *Final Committee Report: The Design, Development & Certification of the Boeing 737 MAX.* Washington, DC: U.S. House of Representatives, 2020.

377. National Transportation Safety Board. *Aircraft Accident Report: Descent Below Terrain Following Loss of Control Involving Boeing 737-8, Lion Air Flight 610.* Washington, DC: NTSB, 2019.

378. Joint Authorities Technical Review (JATR). *Boeing 737 MAX Flight Control System: Observations, Findings, and Recommendations.* Washington, DC: Federal Aviation Administration, 2019.

379. U.S. Environmental Protection Agency. *Coal Combustion Residuals (CCR) Risk Assessment and Damages Case Assessment.* Washington, DC: U.S. Environmental Protection Agency, Office of Resource Conservation and Recovery, 2014.

380. Metabolic. *De Ceuvel: Circular Regeneration of a Contaminated Industrial Site.* Amsterdam: Metabolic Lab, 2014.

381. City of Copenhagen. *Copenhagen Bicycle Account 2021.* Copenhagen: Technical and Environmental Administration, City of Copenhagen, 2021.

382. Australian Sustainable Built Environment National Research Centre (SBEnrc). *Community Attitudes to Construction and Demolition Waste Facilities: Drivers of Opposition and Acceptance.* Perth: SBEnrc, 2022.

383. Public Utilities Board (PUB) Singapore. *Bishan–Ang Mo Kio Park and Kallang River: ABC Waters Design Features.* Singapore: PUB, 2012.

384. U.S. Army Corps of Engineers, Los Angeles District. *Los Angeles River Ecosystem Restoration Feasibility Study.* Los Angeles: USACE, 2015.

385. International Railway Journal. "Tokyo Metro Renews the Ginza Line While Keeping Services Running." *International Railway Journal*, 2016.

386. Tokyo Metro Co., Ltd. *Ginza Line Renewal Project: Track, Signalling and Station Upgrades.* Tokyo: Tokyo Metro Co., Ltd., ongoing programme documentation (2012–present).

387. City of Sydney. *Construction Noise Management Guidelines.* Sydney: City of Sydney, 2019.

388. Transport for New South Wales. *Sydney Harbour Bridge Asset Management and Maintenance Program.* Sydney: NSW Government, 2022.

389. Queensland Government. *Paradise Dam Improvement Project: Detailed Business Case.* Brisbane: Queensland Department of Regional Development, Manufacturing and Water, 2021.

390. Rijkswaterstaat. *Room for the River Programme: Summary and Results.* The Hague: Ministry of Infrastructure and the Environment, 2016.

391. van den Brink, M., et al. "Room for the River: A Paradigm Shift in Dutch Flood Risk Management." *International Journal of River Basin Management* 12, no. 1 (2014): 1–12.

392. Transport for London. *Business Plan 2019–24.* London: Transport for London, 2019.

393. Friends of the High Line. *The High Line: History, Design, and Horticulture.* New York: Friends of the High Line, 2015.

394. O'Neill, Michael J., and Gary W. Evans. "Supportive Design Features in Healthcare Environments and Their Impact on Stress and Wellbeing." *Journal of Environmental Psychology* 20, no. 2 (2000): 145–54.

395. Perth Airport. *Quiet Room and Sensory Space Initiatives for Neurodivergent Travellers.* Perth: Perth Airport Pty Ltd, 2022.

396. Victorian Government. *Burndap Birrarung burndap umarkoo (Yarra Strategic Plan).* Melbourne: Department of Environment, Land, Water and Planning, 2017.

397. Department of Climate Change, Energy, the Environment and Water (DCCEEW). *Climate Risks to Australia's Coastal Infrastructure.* Canberra: Australian Government, 2021.

398. Federal Highway Administration. *National Bridge Inventory (NBI).* Washington, DC: U.S. Department of Transportation. Accessed 2024.

399. Reuters. "Italy Charges Dozens over Genoa Bridge Collapse That Killed 43." *Reuters*, April 14, 2023. https://www.reuters.com/world/europe/italy-charges-dozens-over-genoa-bridge-collapse-that-killed-43-2023-04-14/.

400. Sydney Opera House Trust. *Living Seawalls at Bennelong Point: Restoring Marine Habitat at the Sydney Opera House.* Sydney: Sydney Opera House Trust, 2021. https://www.sydneyoperahouse.com/our-story/environment/living-seawalls.html.

401. Transport for London. *Asset Management Strategy: Maintaining a State of Good Repair.* London: Transport for London, 2019. https://content.tfl.gov.uk/asset-management-strategy.pdf.

402. "Toxic Lead Levels in Children after Flint Water Crisis." *New England Journal of Medicine* 374, no. 21 (2016): 2057–60.

403. Kumakura, Isao. *Kintsugi: The Poetics of Repair.* Tokyo: Kodansha International, 2015.

404. Jencks, Charles. *Maggie's Centres: Architecture for Cancer Care.* London: Frances Lincoln, 2010.

405. de la Mothe, John, and Dominique Foray, eds. *Knowledge Management in the Innovation Process.* Boston: Springer, 2001.

406. Rail Trails Australia. *Rail Trails of Australia: Economic, Social and Health Benefits.* Melbourne: Rail Trails Australia, 2020.

407. Victorian Government, Department of Transport and Planning. *Murray to Mountains Rail Trail: Economic Impact and Visitor Use.* Melbourne: State of Victoria, 2018.

408. Northern Rivers Rail Trail Supporters. *Economic Benefits of Australian Rail Trails: Case Study.* n.d. https://northernriversrailtrailsupporters.org.au/vision/rail-trails-work-2/economic-benefits/case-study-2-australian-rail-trails/.

409. *Brisbane Valley Rail Trail Visitor Research Project Report 2021.* Prepared for The University of Queensland, 2021.

410. Queensland Department of Transport and Main Roads. *Funding Rail Trails Helps Power Queensland Economy.* Brisbane: Queensland Government, 2019.

411. Sunshine Coast Council. *Moreton Central Sugar Mill and Cane Tramway History.* Sunshine Coast Council, Queensland, n.d.

412. Seoul Metropolitan Government. *Cheonggyecheon Restoration Project.* Seoul: Seoul Metropolitan Government, 2005.

413. Gans, Herbert J. *The Urban Villagers: Group and Class in the Life of Italian-Americans.* New York: Free Press, 1962.

414. Brisbane City Council. *Brisbane Powerhouse: History and Redevelopment.* Brisbane: Brisbane Powerhouse Trust, n.d. https://brisbanepowerhouse.org/about/history/.

415. *Reimagining Detroit: Opportunities for Redefining an American City.* Detroit: Wayne State University Press, 2010.

416. Biennale of Sydney. *Cockatoo Island as a Biennale Venue.* Sydney: Biennale of Sydney, n.d. https://www.biennaleofsydney.art/venues/cockatoo-island/.

417. New South Wales National Parks and Wildlife Service. *Cockatoo Island: History, Heritage and Adaptive Reuse.* Sydney: NSW Government, n.d. https://www.cockatooisland.gov.au/about/history/.

418. Sudjic, Deyan. *Tate Modern: Herzog & de Meuron.* London: Tate Publishing, 2000.

419. Dunham-Jones, Ellen, and June Williamson. *Retrofitting Suburbia: Urban Design Solutions for Redesigning Suburbs.* 2nd ed. Hoboken, NJ: Wiley, 2011.

420. Taylor, Rabun. *Public Needs and Private Pleasures: Water Distribution, the Tiber River, and the Urban Development of Ancient Rome.* Rome: L'Erma di Bretschneider, 2000.

421. Frontinus, Sextus Julius. *On the Water Supply of the City of Rome (De Aquis Urbis Romae).* Translated by R. H. Rodgers. Cambridge: Cambridge University Press, 2004.

422. National Audit Office (UK). *The Millennium Dome.* London: The Stationery Office, 2000.

423. Infrastructure New South Wales. *Barangaroo Delivery Authority: Barangaroo Reserve – Design, Culture and Sustainability.* Sydney: NSW Government, 2015.

424. UNESCO World Heritage Centre. *Venice and Its Lagoon.* Paris: UNESCO, 1987. https://whc.unesco.org/en/list/394/.

425. Canterbury Earthquake Recovery Authority (CERA). *Psychosocial Recovery Framework for Greater Christchurch*. Christchurch: New Zealand Government, 2013.

426. U.S. Environmental Protection Agency. *Evaluation of Air Quality Impacts from Large-Scale Demolitions in Detroit*. Washington, DC: U.S. Environmental Protection Agency, 2018.

427. City of Fremantle. *Fremantle Markets and West End Heritage*. Fremantle, WA: City of Fremantle, n.d.

428. Real Commercial. "MANY 6160 Brings New Life to Fremantle's Former Myer Building." *Real Commercial*, 2014.

429. Brisbane City Council. *Legacy Way Project Environmental Impact Statement: Community Consultation Report (Appendix B)*. Brisbane: Brisbane City Council, 2010–2012 (various volumes).

430. Senate Department for Urban Development and Housing (Berlin). *Tempelhofer Feld: Development, Referendum and Civic Use*. Berlin: State of Berlin, 2014.

431. Victorian Auditor-General's Office. *Managing the Latrobe Valley Transition*. Melbourne: Victorian Government, 2018.

432. Latrobe Valley Authority. *Hazelwood Mine Fire Inquiry – Implementation and Community Outcomes*. Morwell: Victorian Government, 2019.

433. Healy, Noel, and John Barry. "Politicizing Energy Justice and Energy System Transitions: Fossil Fuel Divestment and a 'Just Transition.'" *Energy Policy* 108 (2017): 451–59.

434. International Energy Agency. *Managing Coal Plant Retirements: Insights from Advanced Economies*. Paris: IEA, 2020.

435. U.S. Environmental Protection Agency. *Coal Combustion Residuals: Impoundments and Legacy Risks*. Washington, DC: U.S. Environmental Protection Agency, 2019.

436. Tennessee Valley Authority. *Widows Creek Fossil Plant Closure and Ash Management*. Knoxville, TN: TVA, 2016.

437. Victorian Auditor-General's Office. *Rehabilitating Coal Mine and Power Station Sites*. Melbourne: Victorian Government, 2020.

438. Barthes, Roland. *The Eiffel Tower and Other Mythologies*. Berkeley: University of California Press, 1979.

439. New South Wales National Parks and Wildlife Service. *Brewarrina Aboriginal Fish Traps (Baiame's Ngunnhu): Plan of Management*. Sydney: NSW Government, 2006.

440. Kranz, Eugene. *Failure Is Not an Option: Mission Control from Mercury to Apollo 13 and Beyond*. New York: Simon & Schuster, 2000.

441. Infrastructure Victoria. *Infrastructure Strategy 2021–2051*. Melbourne: Victorian Government, 2021.

442. Miller, Roger, and Michael Hobday. "Innovation in Complex Systems Industries: The Case of Aerospace." *Research Policy* 24, no. 3 (1995): 363–84.

443. Seoul Metropolitan Government. *Cheonggyecheon Restoration Project: Environmental, Social and Economic Impacts*. Seoul: Seoul Metropolitan Government, 2005.

444. City of Seattle. *Alaskan Way Seawall Replacement Project: Final Environmental Impact Statement*. Seattle: City of Seattle, 2015.

445. Toft, Jason D., et al. "Shoreline Armoring Effects on Nearshore Ecology: Evidence from the Seattle Seawall." *Ecological Engineering* 91 (2016): 290–301.

446. International Civil Aviation Organization. *Aircraft Accident Report: Collision of Two Boeing 747 Aircraft on the Runway at Los Rodeos Airport, Tenerife, 27 March 1977*. Montreal: ICAO, 1978.

447. Reason, James. *Managing the Risks of Organizational Accidents*. Aldershot: Ashgate, 1997.

448. International Civil Aviation Organization. *Safety Report 2022*. Montreal: ICAO, 2022.

449. National Aeronautics and Space Administration. *Aviation Safety Reporting System (ASRS): Program Overview*. Washington, DC: NASA, 2021.

450. European Union Aviation Safety Agency. *Just Culture in Aviation*. Cologne: EASA, 2016.

451. Presidential Commission on the Space Shuttle Challenger Accident. *Report of the Presidential Commission on the Space Shuttle Challenger Accident*. Washington, DC: U.S. Government Printing Office, 1986.

452. Flint Water Advisory Task Force. *Final Report*. Lansing, MI: State of Michigan, 2016.

453. Auckland Transport. *Te Ara Kotahi – Māori Outcomes Framework*. Auckland: Auckland Transport, 2018.

454. Rozovsky, Julia. "The Five Keys to a Successful Google Team." *re:Work* (Google), 2015. https://rework.withgoogle.com/blog/five-keys-to-a-successful-google-team/.

455. Commissione Parlamentare d'Inchiesta sul MOSE. *Relazione Conclusiva sul Sistema MOSE*. Rome: Italian Parliament, 2014.

456. North Australian Indigenous Land and Sea Management Alliance (NAILSMA). *The North Australian Fire Management Programme: Outcomes and Benefits*. Darwin: NAILSMA, 2015.

457. Russell-Smith, J., et al. "Fire Management and Carbon Abatement in Northern Australian Savannas." *Philosophical Transactions of the Royal Society B* 371, no. 1696 (2016): 20150309.

458. Ministry of Land, Infrastructure, Transport and Tourism (MLIT), Japan. *Metropolitan Area Outer Underground Discharge Channel (G-Cans): Project Overview and Flood Control Function*. Tokyo: MLIT, 2010.

459. Billah, K. Y., and R. H. Scanlan. "Resonance, Tacoma Narrows Bridge Failure, and Undergraduate Physics Textbooks." *American Journal of Physics* 59, no. 2 (1991): 118–24.

460. National Diet of Japan. *The Official Report of the Fukushima Nuclear Accident Independent Investigation Commission*. Tokyo: National Diet of Japan, 2012.

461. Infrastructure Australia. *Assessment Framework*. Canberra: Australian Government, 2018.

462. Ellen MacArthur Foundation. *Completing the Picture: How the Circular Economy Tackles Climate Change*. Cowes, UK: Ellen MacArthur Foundation, 2019.

463. National Infrastructure Commission. *Better Delivery: Improving the Delivery of National Infrastructure*. London: NIC, 2020.

464. National Audit Office. *Lessons from Delivering Major Projects*. London: National Audit Office, 2019.

465. World Bank. *Stakeholder Engagement for Inclusive Infrastructure*. Washington, DC: World Bank, 2017.

466. Task Force on Climate-related Financial Disclosures. *Final Report: Recommendations of the Task Force on Climate-related Financial Disclosures*. Basel: Financial Stability Board, 2017.

467. Engineers Australia. *Engineering Futures 2035: Identifying Engineering Skills for the Future*. Canberra: Engineers Australia, 2018.

468. Noosa Shire Council. *Queensland-First Trial Tests Bushfire Detection Technology*. Media release, 1 December 2022.

469. XPRIZE. *XPRIZE Wildfire: Ending Destructive Wildfires*. Culver City, CA: XPRIZE Foundation, 2023.

470. Swinburne University of Technology. *Swinburne Team Advances in Global XPRIZE Wildfire Challenge*. Melbourne: Swinburne University of Technology, 2024.

471. U.S. Army Corps of Engineers. *Engineering With Nature: An Atlas*. Washington, DC: U.S. Army Corps of Engineers, 2018.

472. National Oceanic and Atmospheric Administration (NOAA). *Natural Infrastructure: Protecting Coastal Communities and Ecosystems*. Washington, DC: NOAA Office for Coastal Management, 2021.

473. Transport for New South Wales. *Asset AI: Using Artificial Intelligence to Improve Road Asset Management Across NSW*. Sydney: Transport for NSW, 2022.

474. Shellharbour City Council. *Smart Road Asset Management Trial Using Asset AI*. Shellharbour, NSW: Shellharbour City Council, 2022.

475. Australian Energy Market Operator. *Engineering Roadmap to 100% Renewables.* Melbourne: AEMO, 2020.

476. Changi Airport Group. *Changi Airport Deploys Autonomous Cleaning Robots to Enhance Operations.* Singapore: Changi Airport Group, 2022.

477. Tokyo Metro Co., Ltd. *Use of AI and Big Data to Optimise Train Operations and Reduce Congestion.* Tokyo: Tokyo Metro, 2021.

478. Rio Tinto. *Autonomous Haulage System in the Pilbara.* London: Rio Tinto, 2023.

479. BBC News. "Ever Given: The Ship That Blocked the Suez Canal." *BBC News*, March 29, 2021.

480. National Transportation Safety Board. *Preliminary Report: Collision of Containership Dali with the Francis Scott Key Bridge, Baltimore, Maryland.* Washington, DC: NTSB, 2024.

481. Engineers Australia. *Engineering Skills and Labour Shortages.* Canberra: Engineers Australia, 2023.

482. Engineers Australia. *The Engineering Profession: A Statistical Overview.* Canberra: Engineers Australia, 2022.

483. Invergowrie Foundation. *Understanding What Works: Evaluation of Programs Supporting Women and Girls.* Melbourne: Invergowrie Foundation, 2022.

Made in the USA
Monee, IL
04 November 2026

67385336R00305